SPRINGER HANDBOOK OF AUDITORY RESEARCH

Series Editors: Richard R. Fay and Arthur N. Popper

SPRINGER HANDBOOK OF AUDITORY RESEARCH

Continued after index

Jochen Schacht
Arthur N. Popper
Richard R. Fay

Editors

Auditory Trauma, Protection, and Repair

 Springer

Editors

Jochen Schacht
Kresge Hearing Research Institute
University of Michigan
Ann Arbor, MI 48109
USA
schacht@umich.edu

Arthur N. Popper
Department of Biology
University of Maryland
College Park, MD 20742
USA
apopper@umd.edu

Richard R. Fay
Parmly Hearing Institute
6525 North Sheridan Road
Loyola University Chicago
Chicago, IL 60626
USA
rfay@luc.edu

Series Editors

Richard R. Fay
Parmly Hearing Institute and
    Department of Psychology
Loyola University Chicago
Chicago, IL 60626
USA

Arthur N. Popper
Department of Biology
University of Maryland
College Park, MD 20742
USA

ISBN: 978-0-387-72560-4   e-ISBN: 978-0-387-72561-1

Library of Congress Control Number: 2007926238

Cover illustration: This image is Figure 8.2 from the text.

Printed on acid-free paper

9 8 7 6 5 4 3 2 1

springer.com

Dedication

This volume is dedicated to Professor Joseph Elmer Hawkins, Jr., scholar, educator, and oto-historian, in recognition of his lifelong contributions to the anatomy and pathology of the auditory system. Every chapter of this book reflects his influence as a scientist and mentor.

Contents

Contributors

RICHARD A. ALTSCHULER
Kresge Hearing Research Institute, Department of Otolaryngology, University of Michigan, Ann Arbor, MI 48109-0506, USA, Email: shuler@umich.edu

KAREN B. AVRAHAM
Department of Human Molecular Genetics and Biochemistry, Sackler School of Medicine, Tel Aviv University, Tel Aviv 69978 Israel, Email: karena@post.tau.ac.il

CAROL A. BAUER
Southern Illinois University School of Medicine, Department of Surgery, Springfield, IL 62794-9662, USA, Email: cbauer@siumed.edu

ERIC BIELEFELD
Center for Hearing & Deafness, SUNY at Buffalo, Buffalo, NY 14214, USA, Email: ecb2@buffalo.edu

THOMAS J. BROZOSKI
Southern Illinois University School of Medicine, Department of Surgery, Springfield, IL 62794-9629, USA, Email: tbrozoski@siumed.edu

ROBERT D. FRISINA
Otolaryngology Department, University Rochester School of Medicine and Dentistry, Rochester NY, 14642-8629, USA, Email: Robert_Frisina@urmc.rochester.edu

QUINTON GOPEN
Division of Otology and Laryngology, Harvard Medical School, Children's Hospital Boston, Brigham and Women's Hospital, Boston, MA 02115, USA, Email: quinton.gopen@childrens.harvard.edu

STEVEN GREEN
Department of Biological Sciences, University of Iowa, Iowa City, IA 52242-1324, USA, Email: steven-green@uiowa.edu

JEFFREY P. HARRIS
Division of Otolaryngology-Head and Neck Surgery,University of California
San Diego Medical Center, La Jolla, CA 92037, USA, Email: jpharris@ucsd.edu

STEFAN HELLER
Departments of Otolaryngology/HNS and Molecular & Cellular Physiology,
Stanford University School of Medicine, Stanford, CA 94305, USA, Email:
Hellers@stanford.edu

DONALD HENDERSON
Center for Hearing & Deafness, SUNY at Buffalo, Buffalo, NY 14214, USA,
Email:donaldhe@acsu.buffalo.edu

BOHUA HU
Center for Hearing & Deafness, SUNY at Buffalo, Buffalo, NY 14214, USA,
Email: bhu@buffalo.edu

JOSEF M. MILLER
Kresge Hearing Research Institute, Department of Otolaryngology, University
of Michigan, Ann Arbor, MI 48109-0506, USA, Email: josef@umich.edu

D. KENT MOREST
University of Connecticut Medical Center, Department of Neuroscience,
Farmington, CT 06030-3401, USA, Email: kentmorest@neuron.uchc.edu

KEVIN K. OHLEMILLER
Department of Otolaryngology, Central Institute for the Deaf at Washington
University Medical School, St. Louis, MO 63110-1010, USA, Email:
kohlemiller@wustl.edu

STEVEN J. POTASHNER
University of Connecticut Medical Center, Department of Neuroscience,
Farmington, CT 06030-3401, USA, Email: SJP9713@neuron.uchc.edu

YEHOASH RAPHAEL
Kresge Hearing Research Institute, University of Michigan, Ann Arbor, MI
48109-0506, USA, Email: yoash@umich.edu

LEONARD P. RYBAK
Southern Illinois University School of Medicine, Department of Surgery/
Otolaryngology, Springfield, IL 62794-9653, USA, Email: lrybak@siumed.edu

JOCHEN SCHACHT
Kresge Hearing Research Institute, University of Michigan, Ann Arbor, MI 48109-0506, USA, Email: schacht@umich.edu

ELLA SHALIT
Department of Human Molecular Genetics and Biochemistry, Sackler School of Medicine, Tel Aviv University, Tel Aviv 69978, Israel, Email: shalitel@post.tau.ac.il

ANDRA E. TALASKA
Kresge Hearing Research Institute, University of Michigan, Ann Arbor, MI 48109-0506, USA, Email: atalaska@umich.edu

PHILINE WANGEMANN
Department of Anatomy & Physiology, Kansas State University, Manhattan, KS 66506, USA, Email: wange@vet.ksu.edu

Series Preface

The Springer Handbook of Auditory Research presents a series of comprehensive and synthetic reviews of the fundamental topics in modern auditory research. The volumes are aimed at all individuals with interests in hearing research including advanced graduate students, postdoctoral researchers, and clinical investigators. The volumes are intended to introduce new investigators to important aspects of hearing science and to help established investigators to better understand the fundamental theories and data in fields of hearing that they may not normally follow closely.

Each volume presents a particular topic comprehensively, and each serves as a synthetic overview and guide to the literature. As such, the chapters present neither exhaustive data reviews nor original research that has not yet appeared in peer-reviewed journals. The volumes focus on topics that have developed a solid data and conceptual foundation rather than on those for which a literature is only beginning to develop. New research areas will be covered on a timely basis in the series as they begin to mature.

Each volume in the series consists of a few substantial chapters on a particular topic. In some cases, the topics will be ones of traditional interest for which there is a substantial body of data and theory, such as auditory neuroanatomy (Vol. 1) and neurophysiology (Vol. 2). Other volumes in the series deal with topics that have begun to mature more recently, such as development, plasticity, and computational models of neural processing. In many cases, the series editors are joined by a co-editor having special expertise in the topic of the volume.

RICHARD R. FAY, Chicago, IL
ARTHUR N. POPPER, College Park, MD

Volume Preface

The past decade has brought great advances in our understanding of the mechanisms underlying auditory pathologies. Molecular biology and genetics primarily have contributed to this enhanced understanding, which in turn has driven the design of novel rational therapeutic interventions. This volume presents recent developments in auditory research and their potential translation to the clinical setting. In particular, the authors address the major entities of peripheral auditory trauma; discuss the underlying mechanisms, the central nervous system consequences, and protective interventions; and finally explore the possibilities to restore cochlear morphology and function.

Two themes pervade the chapters in this book: cellular homeostasis and cell death. In the broadest sense, all auditory pathologies are disorders of cellular homeostasis. The book appropriately starts with a consideration of genetic factors that determine the function and the dysfunction of the auditory organ and predispose an individual to acquired hearing loss. Shalit and Avraham (Chapter 2) review the revolution in genetics that has given us profound insight into the genes that are involved in inner ear disorders. Extending the chapter on genetics from a physiological perspective, Wangemann in Chapter 3 treats disorders of the cochlea with the background of our understanding of cellular metabolism and metabolic regulation. Following a comprehensive assessment of the principles of homeostasis, the author discusses the most prominent or well understood homeostatic disorders.

Tinnitus, a major enigma among hearing disorders, is the topic of Chapter 4 by Bauer and Brozoski. The authors review current theories, potential mechanisms, and promising treatments. Autoimmune inner ear disease is now recognized as a genuine inner ear disorder. In Chapter 5, Gopen and Harris discuss the basic immunology of the inner ear, the pathophysiology of the disease including that studied in animal models, as well as clinical diagnosis and treatment of autoimmune hearing loss. The changes in hearing associated with age-related hearing loss reflect alterations in both the cochlea and the central auditory pathways. In Chapter 6, Ohlemiller and Frisina discuss our current understanding of the various forms of presbycusis derived from clinical observations and animal models. Auditory pathologies resulting from noise or drugs have long been established. Noise-induced hearing loss is discussed in Chapter 7 by Henderson, Hu, and Bielefeld, while in Chapter 8, Rybak, Talaska, and Schacht summarize the latest on drug-induced hearing loss.

While the preceding chapters primarily deal with the peripheral auditory system as the site of the initial lesion of auditory trauma, Morest and Potashner detail the pathophysiology of central auditory pathways and its molecular basis in Chapter 9.

The complexity of the cellular regulation of cell death pathways, as well as endogenous protective mechanisms, are considered in Chapter 10 by Green, Altschuler, and Miller. In addition to a detailed consideration of cell death pathways, the authors outline the state-of-the-art attempts to protect the cochlea against environmental insults, with special attention to the spiral ganglion neurons.

Finally, Heller and Raphael introduce the most recent revolutionary developments in hair cell regeneration and stem cell therapy in Chapter 11. They give us a vision of a future when hearing loss and loss of hair cells may be reversed by genetic or pharmacological manipulation.

As often is the case, earlier volumes of the Springer Handbook of Auditory Research Series also provide background or additional information about material covered in this volume. Related topics in Vol. 7 of the series (*Clinical Aspects of Hearing*, edited by Van de Water, Popper, and Fay) include chapters on molecular genetics (Steel and Kimberling), ototoxicity (Garetz and Schacht), and the psychophysical study of tinnitus (Penner and Jastreboff). An additional discussion of homeostasis is found in a chapter by Wangemann and Schacht in Vol. 8 (*The Cochlea*, edited by Dallos, Popper, and Fay). Supplementing the summary of genetics in this volume are the chapters of Vol. 14 (*Genetics and Auditory Disorders*, edited by Keats, Popper, and Fay) that discuss genes and mutations in hearing impairment (Avraham and Hasson), genetic epidemiology of deafness (Nance and Pandya), and genetic counseling (Arnos and Oelrich). Finally, Vol. 26 (*Development of the Inner Ear*, edited by Kelley, Wu, Popper, and Fay) includes several relevant chapters on the molecular biology of ear development, including a chapter by Herzano and Avraham on developmental genes associated with hearing loss in humans. The current volume thus builds on and expands the information presented in earlier volumes of the Handbook of Auditory Research.

JOCHEN SCHACHT, Ann Arbor, MI
ARTHUR N. POPPER, College Park, MD
RICHARD R. FAY, Chicago, IL

1
Auditory Pathology: When Hearing Is Out of Balance

Jochen Schacht

1. Deafness: Past and Present

Hearing, our most sensitive and exquisitely versatile sense, is also uniquely vulnerable to attacks that range from the brute force of mechanical trauma to the adverse effect of otherwise beneficial drugs and to the subtleties of age-related changes. Today, the World Health Organization (WHO 2006) counts 280 million people with a disabling hearing impairment, of whom 70 million suffer a hearing loss already beginning in childhood. Another 364 million people are estimated to have a mild hearing impairment. (Chapter 2 by Shalit and Avraham includes a concise definition of types and measures of severity of "hearing loss" for the reader unfamiliar with audiological terminology.) The global burden of hearing loss is reflected in the fact that deafness ranks third in the world in "years lived with a disability." An even greater number of people may suffer from other disorders of hearing or balance such as tinnitus or Ménière's disease.

The awareness of, and the public concern for, hearing loss is a relatively recent phenomenon in our medical and sociological history. Deafness due to a variety of causes was undoubtedly present in ancient societies (Schacht and Hawkins 2006), but deafness, or for that matter any disease or disability, was mostly the individual's problem and not that of society. Ancient civilizations frequently deprived the deaf of political, economic, or personal rights or, in the extreme example of the Spartan culture, children with, or suspected of having, birth defects were tossed off the cliffs. In modern societies, the Industrial Revolution brought home the fact that hearing loss was the price to pay for gainful employment or service to the country or its sovereign. "Boilermaker's deafness" became a proverbial cause of hearing loss, first described by St. John Roosa (1873) as "Workmen employed in hammering large iron plates, such as are used in making the boilers of steam engines, are very apt to lose much of their hearing power." Two centuries earlier, Ambroise Paré (1510–1590) had portrayed the "great thunderous noise, large bells and artillery" of warfare and noticed that "one often sees gunners losing their hearing because of the great agitation of the air inside the ear." Undoubtedly, boilermakers and gunners were the predecessors to today's victims of noise-induced hearing loss, which ranks as the most common occupational disease.

In recent times, several events contributed to the growing medical and public awareness of disorders of the inner ear. One of the first was the arrival of chemotherapy and, in particular, the discovery and therapeutic application of aminoglycoside antibiotics (Schatz et al. 1944; Hinshaw and Feldman 1945). These miracle drugs and long-sought cure for tuberculosis presented a novel side effect: an unexpected predilection for killing the sensory cells and devastating the ears and the lives of thousands of patients. Another event was the epidemic of German measles that hit the United States in the early 1960s and brought with it deafness in newborns. The resulting public health campaign for vaccinations became a source of popular knowledge about disabilities. Finally, the follies of self-inflicted hearing loss through modern leisure activities ranging from highly amplified rock concerts to personal music delivery through ear phones produced a prominent casualty in rock musician Pete Townshend of The Who. Prompted by his partial deafness and tinnitus, Townshend supported the formation of the advocacy group H.E.A.R., Hearing Education and Awareness for Rockers, in the late 1980s.

2. The Life and Death of Cells

The chapters in this book provide the reader with an overview of the current knowledge of mechanisms underlying the most common auditory traumas and the present concepts of protection and potential treatment. Nevertheless, the reader will note omissions of some clinically important disorders, such as sudden hearing loss, about which we know too little to discuss their cellular and molecular basis.

Two themes run like a red thread through the chapters: cellular homeostasis and cell death. In the broadest sense, all auditory pathologies are disorders of cellular homeostasis. Under the impact of mutated genes or environmental insults, cells are challenged and eventually overwhelmed and unable to maintain their integrity and function. The importance of this maintenance of the cellular environment was recognized more than a century ago and first formulated as the "milieu interne" by Claude Bernard (1878) in his "Leçons sur les Phénomènes de la Vie." Walter Cannon then coined the term "homeostasis" in his treatise on the "Organization for Physiological Homeostasis" (1929). Every cell expends large resources to the defense of its internal balance through intricate barriers, transport systems, and regulatory pathways. In Chapter 3, Wangemann provides an introduction to these mechanisms and a background to most of the other chapters, touching on such topics as energy metabolism, free radical formation, ionic homeostasis, and regulation of blood flow.

Failing to maintain homeostasis, cells will embark on pathways that lead to their demise. Even in dying, the cell attempts to contain the damage to surrounding partners and preferably invokes apoptosis. Apoptosis is an orderly deconstruction of the cell where debris is internalized and not spilled out, an active process that requires energy and coordinated metabolic signaling networks. Necrosis is the fate of the cell that is unable to plan and execute its own funeral, thereby jeopardizing still intact neighboring tissues. Chapter 10 by Green and colleagues summarizes the essentials of cell death pathways and serves well as background reading for other pathologies described in this book.

3. The Many Facets of Hearing Loss

Although the subsequent chapters in some way build upon one another, each chapter is self-contained and will provide the reader with all necessary information. Inevitably, this attempt to make each chapter complete in itself will expose some redundancy should the reader engage in a cover-to-cover exploration of the book. For example, although the detailed description of homeostasis and cell death pathways was reserved for Chapters 3 and 10, brief discussions of these topics can be found in the appropriate context in other sections.

While the majority of this book is concerned with the many facets of acquired hearing loss, it is appropriate to start with a consideration of genetic factors that determine the function and the dysfunction of the auditory organ and may predispose it to environmental insults. At least one-half of all auditory impairment is of genetic origin, and 3 out of 1000 newborns carry a significant hearing loss. Shalit and Avraham (Chapter 2) review the revolution in genetics that, over the past decades, has given us profound insight into the genes that are involved in inner ear disorders.

In part overlapping with and extending the chapter on genetics from a physiological angle is Chapter 3 by Wangemann. The chapter considers disorders of the cochlea on a background of our understanding of cellular metabolism and regulation. Following an assessment of the principles of homeostatic mechanisms, Chapter 3 discusses the most prominent or well understood homeostatic disorders including connexin-related deafness, Pendred syndrome, Jervell Lange Nielsen syndrome, renal tubular acidosis with sensorineural deafness syndrome, Usher syndrome type 2B, Alport's syndrome, and Ménière's disease.

Tinnitus, a major enigma among hearing disorders, is the topic of Chapter 4. Most adults experience transient "ringing in the ears" as a minor discomfort. However, about 15% of the population are chronic sufferers, some of them to the extent of severe impairment of daily activities. Studies in a variety of animal models are now giving us insight into possible etiologies of the different forms of tinnitus, and even the human model is becoming accessible to objective research techniques. Bauer and Brozoski review current theories, potential mechanisms, and promising treatments.

During the last three decades, autoimmune inner ear disease has not only gained recognition as a genuine inner ear disorder but has also been an expanding subject of experimental research. The immune response in the inner ear, which can be beneficial in protecting from invading pathogens, can cause specific and significant pathology. In Chapter 5, Gopen and Harris discuss the basic immunology of the inner ear, the pathophysiology of the disease including animal models, as well as clinical diagnosis and treatment of autoimmune hearing loss.

Age-related hearing loss is a condition that will eventually affect all humans who live long enough. It represents the most prevalent neurodegenerative disease of aging, affecting 40% of the population by age 65 and 90% by age 80. The changes in hearing associated with presbycusis reflect alterations in both the cochlea and the central auditory pathways. In Chapter 6, Ohlemiller and Frisina discuss our current understanding of the various forms of age-related hearing impairment derived from clinical observations and animal models.

Noise-induced hearing loss and drug-induced hearing loss are probably the two forms of auditory trauma that have historically been most amenable to basic research because suitable animal models have long existed. Pathology akin to that seen in the human can be induced by noise exposure or the injection of ototoxic drugs, and it was a century ago that experimental studies of noise trauma in animals were initiated by Wittmaack (1907; see also Hawkins and Schacht 2005). Recent insights into the underlying mechanisms of noise trauma are reviewed by Henderson, Hu, and Bielefeld in Chapter 7.

Rybak, Talaska, and Schacht summarize the latest on drug-induced hearing loss in Chapter 8. Although more than 100 drugs are potentially "ototoxic," two classes in particular demand attention today. Antineoplastic compounds such as cisplatin are indispensable tools against certain cancers, and aminoglycoside antibiotics are commonly used worldwide because of their efficacy paired with low cost. Promising clinical prevention to attenuate the toxic side effects is now emerging from basic studies on the underlying mechanisms by which these drugs kill the auditory sensory cells.

While these chapters primarily deal with the peripheral auditory system as the site of the initial lesion of auditory trauma, we must not underestimate the role that central defects can play in various pathologies, as mentioned in Chapter 4 on tinnitus by Bower and Brozoski and in Chapter 6 on presbycusis by Ohlemiller and Frisina. Morest and Potashner further detail the pathophysiology of central auditory pathways and its molecular basis in Chapter 9.

The complexity of the cellular regulation of cell death pathways and endogenous protective mechanisms are the topics of Chapter 10 by Green, Altschuler, and Miller. This chapter is not only a primer on the multifaceted nature of cell death, but the authors also outline the state-of-the-art attempts to protect the cochlea against environmental insults. In addition, this chapter contributes a detailed deliberation of the spiral ganglion neurons, which are affected by so many of the pathologies introduced in the other chapters.

Finally, in Chapter 11 Heller and Raphael tackle the question of how the deaf ear can be restored. This chapter is based on the most recent revolutionary developments in hair cell regeneration and stem cell therapy in the mammalian auditory system. It gives us the vision for the future when an already existing loss of hair cells and auditory function may be amenable to genetic and pharmacological manipulation.

4. Tomorrow

Despite the impressive progress in elucidating auditory pathologies and establishing the principles of intervention and repair, there still are major challenges ahead. The various chapters of this volume touch on some of the unresolved questions, which may range from the choice of an appropriate experimental model (e.g., in tinnitus, autoimmune hearing loss, homeostatic disorders) to the most effective therapies for prevention or reversal of hearing loss.

One overarching motif is the delineation of the basic principles of sensory cell death. While individual pathways to cell suicides or funerals are already beginning to be elucidated, we must realize that such pathways cannot be viewed in isolation, but constitute a complex and infinitely interrelated network. This network operates in a continuous fashion. The initial death throes of a cell are balanced by attempts at restoring homeostasis until one outweighs the other. Therefore, we cannot expect "only" apoptotic or necrotic signaling in a cell undergoing a challenge to its integrity and, even when a cell is committed to its demise, not a single but a plethora of intimately connected pathways will be activated. Once we know more about the intricacies of such networks, it may also become apparent whether different noxious stimuli act by distinctly different mechanisms or whether they provide trigger points through which eventually the same networks of cellular responses are invoked. As a case in point, drug-induced, noise-induced, and age-related hearing loss all cause oxidative stress in the sensory cells and current research presents both differences and commonalities in the pathways of their death. With further understanding, the networks of cell death may turn out to be indeed distinct or essentially identical, which would have considerable consequences for any attempts at preventing cochlear pathologies.

In the context of cell death and its prevention, two problems will continue to demand our attention. One is the question why, among the different cell types of the cochlea and the vestibular system, the hair cells are so singularly sensitive to environmental insults. A few suggestions have been made, such as a heightened intrinsic susceptibility to oxidative stress (Sha et al. 2001), but a more thorough understanding is needed. Is there a difference in inducible survival pathways? Is there a difference in the readiness for apoptotic pathways? The answer to this question has direct bearing on our ability to design effective treatments for protection which may include pharmacological or genetic manipulations to confer improved resistance to hair cells. The other issue, already beginning to be addressed, is the "window of opportunity" for rescue, the time frame following an insult during which the damage to sensory cells or the spiral ganglion neurons can still be contained. We know that, depending on the severity of the insults, hair cells may be destroyed instantaneously, such as by an aggravated noise exposure, but this is only the initiation of a slow cascade of death that expands from the point of original insult to adjacent sites. Likewise drug-exposed and, in particular, aging hair cells die over an extended period of time, and it remains to be seen how attempts at rescue can capitalize on such delayed time courses of cell death.

Even though we lack detailed knowledge of some of the basic pathological mechanisms, the next decade will likely see effective pharmacological protection against a number of the disorders that are discussed in this volume. Animal experimentation has clearly shown the principles of intervention for noise-induced hearing loss, drug-induced hearing loss, and to a lesser extent for other pathologies. Although successful animal experimentation does not guarantee a translation into the clinic, initial trials on the prevention of gentamicin-induced hearing loss are highly encouraging by showing that a mechanism-based intervention can indeed have clinical utility. Formulation of protective medications should thus be possible based on already existing results. Additional drugs will, of course, be discovered and powerful models for high-throughput screening would be helpful in this context. Such high-throughput screening requires appropriate models for different disorders, and at the moment seems plausible only for drug-induced hearing loss. In vivo experimentation, although tedious, time consuming, and costly, will remain indispensable in studying most pathologies and will remain the ultimate test before any treatment can be translated into the clinic.

The future of clinical treatment will also go beyond pharmacological intervention. For the spiral ganglion, a protection by neurotrophic factors can be coupled with physiological stimuli such as sustained depolarization to activate survival mechanisms. Such a treatment may be integrated into cochlear implants that then serve both the delivery of sound sensation and survival or regenerative factors. Even deafness-induced plastic changes in central synaptic activities and signaling pathways may similarly come under pharmacological control.

Restoration of cochlear hair cells is currently considered the final frontier, and new tantalizing results may be boosted by developments in the areas of stem cell therapy and gene therapy. Thus, genetic hearing loss may become amenable to intervention based on success in correcting hearing function in mouse models (Probst et al. 1998). The problems of restoring hearing after the loss of hair cells, however, surpass those posed for intervention strategies. Even if such issues as cell-specific transfection and guided differentiation are being resolved, the auditory system poses exquisite challenges that go beyond simple restoration of cells. Hair cells must not only be restored and functional, they also must be arranged in the appropriate tonotopical organization with neural connections. Yet, we know very little at the moment about the molecular control of the cytoarchitecture of the mammalian cochlea.

While the challenges to resolve mechanisms of auditory trauma, design protection and accomplish repair are formidable, they are not—given time and resources—insurmountable. The unprecedented progress of the last decades that is documented in this volume is encouraging for the future of our field.

References

Bernard C (1878) Leçons sur les phénomènes de la vie. Paris: Ballière.
Cannon W (1929) Organization for physiological homeostasis. Physiol Rev 9:399–431.
Hawkins JE, Schacht J (2005) Sketches of otohistory. Part 10: Noise-induced hearing loss. Audiol Neurotol 10:305–309.

Hinshaw HC, Feldman WH (1945) Streptomycin in treatment of clinical tuberculosis: a preliminary report. Proc Mayo Clinic 20:313–318.

Probst FJ, Fridell RA, Raphael Y, Saunders TL, Wang A, Liang Y, Morell RJ, Touchman JW, Lyons RH, Noben-Trauth K, Friedman TB, Camper SA (1998) A novel unconventional myosin in a BAC transgene corrects deafness in *shaker-2* mice. Science 280:1444–1447.

Schacht J, Hawkins JE (2006) Sketches of otohistory. Part 11: Ototoxicity: drug-induced hearing loss. Audiol Neurotol 11:1–6.

Schatz A, Bugie E, Waksman SA (1944) Streptomycin, a substance exhibiting antibiotic activity against gram-positive and gram-negative bacteria. Proc Soc Exp Biol Med 55:66–69.

Sha S-H, Taylor R, Forge A, Schacht, J. (2001) Differential vulnerability of basal and apical hair cells is based on intrinsic susceptibility to free radicals. Hear Res 155:1–8.

St. John Roosa DB (1873) Treatise on the Diseases of the Inner Ear. New York: William Wood and Company.

Wittmaack K (1907) Über Schädigung des Gehörs durch Schalleinwirkung. Z Ohrenheilk 54:37–80.

World Health Organization (2006) Deafness and hearing impairment. Fact sheet no. 300. http://www.who.int/mediacentre/factsheets/fs300/

2
Genetics of Hearing Loss

ELLA SHALIT AND KAREN B. AVRAHAM

1. Introduction

The revolution in genetics in the past decades has enabled identification of many of the genes associated with human hereditary diseases, and hearing loss is no exception. These discoveries have a profound impact on knowledge about inner ear function and the pathology caused by mutations in these genes, which becomes clinically and socially relevant because a significant proportion of hearing loss is caused by hereditary mutations in genes. The identification of these genes and the proteins they encode allow for molecular diagnostics and genetic counseling for patients, and will help devise new ways to diagnose, treat, and prevent disorders of the auditory system.

1.1 Hearing Loss

Sound can be described in terms of frequency (pitch), measured in hertz (Hz), and in terms of intensity (loudness), measured in decibels (dB). Hearing is considered within the normal range if a person can process sound frequencies between 20 and 20,000 Hz and decibel levels from 0 to 140 dB. Hearing loss or hearing impairment occurs when all or a part of the normal range of hearing is lost (Newby 1992).

In brief, the mechanism of hearing is the transduction of sound, turning it into neural impulses and interpreting these impulses by the central nervous system. A defect at one or more of the levels in this system can lead to hearing loss. Occupying a wide spectrum of decreased hearing, "hearing loss" is defined as the reduced or absent ability to perceive or process auditory information. The World Health Organization (WHO) defines the term "deaf" only when referring to severe cases in which alleviation by hearing aids and cochlear implants are not optimal. For educational placement and integration purposes, children with hearing loss over 90 dB are considered deaf (National Dissemination Center for Children with Disabilities, Table 2.2).

Hearing loss is partitioned into several categories according to different factors. Hearing loss can affect either one or both ears, designated as unilateral hearing loss or bilateral hearing loss, respectively. Hearing loss can also be sorted into stable hearing loss (permanent year after year), progressive, fluctuating (worsening and improving alternately, such in cases of inflammation of the tympanic membrane or fluid in the middle ear), or even transient (due to wax buildup in the ear canal, head injuries, ear infections, and reactions to medications, for example, aspirin). Based on the onset of hearing loss, it is possible to distinguish between congenital (present at birth), early-onset (commence at childhood), and late-onset loss (begins at adulthood). An additional discrimination relates to the development of language acquisition, as either prelingual or postlingual (before and after acquisition of language and speech, respectively).

Hearing loss severity is divided into five groups organized according to gradual deterioration of hearing: mild (thresholds of 21–40 dB), moderate (41–60 dB), moderately severe (61–80 dB), severe (81–100 dB), and profound (>100 dB). The range of hearing loss is generally grouped into three categories: low-frequency hearing loss (<500 Hz), middle-frequency hearing loss (501–2000 Hz), and high-frequency hearing loss (>2000 Hz).

Yet another classification subdivides hearing loss into four classic groups according to underlying pathologies. The first, sensorineural hearing loss, is due to malfunction of the inner ear or along the neuronal pathway between the inner ear and the brain. Such damage to the delicately correlated system of the transmission of sound waves from the hair cells to the supporting nervous tissue often causes hearing loss. In general, it is a permanent disturbance that cannot be cured by medical or surgical intervention.

The second group, conductive hearing loss, represents hearing obstructions present in the conduction canal leading to the inner ear, consisting of the external and middle ear. Common factors in this kind of hearing loss are wax in the ear canal, a perforation in the eardrum, infections, fluid in the middle ear, and fixation of the ossicles as occurs in otosclerosis, and erosion of the ossicles as in cholesteatomas, epithelial tumors of the middle ear that can be congenital or result from chronic infection. Conductive losses generally affect all frequencies and in many cases are surgically treatable. The third group, mixed background hearing loss, results from combined sensorineural and conductive factors. Finally, the fourth group, central auditory dysfunction, results from damage at the level of the eighth cranial nerve, auditory brain stem, or cerebral cortex.

1.2 Genetic Hearing Loss and Its Prevalence

A significant difference in the cause of hearing impairment is whether its origin is genetic or nongenetic. Genetic hearing losses are due to single or multiple lesions throughout the genome that may be expressed at birth or sometime later in life. Nongenetic "acquired" hearing loss, on the other hand, is a consequence of environmental factors that result in hearing impairment, with no regard to inheritance. Such factors might include infections such as meningitis and otitis media, traumatic injuries such as perforation of the eardrum, skull fractures and acoustic trauma (see Henderson et al., Chapter 7), and use of toxic drugs such as aminoglycoside antibiotics or cisplatin (see Rybak et al., Chapter 8). However, even when speaking of environmental causes, genetic factors may be involved as modifying genes (Table 2.1) that may have an impact on onset, severity, and progressiveness of nongenetic hearing loss.

TABLE 2.1. Genetic definitions.

Genetic term	Definition
Recessive inheritance	Normal individual holds within its DNA two copies of each gene. A recessive characteristic or trait is apparent only when two copies of the gene encoding it are present. If a mutation in a gene is inherited in a recessive pattern, the mutated trait or characteristic will be phenotypically expressed only in a homozygous mode.
Dominant inheritance	A dominant pattern of inheritance of characteristic or trait in which one copy of an allele is sufficient to phenotypically confer this feature. In the case of a dominant mutation in a gene, it will be visually expressed in either a heterozygous or a homozygous condition. As a result, a dominant disorder is apparent when inherited from only one of the parents and will appear in each generation.
X-linked inheritance	Out of the 23 pairs of chromosome that a normal individual carries in his DNA, one pair of chromosomes is discriminated as the sex chromosomes (X and Y chromosomes) that contain genes that determine gender development and function, as well as other genes that are non related to sex configuration. X-linked is descriptive of an allele located on the X-chromosome.
Mitochondrial inheritance	Mitochondria are microscopic rod-like subcellular structures that are the principal energy source of the cell, by metabolizing nutrient molecules into available energy. A mutation in mitochondrial gene is maternally passed on to all of her children with no regards to their gender, while only her daughters will pass it to the next generations.
Expressed sequence tags (ESTs)	Expressed sequence tags are short DNA sequences from the expressed regions of a gene only (called 'exons'). These pieces of DNA (several hundred base pairs of length) can serve for rapid identification ('tagging') of the full length sequencing of the expressed genes (called cDNAs) and in developing DNA markers.
Penetrance	The likelihood or probability that a characteristic or a trait will be expressed as a phenotype as a result of a specific given phenotype.
Transcriptional regulators	Specific proteins that are required for the initiation of transcriptional process, thereby regulating it.
Heterozygous	A genotype that possesses two distinct copies of an allele of a certain characteristic or a trait.
Homozygous	A genotype that possesses two identical copies of an allele of a certain characteristic or a trait.

TABLE 2.1. (continued)

Genetic term	Definition
Bacterial artificial clones	In order to artificially transport fragments of DNA into cells, these fragments need to be introduced first into a molecule which is able to carry them into the host cell in vitro and facilitates their multiplication within the host. Bacterial artificial chromosomes are DNA vectors derived from genetically engineered bacteria *E. coli* chromosome, used to incorporate large fragments of DNA (100 to 300 kb).
Transgenes	DNA fragment or a gene which is artificially inserted into the germ line of another organism, which is then called a transgenic organism. These fragments of DNA can integrate into the host genome and alter its genotype.
In situ hybridization	An assay testing the hybridization ability of DNA/RNA probes which are applied on an intact tissue, in order to detect presence of the complementary DNA sequence
Hypomorphic allele	An allele that disrupts and diminishes the function of a gene, but does not absolutely abolish its activity.
Modifying genes	A gene that can influence the expression of another gene. Thus, modifying genes can alter the phenotype of a feature associated with a particular gene, and can result in multiple phenotypes for the same genotype.
Allele	One of the two or more alternative forms of a certain gene. A normal individual caries two alleles for each of his genes (one from each parent), which can either be two identical alleles (homozygous) or two distinct alleles (heterozygous).
Organogenesis	The process of organ formation during embryonic stages.
Monogenic	A monogenic disorder is a condition caused and controlled by a single gene.
Homologue	A gene or a locus from one species that shares high similarity in sequence to a gene or a locus of another organism. Homology between genes suggests a common origin and function of the gene in question.
Gene	A sequence of DNA that is the basic unit of inheritance. Most of the genes encode for proteins while their minority are non coding genes.
Phenotype	The observable or measurable expression of a gene coding for a certain characteristic. The phenotype can be the consequence of numerous factors including genotype, environmental factors, age, presence of modifier genes or interaction among several genes.
Genotype	The genetic identity of a certain gene of locus in the DNA. The genotype is not necessarily expressed visually.
Autosomal	Any of the chromosomes that are not the sex chromosomes (X and Y chromosomes).
Logarithmic odds (LOD)	A statistical estimate of whether two loci are likely to lie near each other on a chromosome and are therefore likely to be inherited together. A LOD score of three or more is generally taken to indicate that the two loci are close.

At least 50% of all hearing impairment are due to genetic factors (Skvorak-Giersch and Morton 1999; Nadol and Merchant 2001). About 3 out of 1000 newborns have a significant hearing impairment, and one half of the population older than 70 years of age develops some degree of hearing impairment. These numbers are turning hearing loss into one of the most common birth defects and into the most inherited sensory defect among adults worldwide (Hone and Smith 2003).

When relating to symptom manifestations, hearing loss is classically subdivided to syndromic hearing loss (SHL) and nonsyndromic hearing loss (NSHL). Hearing impairment accompanied by additional clinical symptoms is referred as SHL. Hearing dysfunction as the only observed phenotype is defined as NSHL. Worldwide, the incidence of nonsyndromic hearing loss (NSHL) is estimated to be responsible for 70% of genetic hearing loss. While the syndromic component (30% of all hearing loss) is almost purely of genetic etiology, NSHL can include nongenetic factors as well. Hereditary NSHL is most often autosomal recessive (75% to 80%; Table 2.1), with 18% to 20% attributed to autosomal dominant inheritance (Table 2.1) and the remaining percentage related to X-linked or mitochondrial inheritance (Table 2.1; Hertzano and Avraham 2005). Although the exact proportions can differ over time and place, current estimates ascribe greater than 50% of all hearing loss cases as originating from a mutation in a sole gene (a monogenic trait). As might be expected of a highly sophisticated and tightly coordinated mechanism such as the hearing apparatus, a large portion of the 30,000 genes in the human genome are dedicated to participate in this task, many of which were identified in the Human Genome Project (Collins and McKusick 2001). Sixty-five deafness-related genes have been already cloned (Hereditary Hearing Loss Homepage, Table 2.2), meaning that they were identified as hearing loss causative factors and were characterized. Moreover, 120 loci have been mapped for hearing loss so far, meaning that the region within the genome that contains the defective gene has been linked to hearing loss, though the specific gene has not yet been identified.

The first deafness locus to be mapped was *DFN3* (Brunner et al. 1988). The first autosomal dominant gene to be mapped was *DFNA1*. This form of NSHL was discovered in a Costa Rican kindred (Leon et al. 1992); the gene responsible was subsequently identified in 1997 as diaphanous (Lynch et al. 1997). No additional diaphanous mutations have been reported. The first autosomal recessive gene to be mapped was *DFNB1* (Guilford et al. 1994). Ironically, this locus has turned out to be responsible for the most prevalent form of NSHL, those associated with connexin 26*GJB2* mutations (Denoyelle et al. 1997; Kelsell et al. 1997; Zelante et al. 1997). Overall, the plethora of genes associated with NSHL has turned out to be fascinating, being involved in every aspect of auditory function. To add to the complexity, there are a number of genes that are associated with both NSHL and SHL, as well as both recessive and dominant hearing loss.

The comprehensive and diverse selection of cloned genes includes genes encoding a variety of different proteins with known functions. A few examples include genes that encode either components constituting the synaptic apparatus

TABLE 2.2. Web sites used for the genetics of hearing loss.

Web site	URL
National Dissemination Center for Children with Disabilities	http://www.nichcy.org/pubs/factshe/fs3txt.htm
Hereditary Hearing Loss Homepage	http://webhost.ua.ac.be/hhh/
Human Genome Project	http://www.ornl.gov/sci/techresources/Human_Genome/home.shtml
UCSC Genome Browser	http://genome.ucsc.edu/
Genetic Linkage Analysis	http://linkage.rockefeller.edu/
The Connexin-Deafness Homepage	http://davinci.crg.es/deafness
The Jackson Laboratory Hereditary Hearing Impairment in Mice: Mouse Models of Human Hearing disorders	http://www.jax.org/hmr/models.html
National Institutes of Health Office of Human Subjects Research	http://ohsr.od.nih.gov/
The German Mouse ENU Project	http://www.gsf.de/isg/groups/enu-mouse.html
Harwell Mutagenesis Program	http://www.mut.har.mrc.ac.uk/
Epitope Prediction	http://www.sbc.su.se/~pierre/svmhc/new.cgi
Pubmed	www.pubmed.gov
Gene Expression Omnibus	http://www.ncbi.nlm.nih.gov/geo/
Unigene	http://www.ncbi.nlm.nih.gov/UniGene
SOURCE	http://source.stanford.edu

of the hair cells of the inner ear and the nerves (PMCA2 and otoferlin), cytoskeleton proteins (myosin VI, myosin VIIA, and myosin XVA), proteins implicated in structural integrity of the organ of Corti (α-tectorin, COL11A2, and COCH), and proteins significant for potassium recycling in the organ of Corti (connexin 26, connexin 31, KCNQ4, Pendrin, and Claudin 14) (Steel and Kros 2001; Tekin et al. 2001). Up to 50% of genetic hearing impairment is estimated to derive from mutations in the connexin 26 (*GJB2*) gene (Marazita et al. 1993; Steel and Kros 2001) that encodes the protein connexin responsible for the proper function of gap junctions between cells.

Chapter 3 by Wangemann complements the genetic information in this chapter by describing the consequences of several of the mutations in the context of cochlear homeostatic mechanisms.

2. Syndromic Hearing Impairment

Hearing impairment is denoted as an integral clinical phenotype in more than 400 genetic syndromes (Gorlin 1995; Steel and Kros 2001; Nance 2003). The presence of clinical features accompanying hearing impairment can vary on a wide scale, while hearing abnormalities are often mild, unstable, or a late-onset trait in these syndromes (Friedman et al. 2003; Nance 2003). Syndromic forms of hearing loss are estimated to be responsible for up to 30% of prelingual deafness, although in general, it endows only a small portion of the broad spectrum of hearing loss. The prominent portion of these disorders are monogenic (Friedman et al. 2003), meaning that their hereditary component is derived from one mutated gene throughout the genome. A glance at some of the major SHL types that have been studied so far is given in the paragraphs that follow.

2.1 Waardenburg Syndrome

In 1947 P.J. Waardenburg, a Dutch eye physician, was the first to notice a link between retinal pigmentary differences, found in one of his patients, and congenital hearing loss. Four years later, after tracing other patients with similar symptoms, he described a new syndrome (Waardenburg 1951), involving congenital hearing loss; lateral displacement of the inner canthi, the inside corner of the eye (dystopia canthorum); and retinal pigmentary differences, referred to today as Waardenburg syndrome (WS) type 1 (WS1).

WS is mostly a genetic autosomal dominant disorder, evident at birth. It is considered to be the most frequent autosomal dominant form of syndromic hearing loss, constituting approximately 2% of all congenital hearing loss (Apaydin et al. 2004). The primary phenotypes observed in this syndrome may include irregular skin and eye pigmentations, seen as white lock patches of hair above the forehead, premature gray hair, light pigmented zones in the skin (leukoderma), two different colored segmented eyes (heterochromia irises), or as extraordinary brilliant blue eyes. In addition, one of the visible phenotypes present in WS patients is an unconventionally wide distance between the inner corners of the eyes and sometimes also confluent connected eyebrows and a high nasal root. The wide scale of symptoms and severities can vary greatly from one person to another and hearing loss in WS can fluctuate from moderate to profound among different WS patients.

Four different types of WS have been described, grouped by distinct physical characteristics. All four types share common sensorineural hearing loss and pigmentary abnormalities expressed at variable degrees. The majority of WS1 individuals display dystopia canthorum, a feature in which the inner corners of the eye are spaced farther apart than normal. WS type II (WS2) is not associated with this phenotype and is caused by mutations in a different gene. Type III and type IV (WS3 and WS4, respectively) are much rarer subforms of WS. WS3 is ascribed to malformations of the upper limbs. WS4 is associated with Hirschsprung disease, a digestive disorder typified by diminished motility in segments of the bowel caused by lack of nerve cells, and is therefore known as Waardenburg-Hirschsprung disease as well.

WS hearing and pigmentary deformities can both be caused due to a failure of proper melanocyte differentiation during embryonic development. Deficiency or lack of melanocytes can affect pigmentation in the skin, hair, and eyes and hearing as well. Melanocytes exist as intermediate cells of the stria vascularis in the organ of Corti, where they play a vital role in creating the endocochlear potential, positive voltage of 80–100 mV seen in the endolymphatic space of the cochlea, which is necessary for normal hair cell function.

All of WS1 and some of WS3 cases are associated with mutations in the *PAX3* gene. WS2 is caused by mutations in the transcription factors *MITF* and *SNAI2*. WS4 is due to alterations in either *EDNRB*, *EDN3*, or *SOX10* genes. *PAX3* is a member of the *PAX* paired-domain proteins family that function during embryonic development at the level of transcription. *PAX3* was found to strongly activate MITF protein expression, in synergy with *SOX10*, in vitro (Bondurand et al. 2000). Watanabe et al. (Watanabe et al. 1998) showed that *PAX3* regulates MITF expression through the MITF promoter. MITF was shown in 1996 (Tachibana et al. 1996) to act as a transactivator of the tyrosinase gene, a crucial key enzyme in melanocyte differentiation. MITF transactivates the *SNAI2* promoter (Sanchez-Martin et al. 2002). These interactions demonstrate that there is a common pathway that links several forms of this disease.

2.2 *Branchiootorenal Syndrome*

After WS, branchiootorenal syndrome (BOR) is the second most prevalent autosomal dominant syndromic type of hearing loss, responsible for approximately 2% of all profoundly deaf children (Steel and Kros 2001). The name branchiootorenal refers to three terms concerning its common symptoms: disturbances in the neck (*branchio*), ear disorders (*oto*), and irregular kidney formations (*renal*). The major clinical features of BOR syndrome are branchial cysts or fistulae (abnormal connections between organs/vessels); renal malformations ranging from asymptomatic hypoplasia (underdevelopment of a tissue) to entirely absence (agenesis; Melnick et al. 1976; Fraser et al. 1978); and deformities in formation of the external, middle and inner ear leading to either conductive, sensorineural, or mixed background hearing loss. Manifestations of BOR symptoms can differ in their presence and severity between individuals, and even within the same person, between the two body sides. A very mildly affected BOR parent can have a severely affected child and vice versa, a situation referred to as "variable expressivity." Although extremely variable, BOR symptoms are of high penetrance (Table 2.1).

Mutations in the *EYA1* gene were found in about 40% of families segregating BOR features. Mutations within the *Drosophila* fly gene *eyes absent* (*eya*), a known homologue of the human *EYA1* gene (Bonini et al. 1998), result in the complete absence of the fly eyes or a reduced eye phenotype. The *EYA1* particular mechanism of action has yet to be uncovered, although it is clear that this is a transcriptional regulator implicated in the development of numerous tissues and organs including the eye, ear, and the branchial arches (tissues that are involved in face and neck creation during early embryonic stages) formation.

A minority of BOR families were characterized as carrying a mutation in the *SIX1* gene (Ruf et al. 2004). EYA1 and SIX1 proteins are known to cooperate together with the *PAX* genes as transcriptional regulators (Table 2.1) in the organogenesis (Table 2.1) patterning (Ruf et al. 2004). Besides the two genes mentioned above, mutations in other genes, which have yet to be discovered, are likely to occur as well.

2.3 Usher Syndrome

Usher syndrome (USH) is the most frequent autosomal recessive syndromic form of hearing loss. Dual sensory defects, involving hearing and sight, are obvious in all USH-affected individuals. Sensorineural hearing loss is congenital, whereas vision is usually adversely affected after the first decade of life and worsens over time, as part of a retinitis pigmentosa disorder (an eye impairment leading to blindness). More than 50% of the deaf–blind community in the United States is affected by USH, making it the most prevalent condition of genetic hearing and vision deformities. In addition to auditory–vision symptoms, some patients suffer from balance problems, originating from internal ear defects in the vestibular apparatus, the center of equilibrium and balance. Cataract (cloudiness forming on the lens inside the eye that may cause blurred vision) can be an additional feature on top of those mentioned in each of the different types of USH.

The USH manifestations are variable and therefore are split into three distinct types of USH (USH1, USH2, and USH3), based on the degree of hearing and vestibular features. USH1 and USH2 subtypes are widespread, while USH3 account for 2% to 5% of USH cases. Each subtype of USH syndrome is heterogeneous, and is further subdivided according to their genetic cause.

USH1 individuals are born with severe to profound sensorineural hearing loss and suffer from improper vestibular function. Their balance and communication dysfunction comes to fruition in early infancy, as their motor capabilities (such as sitting and walking on their own) fall much behind normal. Prepubertal retinitis pigmentosa in USH1 commences with diminished sight at night and is rapidly followed by progressive reduction in the visual field that deteriorates until absolute blindness.

The hearing impairment characterizing USH2 is usually mild to severe, with progressive visual problems and no visible vestibular deficits. In contrast to USH1, USH2 patients can benefit from hearing aid amplification of sounds and their verbal communication is usually normal. Decreased eyesight first begins with blind spots appearing in the early teenage years and usually degenerates until complete loss of sight.

USH3 patients experience progressive hearing, vestibular, and vision deterioration. At birth, their hearing appears normal and their balance abilities are normal or near normal. Noticeable hearing and vision hypoactivity signs develop most often by the second decade of life, though they can appear at varying ages even between siblings of the same pedigree (Karjalainen et al. 1983; Pakarinen et al. 1995). Retinitis pigmentosa symptoms progress from night blindness that usually emerges along with puberty through loss of central visual acuity that begins after the age of 20, eventually resulting in blindness by mid-adulthood.

Twelve genetic loci comprising USH genes have been mapped, eight of which have been identified. Much of the genetic identification of USH genes relied on the study of mouse models. Despite the fact that the precise involvement in hearing and vision has yet to be elaborated, it is evident that mutations in USH genes result in anomalous inner ear hair cell function and in the retinal sensory cells. Interestingly, a considerable portion of USH loci overlap with genetic loci identified in NSHL.

The *USH1B* gene isolation was achieved by the integration of a genetic study of human USH pedigrees and of the genetic mapping of the shaker 1 (*sh1*) mouse model. The *USH1B* locus was first defined as a result of research of numerous families from around the world. The identification of the myosin VIIa (*Myo7a*) gene as the causative mutation in the deaf and vestibular homozygous (Table 2.1) *sh1* mice rapidly led to the recognition that its corresponding human gene was the mutated gene in USH1B. The human orthologue, *MYO7A*, encodes for the myosin VIIa protein, which is a member of the myosin superfamily. Myosins are typified by the ability to move along actin filaments via hydrolysis of ATP molecules. Evidence gleaned from several myosin VIIa studies show that this protein is expressed in hair cell cytoplasm and stereocilia within the cochlea and that myosin VIIa may have a vital role in regulating the development and function of the hair bundle. When defective, this protein causes disturbances in the stereocilia distribution on top of the hair cells, which adversely impacts hearing perception. More than 80 *MYO7A* distinct mutations have been reported, some of which were found as the underlying genes in autosomal recessive NSHL DFNB2 and in autosomal dominant NSHL DFNA11.

The underlying gene of another form of USHI, named *USH1C*, codes for harmonin, a protein containing a PDZ domain, which specializes in protein–protein interactions (Verpy et al. 2000). Harmonin is probably an essential player in the gathering of the USH1 complex of proteins.

In 2000, mutations in the gene *CDH23/USH1D* encoding the cadherin-like protein were linked to the USH1D form of USH and were found a year later as the mutation responsible for the deaf mouse *waltzer* phenotype that exhibits disorganized stereocilia bundles. Cadherin proteins promote intercellular adhesion activities, from which they receive their name. Evidence of interaction between cadherin-like protein and harmonin connects these proteins to the evolving stereocilia bundles (Adato et al. 2005).

In addition to *CDH23*, an additional cadherin-like protein named protocadherin was revealed to be mutated in the USH1F subform (Bolz et al. 2001). The protocadherin gene, *PCDH15*, was also reported to affect hearing in the Ames *waltzer* deaf mouse model, which again displays hair bundle disorganization similar to those seen in the *waltzer* and *shaker1* mice. Autosomal recessive NSHL in the *DFNB23* locus resides in the *USH1F* interval. The transmembrane protocadherin 15 protein is localized to stereocilia as well as to the photoreceptor cells of the retina. It is thought to act as a significant mediator in the adhesion process related to stereocilia morphogenesis by means of protein–protein interactions.

The *USH1G* underlying mutation was detected in the *SANS* gene (Mustapha et al. 2002). The corresponding homologous gene in mice was found to be mutated in a deaf mutant mouse model named Jackson shaker (Kikkawa etal. 2003). The *Sans* gene codes fora scaffold-like protein that was shown to interact with harmonin. Similar to the other USH mice models, the hair bundles demonstrated abnormal arrangement of the stereocilia.

Two isoforms of a novel protein were isolated for USH2A, namely USH2 isoform a (previously called usherin) and USH2A isoform b (Reiners et al. 2006). The latter is a much larger isoform, which weighs more than three times that of isoform a. Isoform a carries several interesting domains: a transmembrane region enabling its anchoring to membranes, two laminin (glycoproteins found in the basal lamina) domains that may point to involvement in the construction of the lamina (the basement membrane, a glycoprotein sheet secreted by cells to form the extracellular matrix) and a PDZ-binding domain allowing protein–protein interconnections. Moreover, it shares a great similarity with extracellular matrix and cell–cell adhesion proteins. Both isoforms of USH2A localize to the basement membrane of the cochlea and retina; specifically USH2A is expressed in stereocilia and within the hair cell–nervous synaptic junctions and in the synaptic terminals of photoreceptor cells of the retina (Bhattacharya et al. 2002). These findings are providing evidence for the hypothesis that USH2A functions in the adhesion of pre- and postsynaptic membranes, which may fit into the USH network of proteins.

2.4 Pendred Syndrome

Recent estimations predict that 4% to 10% of prelingual genetic hearing losses are ascribed to a syndromic form of hearing loss named Pendred syndrome (Park et al. 2003). Reliant on these evaluations, Pendred syndrome is the far most common syndromic configuration of deafness. The two hallmarks characterizing this autosomal recessive syndrome are neurosensory deafness and enlargement of the thyroid gland, a feature defined as thyroid goiter.

The auditory failure in Pendred syndrome is usually in the form of severe to profound deafness, while an uncommon form of Pendred hearing loss may appear as mild to moderate. Hearing loss in Pendred syndrome is most frequently sensorineural, but in rare cases can manifest itself as mixed. In 40% of the cases, hearing loss is concomitant with vestibular hypofunction that can vary among patients. Auditory deformities are mainly associated with Mondini dysplasia, a situation of incomplete cochlear formation, whereby instead of 2.5 turns of a cochlear spiral, a diminished number of turns are present (usually 1.5 turns). Likewise, vestibular dysmorphologies in Pendred syndrome are often a direct consequence of a dilation of the vestibular aqueducts (very narrow channel connecting the inner ear "vestibule" and the skull) and their internal substances.

Malfunctions or blocks in thyroid hormone production, which are prominent characteristics of Pendred syndrome, usually lead to goiter of the thyroid. Extended size of the thyroid gland is not congenital and onset can range from early puberty to adulthood. Swelling in the front area of the neck can emanate from the goiter defects. Sometimes the thyroid gland goiter cannot be clinically diagnosed and thus remains inapparent.

Originally, Pendred syndrome was clinically documented in 1896 by the British physician Vaughan Pendred. A century of scientific research led to the isolation of the Pendred gene in 1997, named *PDS* or *SLC26A4*, encoding the pendrin protein (Everett et al. 1997). Pendrin was defined as an anion transporter protein operating in the inner ear, thyroid, and kidney. It is considered to be the causative mutation in most Pendred syndrome cases, as well as the underlying gene in DFNB4 nonsyndromic deafness.

The phenotype–genotype correlation between the goiter and *PDS* function has been elucidated. Normal pendrin permits passage of iodide in the thyroid gland, which is then used for the assembly of the hormone thyroxine. When pendrin function is disrupted, a block or insufficient production can lead to the thyroid symptoms associated with Pendred syndrome. Little is known regarding the exact association of pendrin function and deafness. Based on the *Pds* null mouse model, it was postulated that pendrin upholds a function in the maintenance of the endocochlear potential in the intermediate cells of the stria vascularis (Royaux et al. 2003).

2.5 Alport Syndrome

Alport syndrome is a paradigm of SHL that can be inherited in either an X-linked mode (85% of the cases; Table 2.1), in an autosomal recessive mode (about 14% of the cases), or in autosomal dominant mode (less than 1%) (Endreffy et al. 2005). A triad of medical symptoms depicts Alport syndrome: sensorineural hearing loss, renal disorders, and visual problems. Hearing impairment usually begins in the second decade of life, while severity is progressive and can differ between patients. Kidney abnormalities are also progressive and ranges from glomerulonephritis (inflammation of the glomerulus in the kidney), through hematuria (presence of blood in the urine), to end-stage renal failure. Common ophthalmologic symptoms characterizing Alport's syndrome are lenticular abnormalities such as congenital cataract, anterior lenticonus (conical or spherical lens protrusion into the anterior chamber), and fleck retinopathy (thickening of the basement membrane).

The genetics in Alport syndrome is dictated by three genes: *COL4A3*, *COL4A4*, and *COL4A5*. As discussed, Alport syndrome can be the result of recessive, dominant or X-linked inheritance. Mutations in *COL4A3* and *COL4A4* accounting for recessively inherited Alport syndrome, mutations in *COL4A5* account for an X-linked mode of inheritance, and roughly 1% of Alport syndrome families have dominant mutations in *COL4A3* and *COL4A4* .

The *COL4* gene family encodes collagen IV, the primary structural protein constituent of the basal lamina (basement membrane). Type IV collagen protein is composed of three α chains encoded by three separate genes, which assemble to form a triple-helical molecule. The *COL4A3*, *COL4A4*, and *COL4A5* genes give rise to the α3, α4, or α5 chains of collagen IV, respectively. Collagen IV trimers that aggregate from the three α chains (α3–5) are constrained in their expression to the kidney, inner ear, and eye basal lamina, whereas α1 and α2 are universally expressed in body tissues (Zehnder et al. 2005).

In the kidney, and most probably in the cochlea as well, a change in the expression of collagen IV α subunits occurs during embryonic development from α1 and α2 Type IV collagen in early development to α3–6 during maturity, called an "isotype switching" process (Zehnder et al. 2005). Alport syndrome renal pathologies are the result of glomerular basement membrane dysmorphologies in the kidney, bringing on a defective glomerular function that gradually induces hematuria or proteinuria (presence of an excess of protein in the urine), up to renal failure. The absence of α3–5 renal type IV collagen chains in most of Alport syndrome males implies that isotype switching does not take place in the patients' kidneys.

Recently, similar findings regarding the inner ear have been published (Zehnder et al. 2005). The first substantial finding was that α3 and α5 chains expression in the cochlea is selectively limited to the basilar membranes of the spiral ligament (the attachment of the basilar membrane to the outer bony wall in the cochlea) and spiral limbus (the thickened connective tissue membrane of the osseous spiral lamina at the attachment of Reisner's membrane), while α1 exists ubiquitously in the basilar membrane of the inner ear. The second finding was that patients with X-linked Alport syndrome α3 and α5 chains were completely missing while α1 sustained its regular expression. This suggested that the isotype switching process also does not appear to occur in the cochlea in Alport syndrome, further supporting the longstanding theory that Alport syndrome hearing loss is associated with functional disruption in cochlear micromechanics or the spiral ligament (Gratton et al. 2005). Deformities of the extracellular matrix due to alterations in the *COL4* genes, which lead to failure of the interaction between cells and their surroundings, are the main cause of Alport conditions in the kidney and the inner ear.

3. Nonsyndromic Hearing Loss

Thus far, more than 100 forms of NSHL have been discovered (Hereditary Hearing Loss Homepage). Most of these are still identified as loci: only the chromosomal location of the defective gene is known. These are designated as DFNA for autosomal dominant inheritance, DFNB for autosomal recessive inheritance, and DFNX for X-linked inheritance. Each locus represents a family or number of families with hearing loss inherited in the relevant mode. In some cases, the associated gene has been cloned and in others, the symbol only reserved. There are almost 100 loci for dominant and recessive forms and fewer than 10 for X-linked, though it appears that most of these are actually forms of SHL. The most recent additions are DFNY for Y-linked inheritance, DFNM for modifiers, OTSC for otosclerosis, and AUNA for auditory neuropathy. In these cases, only a handful of loci have been identified. The loci are numbered in the order in which they were discovered. The information is updated regularly on the Hereditary Hearing Loss Homepage (Table 2.2).

Genes for NSHL have been categorized by the function of the proteins each gene encodes. Several superfamilies are represented, including gap junctions, myosins, adhesion proteins, and ion channels. The most prevalent protein families are discussed in the subsections that follow.

3.1 Gap Junction Proteins

A specialized intersection zone connecting the cytoplasms of two adjoining cells, which permit a free passage of ions and molecules, is defined as a gap junction. In vertebrates, the protein constructing unit of this junction is a connexin. Two connexons form hemichannel structures, each made of six subunits of connexins, resting within the plasma membrane of neighboring cells adjacent to each other to create the gap junction channel (Sabag et al. 2005). In the inner ear, a network of these junctions is situated between the ephithelial supporting cells.

Thus far, three connexin genes have been implicated in NSHL: *CX26*, *CX30*, and *CX31*, corresponding to the proteins connexin 26, connexin 30, and connexin 31, which are designated by their molecular mass (connexin 26, for example, has a mass of 26 kDa; Gerido and White 2004; Sabag et al. 2005). An additional connexin, *CX32*, is involved in a syndromic hearing loss condition named Charcot–Marie–Tooth syndrome.

3.1.1 Connexin 26

Connexin 26 is by far the foremost prominent hearing impairment gene. The discovery that *GJB2*, located at the *DFNB1* locus, is a causative gene for hearing impairment (Kelsell et al. 1997) led to the unexpected revelation that more than half of the recessive NSHL and approximately 30% to 50% of innate hearing loss are due to *GJB2* mutations (Denoyelle et al. 1997; Kelsell et al. 1997; Zelante et al. 1997; Kelley et al. 1998). Even more surprising was the finding that among the 101 mutations documented worldwide up to now (The Connexin-Deafness Homepage; Table 2.2), one particular mutation, 35delG, is responsible for 70% of all the *GJB2* mutations (Maw et al. 1995; Snoeckx et al. 2005). This finding made an enormous impact in the hearing loss field, as it facilitated diagnostic and genetic screening of NSHL patients. Another mutation imperative to note is 167delT, which is most frequent among Ashkenazi Jews and is also present in Palestinians (Sobe et al. 1999; Shahin et al. 2002).

The most accepted theory regarding the role of connexin 26 in the inner ear is that it propagates the recycling of potassium ions necessary for the transmission of the auditory signal within hair cells, through gap junctions situated between supporting cells. This way, rapid circulation of potassium ions back to the endolymph through the stria vascularis (Kikuchi et al. 1995; Forge et al. 1999), allows maintenance of a high cochlear potential. The above hypothesis is based on the expression profile of connexin 26 in the cochlea, which is markedly noticeable in the bulky gap junctions of supporting cells of the sensory hair cells of the cochlea, the spiral ligament and in the spiral limbus (Kikuchi et al. 1995; Lautermann et al. 1998).

3.1.2 Connexin 30

Connexin 30 is the protein product of the *GJB6* gene, which lies within the same genetic interval of *GJB2* gene, *DFNB1*. It presents a similar expression configuration in the inner ear as connexin 26. A 342-kb deletion in the *GJB6* gene is directly linked to recessive NSHL (Lerer et al. 2001; del Castillo et al. 2002). The deletion was initially identified when an unusually larger number of profoundly deaf individuals with only one *GJB2* mutation were documented, more than expected given the carrier rate of *GJB2* mutations in the general population. Linkage analysis suggested that the deafness was due to a defect from the same chromosomal region. Further investigation of the chromosomal region uncovered this large deletion, which is extremely rare in the homozygous form. The *GJB6* mutation is usually seen on one allele only, in heterozygous form and in conjugation with a heterozygous *GJB2* mutation, creating a recessive effect, which indicates that these two proteins act together and/or share similar redundant functions in the ear.

3.2 Myosins

A substantial number of members of this superfamily can bind to different types of molecules and transfer them across the actin filament network. As a direct consequence of the above activities, the myosin molecules participate in cell functions including muscle contraction, chemotaxis, cytokinesis, pinocytosis, targeted vesicle transport, membrane traffic, and signal transduction (Mermall et al. 1998; Frank et al. 2004).

The entire superfamily of myosins shares a few common domains reviewed in (Sellers 2000). In the N-terminal, there is the highly conserved head domain (also called the motor domain) which contains two subdomains, an actin binding domain and an ATP binding domain, both enabling the head domain to associate and dissociate from actin fibers, effecting movement along them. Next, there is the neck domain, which contains 1–6 IQ motifs (a consensus sequence serves as a binding site for calcium-binding domains). Following the neck region, at the C-terminus, comes the tail domain, the most diverse domain (in length and in sequence) in the myosin superfamily. The myosin tail domain binds cargo molecules and has been predicted to target particular myosins to specific cellular compartments, therefore determining the whole molecule function and specificity of each myosin type.

Numerous types of unconventional myosins have been identified as NSHL-causing genes: *MYO1A*, *MYO3A*, *MYO6*, *MYO7A*, and *MYO15A* (encoding myosin I, myosin IIIA, myosin VI, myosin VIIA, and myosin XVA proteins, respectively). The term "unconventional" myosins refers to all nonmuscle myosin subclasses (in contrast with the "conventional" myosins that include only the myosin II subclass). Several of these myosins are described in the paragraphs that follow.

3.2.1 Myosin VI

The myosin VI protein was first implicated in the hearing mechanism when it was found to be mutated in the spontaneously deaf mouse Snell's *waltzer* (*sv*) (Avraham et al. 1995), with a deletion in the myosin VI (*Myo6*) transcript. The deletion results in a truncated protein, leading to a frameshift and subsequent premature stop codon. Mutations in the corresponding human *MYO6* gene, which shares great similarities with the mouse gene, were also identified; a missense mutation (DFNA22) in the motor domain causing dominant NSHL (Melchionda et al. 2001), and three recessive mutations found in DFNB37, also related to profound congenital NSHL (Ahmed et al. 2003a). A form of cardio–auditory syndrome may also be due to a myosin VI mutation, as a dominant missense mutation was discovered in a family segregating both sensorineural deafness and familial hypertrophic cardiomyopathy (Mohiddin et al. 2004).

Snell's *waltzer* mice possess vestibular pathologies such as head tossing, circling, and hyperactivity (Deol and Green 1966). Likewise, their cochlear hair cell phenotype is abnormal; their hair cells appear normal after birth but become disorganized and fuse to form massive stereocilia by the 20th day after birth. Further study led to the revelation that the *sv* mice do not bear any detectable myosin VI in their hair cells (Avraham et al. 1995).

Two interesting facts suggest that myosin VI has a fundamental role in the inner ear; this is the only myosin known to move "backwards" toward the minus end of actin filaments (Wells et al. 1999), and in the inner ear it is expressed solely within the hair cells (Hasson et al. 1997). Within the hair cell, myosin VI is normally expressed in the cytoplasm, and mainly at the cuticular plate, the apical region of the hair cell at the base of the stereocilia, which serves to anchor the stereocilia to the intracellular cytoplasm, and at the pericuticular necklace (the intacellular zones surrounding the cuticular plate, from both sides) at the apical side of these cells (Avraham et al. 1997; Hasson et al. 1997).

The specific function of the myosin VI protein in the inner ear remains to be discovered. Suggestions of its potential roles came mainly from mouse and zebrafish myosin VI mutant models and from cultured cell lines. Among the more popular theories is that myosin VI functions as an anchoring protein that stabilizes the interface connection between the apical membrane to the actin mesh in the cuticular plate of hair cells (Hasson et al. 1997) or that myosin VI plays a role in endocytosis or membrane trafficking in hair cells, thereby involved in the development and maintenance of stereocilia structure.

3.2.2 Myosin VIIA

Hearing loss–related mutations in *MYO7A* were originally exposed in the *shaker1* (*sh1*) mouse, which was proposed as a model for Usher syndrome. Until now, 10 distinct mutations have been associated in different *sh1* alleles (Libby and Steel 2001), carrying vestibular defects pronounced as head tossing, circling, and hyperactivity (Gibson et al. 1995). The mutant cochlear phenotype is accompanied by progressive degeneration of the organ of Corti, resulting eventually in profound deafness (Libby and Steel 2001). By virtue of using these models, mutations in the *MYO7A* human homologue were identified as the genetic cause for Usher syndrome type 1B (Weil et al. 1995, 1996) and Usher syndrome type 2A (Maubaret et al. 2005). Contiguous to this discovery, human *MYO7A* was also linked to NSHL in two distinctive loci: *DFNB2* (Weil et al. 1997) and *DFNA11* (Liu et al. 1997).

Myosin VIIa operates in a wide range of tissues but is highlighted in its appearance within the cochlea in hair cell epithelia, where it is expressed in the cytoplasm, the cuticular plate and along the length of the stereocilia (Hasson et al. 1997). A breakthrough in the elucidation of myosin VIIa function in the ear was achieved by identifying some of its protein interactors via yeast two hybrid screens. One possible destination for myosin VIIa in the inner ear is the anchoring of stereocilia crosslinks, in which an interactor transmembranal protein named vezatin can act as a mediator (Petit et al. 2001). Another possibility that myosin VIIa takes part in the development of the stereocilia bundle arose from observations that myosin VIIa binds to cadherin 23 and harmonin (Boeda et al. 2002).

3.2.3 Myosin XV

An interesting series of events led to the discovery of the *MYO15* gene as a causative deafness gene. Initially, a broad-scale mapping strategy in a Balinese community of 2200 peoples prompted the identification of the recessive NSHL locus, referred to as *DFNB3* on the long arm (p11.2) of chromosome 17 (Friedman et al. 1995; Liang et al. 1998). Based on homology of the *DFNB3* locus region to the mouse chromosomal region, the *shaker2* mouse mutant was a candidate mouse model. *Shaker2* is a profoundly deaf mutant presenting classical vestibular abnormal behavior with short hair cell stereocilia in the cochlear and vestibular systems (Probst et al. 1998; Beyer et al. 2000). A molecular genetic strategy was undertaken to prove a correlation between the mouse locus and its human corresponding locus. Bacterial artificial clones (BAC) (Table 2.1) containing candidate genes from the *DFNB3* interval were injected into *shaker2* embryos and their influence on the phenotype was assessed. One of these transgenes (Table 2.1) that managed to "rescue" the mutant *shaker2* contained the *Myo15a* gene (Probst et al. 1998). Sequencing of the *Myo15a* gene in the mutant *shaker2* mouse revealed a missense substitution of a highly conserved residue in the myosin XV head domain, clarifying that this is the deafness causing gene in these mice. *MYO15A* sequencing analysis in DFNB3 families confirmed that this unconventional myosin is implicated in human recessive NSHL (Wang et al. 1998; Liburd et al. 2001).

Myosin XVa was first documented to be expressed in inner and outer hair cells of the cochlea in an in situ hybridization analysis (Table 2.1; Lloyd et al. 2001). Further observations using an antibody raised against a portion of the myosin VIIa protein refined the localization of myosin XV within the hair cells, and proved that this protein is situated in the tip of each hair cell stereocilia (Belyantseva et al. 2003). Tip links, formed by microscopic filaments interconnected between adjacent stereocilia, are speculated to be the position where the mechanically gated transduction channels are situated (Hudspeth and Corey 1977; Pickles et al. 1984). Therefore localization of a protein near this region implies its significance in hearing process.

One of the roles postulated for myosin XV is that it is a critical factor in the evolvement of the hair bundle and in the staircase formation of the bundle (Belyantseva et al. 2003), as the appearance of myosin XV expression coincides with the appearance of the graduated patterning of the stereocilia. Moreover, based on observations showing that myosin XV binds another deafness molecule, whirlin, at the tip link zone, it was suggested that the two molecules might control stereocilia elongation patterning during development and may be implicated in stabilizing connections between stereocilia (Delprat et al. 2005).

3.3 Adhesion Proteins

Adhesion proteins are a diverse group of distinct protein families including integrins, selectins, members of the immunoglobulin superclass family, and cadherins. They protrude upon the cell surface, where they facilitate cell to cell or cell–extracellular matrix interactions and binding. The common denominator for this set of molecules seems to be that they are all glycoproteins involved in adhesion, recognition, activation, and migration activities.

3.3.1 Cadherin 23

Cadherins are a subset of the adhesion molecule superfamily comprising cadherins, protocadherins, desmogleins, and desmocollins. They promote cell binding activities, and rely on the presence of calcium ions for their proper functions. Cadherin-specific calcium binding motifs are repetitively present along the molecule structure called EC domains (Suzuki 1996; Nollet et al. 2000). They are also predicted to share a common cytoplasmic motif enabling their connection to the cytoskeleton network (Gumbiner 1996).

As mentioned in Section 2.3, mutations in the CDH23 gene, coding for the cadherin 23 protein, are associated with USH1D. The USH1D locus was found to coincide with the recessive NSHL locus DFNB12, which is associated with prelingual moderate to profound hearing loss. These data suggested that DFNB12 and USH1D in fact are mutually derived from CDH23 allelic mutations, both resulting in syndromic and nonsyndromic forms of hearing loss (Bork et al. 2001; Astuto et al. 2002).

Cadherin 23 appears as early as P0 in cochlear hair bundles and in Reissner's membrane, but later on, by P42, cadherin 23 expression is restricted to stereocilia tip links exclusively (Siemens et al. 2004), where it binds to myosin 1c, a predicted component of the mechano-transduction complex. Therefore, cadherin 23 was hypothesized to participate in the activation of this process. Interestingly, data from research in the Usher syndrome field has proven that cadherin 23 forms a complex with myosin VIIa and harmonin b, indicating that this molecule is a vital member of a specific unit designated for maintaining stereocilia inter-cohesion (Boeda et al. 2002). The fact that mice cochlea deficient in *Cdh23* display disorganized stereociliary bundle patterns (Di Palma et al. 2001) emphasizes that this protein is a central factor in stereocilia maintenance.

3.3.2 Protocadherin 15

Another member of the cadherin superfamily of molecules is protocadherin 15, the protein product of the *PCDH15* gene. As discussed in the section dealing with Usher syndrome, the *PCDH15* gene was originally implicated as a hearing loss causing gene via the deaf and vestibular deformed Ames *waltzer* mouse mutant (Alagramam et al. 2001). Thereafter the human orthologous gene was cloned and identified as one of the genes responsible for Usher syndrome, USH1F (Ahmed et al. 2001). Only two years later, *PCDH15* was documented as a NSHL causing gene (Ahmed et al. 2003b). This case of a gene locus causing both NSHL and SHL leading to Usher syndrome is similar to the case of the *CDH23* gene.

An explanation of why certain mutations in *PCDH15* cause NSHL, while others bring on a set of additional symptoms and cause SHL, was suggested (Ahmed et al. 2003b). They demonstrated a genotype–phenotype correlation between distinct mutations and their derived phenotype, from which it was deduced that more acute *PCDH15* mutations result in USH1, whereas hypomorphic (Table 2.1) mutations render the phenotype into NSHL (DFNB23) only.

Clues regarding the function of protocadherin 15 in the hearing process emerged from the study of wild type and Ames *waltzer* mice. Normal expression of protocadherin 15 was seen at embryonic day 16, overlapping the period during which stereocilia begin their formation. On the other hand, the stereocilia of Ames *waltzer* were remarkably disoriented, and displayed degeneration. Clearly, this indicated that protocadherin 15 plays a vital role in the formation and maintenance of the stereocilia. Recent data support this theory, whereby protocadherin 15 binds to myosin VIIa protein in the hair bundle, where they seem to collaborate in the regulation of the bundle development (Senften et al. 2006). In addition, due to the expression pattern of protocadherin 15 seen along the length of the stereocilia, the cuticular plate and the cytoplasm of auditory hair cells, it was suggested that protocadherin 15 contributes to evolvement and maintenance of the stereocilia lateral links (Ahmed et al. 2003b).

3.4 Ion Channels

Opening or closing of ion channels regulates the diffusion of ions through membranes. In the inner ear, ion channels have a tremendous weight in the recycling and maintaining of endolymph ionic homeostasis, a crucial condition for normal auditory transduction. The mechano-transduction theory, in which ion channels at the stereocilia tip links open when the stereocilia bundle is deflected, thus permitting an ionic potassium current to alter the hair cell voltage, emphasizes the imperative central role of ion channels in auditory hair cells. A wealth of evidence that has accumulated over the years has identified several different genes, both for syndromic and nonsyndromic hearing loss, which encode ion channels and transporters.

3.4.1 *KCNQ4*

The *KCNQ4* gene codes for one of the voltage-gated potassium channel proteins. Its ultrastructure and mechanism of action were precisely deciphered by X-ray three-dimensional modeling (Doyle et al. 1998); it is composed of four protein units forming a homo/heterotetramer. Six transmembrane domains and a one pore-loop segment in each of these proteins create the specific transformational structure of the whole channel.

In 1994, a new locus of dominant progressive NSHL, *DFNA2*, was defined (Coucke et al. 1994). The *KCNQ4* gene, which is located at the *DFNA2* interval, was a natural candidate as the causative gene for hearing loss in this locus. A few years later, this was indeed proven to be the case (Coucke et al. 1999; Kubisch et al. 1999). The *KCNQ4* mutations result in late adult-onset progressive hearing loss, commencing in the second decade of life, with a comparatively preserved lingual abilities (Coucke et al. 1999; Bom et al. 2001), while it deteriorates to profound hearing loss within 10 years.

Hypotheses regarding the role of the KCNQ4 protein in the inner ear are mostly derived from its cellular and subcellular pattern of expression in the auditory apparatus. The KCNQ4 protein is localized in the mouse cochlea in outer hair cells only, where it is bound to the basal membrane (Kharkovets et al. 2000), and appears also to express in a base to apex gradient in the spiral ganglion sensory neurons (Beisel et al. 2000). The gradient configuration of expression may be the underlying explanation of high-frequency hearing loss in DFNA2 families. This was the first ion channel noted to be specifically expressed in a sensory pathway, as KCNQ4 was demonstrated to be expressed in numerous nuclei of the central auditory pathway in the brain stem. Thus, KCNQ4 might stimulate a change in the electrical characterizations of outer hair cells, enable a path of exit for potassium ions out of the hair cells, and potentially operate in both functions (Kharkovets et al. 2000).

4. How Genes are Identified

Recombinant DNA and molecular biology techniques developed in the 1970s and 1980s made the discovery of human genetic disease genes possible. Several techniques revolutionized the way disease genes are found, including restriction fragment length polymorphism (RFLP) analysis using radioactive isotopes and Southern blotting, the polymerase chain reaction (PCR), and sequencing. For details on cloning techniques, refer to the *Cold Spring Harbor Laboratory Manual* (Sambrook and Russell 2001).

In brief, a family with inherited hearing loss is first ascertained by a medical geneticist or otolaryngologist. Close collaboration with audiologists is crucial for characterizing the auditory phenotype. A medical history is taken, with particular emphasis given to subjective degree of hearing loss, age at onset, evolution of hearing loss, symmetry of the hearing impairment, hearing aids, presence of tinnitus, medication, noise exposure, pathologic changes in the ear, and other relevant clinical manifestations. Blood samples for the extraction of genomic DNA are drawn from participating individuals after informed consent is obtained in accordance with guidelines of a Helsinki or Institutional Review Board (IRB) Committee (see National Institutes of Health Office of Human Subjects Research, Table 2.2).

Owing to the prevalence of connexin 26 and 30 mutations and the ease of screening, PCR is first performed on the proband to rule in or out connexin involvement. Several diagnostic algorithms have been suggested for clinical practitioners caring for the deaf when searching for the etiology of the hearing loss (Greinwald and Hartnick 2002; Brownstein et al. 2004). Unfortunately, the other genes are either less prevalent or cumbersome and costly to screen, making fast diagnostics impractical. Several approaches have been taken. When examining a particular ethnic group, one may examine for mutations already identified in this group. For example, in the case of Jewish Ashkenazi children diagnosed with profound congenital deafness, the R245X mutation in *PCDH15* is also examined (Brownstein et al. 2004). In the Palestinian population, mutations in several genes have already been identified, including *TMPRSS3* (Bonne-Tamir et al. 1996; Scott et al. 2001), *stereocilin* (Campbell et al. 1997; Verpy et al. 2001), *otoancorin* (Zwaenepoel et al. 2002), *whirlin* (Mustapha et al. 2002; Mburu et al. 2003), *DFNB33* (Medlej-Hashim et al. 2002), *pendrin* (Baldwin et al. 1995), *DFNB17* (Greinwald et al. 1998) and most recently, *TRIOBP* (Shahin et al. 2006). An alternative approach was taken to search for mutations in 156 Palestinian deaf probands and their families (Walsh et al. 2006). Initially, connexin 26 mutations were found in 17% of the group. Thereafter, a hearing-loss-targeted genome scan was performed on 10 families, using microsatelite markers flanking 36 loci associated with hearing loss. In the cases where linkage was found, the genes lying in the region were examined, leading to identification of *TMPRSS3*, *pendrin*, and *otoancorin* mutations.

When DNA from a large family can be obtained, the method of choice is to perform a whole genome scan, using evenly distributed microsatellite markers spanning the genome. Usually a set of 364 markers are used to identify a chromosomal location. To perform such a scan, multiply affected members (as well as unaffected members) are required from one extended family. The minimum number is approximately five for a family with recessive inheritance and 10 for a family with dominant inheritance; more individuals provide a greater chance of identifying a significant chromosomal location. Once such a region is defined by a logarithmic odds (LOD) score (Table 2.1) of at least three, the critical region is further defined by typing additional markers in the region. Statistical analyses are essential for human genetic mapping and programs are available on the Internet (for examples, see Genetic Linkage Analysis, Table 2.2). Eventually, a map of the genes in the region of linkage is examined and mutation analysis by sequencing is performed. The most widely used resource for identifying the genes in the region is the UCSC Genome Browser (Table 2.2), which lists known annotated genes as well as expressed sequence tags (ESTs) (Table 2.1) that might represent additional splice forms. For example, while the *TRIOBP* gene was known to reside in the chromosome 22-linked region of the DFNB28 Palestinian families, no mutations were found. Additional ESTs revealed the presence of a longer isoform of TRIOBP; subsequent mutation analysis uncovered novel mutations in this novel isoform in the Indian and Pakistani population as well (Riazuddin et al. 2006; Shahin et al. 2006).

5. Animal Models to Study Human Genetic Hearing Loss

The study of the genetics of hearing loss in humans is fraught with difficulties, mostly due to the shortage of large affected families and heterogeneity of genetic NSHL in humans. On the other hand, utilization of the mouse as a research tool brings on several significant advantages that answer the needs of biological studies of genetic hearing loss: vestibular abnormalities are more apparent in mice and are often linked to deafness, making it easier to detect deaf mice. Moreover, the accessibility to the ear organs and the ability to introduce defined genetic changes through the generation of transgenic or knockout mouse models (Ahituv and Avraham 2002) have proved an exceptional tool for studying the mechanisms of genetic hearing loss.

The large reservoir of mouse models currently available are divided into three main groups: spontaneous mutants, transgenic or knockout mice, and radiation or chemically induced mutants. The main goal of inducing mutations by radiation or chemicals is to create random mutations throughout the genome, in order to produce new mouse mutants, which then could be screened according to their different mutant phenotypes. The *N*-ethyl-*N*-nitrosourea (ENU) mutagen is a potent alkylating reagent that primarily induces point mutations and mutagenizes mouse spermatogonial stem cells efficiently (Balling 2001). It produces single locus mutation frequencies of 6×10^{-3}–1.5×10^{-3}, which is equivalent to obtaining a mutation in a single gene of choice (Popp et al. 1983). The phenotype-driven screening approach is used to select the mutants one needs from G1 (first generation) offspring (Nolan et al. 2002). ENU mutagenesis is an efficient method for generating auditory and vestibular mouse mutants. First, in the context of the genetics of hearing research, several simple behavioral phenotypic tests are available that can easily indicate an auditory or a vestibular dysfunction, such as the Preyer reflex test (ear flick), which examines the response of the mouse to a calibrated sound burst (Balling 2001). Second, statistically, the ENU mutagen causes a point mutation in a single gene only, which simplifies the isolation of the mutated gene afterwards. Several examples demonstrate the discovery of mouse models for human deafness loci and syndromes, including Beethoven for

DFNA36 (Kurima et al. 2002; Vreugde et al. 2002), Headbanger for *DFNA11* (Rhodes et al. 2004; Street et al. 2004), and 9 mutant alleles mapping to mouse chromosome 4 for CHARGE syndrome (Bosman et al. 2005). Databases of ENU mutants with inner ear and other defects are available (The German Mouse ENU Project and The Harwell Mutagenesis Program, Table 2.2).

While using ENU or radiation-induced mutagenesis is a phenotype-driven approach (the phenotype is known, while the genotype has yet to be discovered), transgenesis or gene-targeted knockouts are part of a genotype-driven approach. By injecting a known gene into the single cell embryo (for producing trans-genics) or gene-targeted embryonic stem cells into blastocysts (for generating knockouts), one discovers the role of the gene by observing the phenotype that arises from overexpressing the relevant gene or removing its function altogether from a mouse. The ability to create these mouse models has revolutionized the study of human genetic disease, since it opened up the way for creating valuable mouse models for disease. A knockout for the most prevalent form of deafness, connexin 26, was created by removing connexin 26 from specific portions of the inner ear only using an otogelin promoter (Cohen-Salmon et al. 2002); a gene knockout in mice is lethal due to species-specific differences between humans and mice (Gabriel et al. 1998). Other examples of mouse knockouts for human deafness loci are *DFNA2* and *KCNQ4* mutant mice (Kubisch et al. 1999; Kharkovets et al. 2006). An analysis of the mice demonstrated that *KCNQ4*-associated hearing loss is predominantly caused by a slow degeneration of outer hair cells (OHCs) resulting from chronic depolarization.

A comprehensive list of gene-targeted mice with inner ear defects is provided in a thorough review (Anagnostopoulos 2002) and several regularly updated web sites (Table 2.2).

6. Approaches to Understanding Protein Function

The gene "hunting" research commences with genotyping DNA derived from a family or families with hearing loss with microsatellite markers, followed by sequencing for mutations in the linked regions. This phase of the research is a time-consuming process and may take years. If a candidate gene is subsequently found to bear a mutation, the researcher is compelled to study the protein encoded by the deafness gene in order to understand how auditory function is compromised.

Expression studies make it possible to follow the localization of the gene and protein and often help provide a hypothesis regarding function based on the temporal and spatial expression. For expression of genes, PCR can be performed to determine whether the gene is expressed in the inner ear and at what stage. An example for the *Myo6* gene is shown in Fig. 2.1 (see Section 3.2.1). For evaluation of quantity, real-time reverse transcriptase (RT)-PCR should be performed. An example for the *Tmc1* gene is shown in Fig. 2.2. For examination of the temporal and spatial expression of a gene, mRNA in situ hybridization

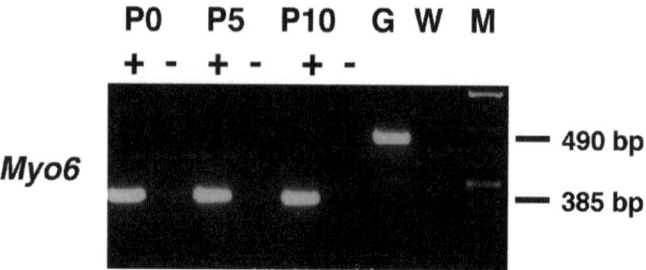

FIGURE 2.1. Expression of myosin VI (*Myo6*) in the mouse cochlea. RNA from postnatal (P) day stages P0, P5, and P10 was used to amplify the gene by RT-PCR. Amplifications were carried out with and without reverse transcriptase (+/–), by using cochlear RNA, genomic DNA (G), and a water control (W). (Modified from Walsh et al. 2002.)

is performed. An example for the *Ush3* gene is shown in Fig 2.3 (see Section 2.3). To follow the expression pattern of the protein, a suitable antibody directed against the protein is essential. This requires identifying an epitope that will be recognized uniquely by the protein, which can be determined using bioinformatic analyses (for example, Epitope Prediction, Table 2.2). The localization of Pou4f3, myosin VI and Lhx3 expression in the inner ear was revealed using specific antibodies against each protein, shown in Fig. 2.4. In many cases, the work has already been done by other investigators and a search through Pubmed (Table 2.2) and expression data sites will reveal where the gene and/or protein is expressed (for examples see NCBI sites: Gene Expression Omnibus, Unigene,

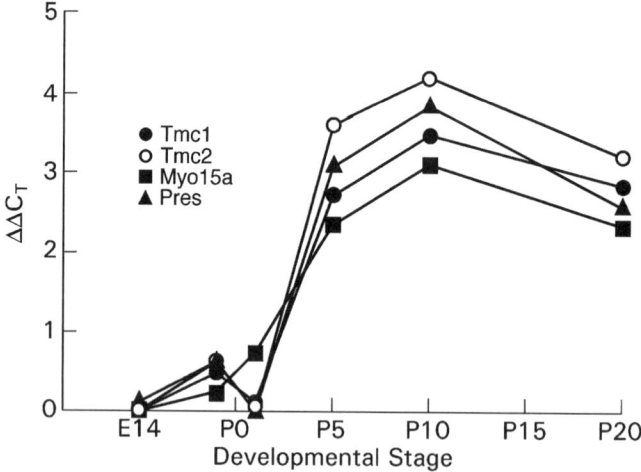

FIGURE 2.2. Real-time RT-PCR analysis of *Tmc1*, *Tmc2*, myosin XVa (*Myo15a*), and prestin (*Pres*) mRNA levels in mouse temporal bones at embryonic and postnatal stages. (From Kurima et al. 2002.)

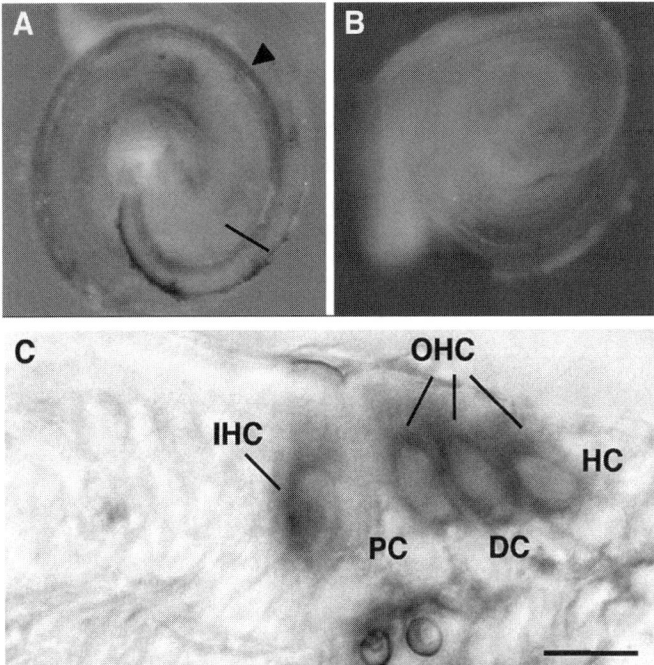

FIGURE 2.3. Detection of *Ush3a* mRNA expression by in situ hybridization. (**A**) Whole mount of a cochlea at embryonic (E) day 16, with staining along the region of the hair cells (*arrowhead*). (**B**) Sense probe demonstrates that staining in **A** is specific. (**C**) Sectioning through the cochleae (line in **A**) revealed specific hybridization in the inner (IHC) and outer hair cells (OHC) of the organ of Corti, but not in the Deiters cells (DC), pillar cells (PC), or the Hensen cells (HC). (Modified from Adato et al. 2002.)

and SOURCE, Table 2.2). There are fewer data about inner ear expression in these general expression databases.

Experiments using cell culture techniques may also reveal the function of a protein. One can overexpress the wild-type and mutant form of the gene that corresponds to the deafness phenotype. For example, to determine the mechanism for *POU4F3 DFNA15*-associated hearing loss in an Israeli kindred, the human gene was cloned into an expression vector and overexpressed in HEK293, COS-7, and the established cochlear cell line UB/OC-2 cells (Weiss et al. 2003). While the wild-type form of the gene was localized to the nucleus, as expected for a transcription factor, the mutant form was also expressed in the cytoplasm. Subsequent bioinformatics and experimental analysis revealed that a bipartite nuclear localization signal (NLS) was removed due to the truncation caused by the 8-bp deletion, leading to partial loss of nuclear localization (Fig. 2.5). To examine connexin mutations and proper localization of gap junction formation, the gene has been cloned in expression vectors and fused to a reporter, GFP, to enable localization of the wild-type and mutant proteins. In this way, many

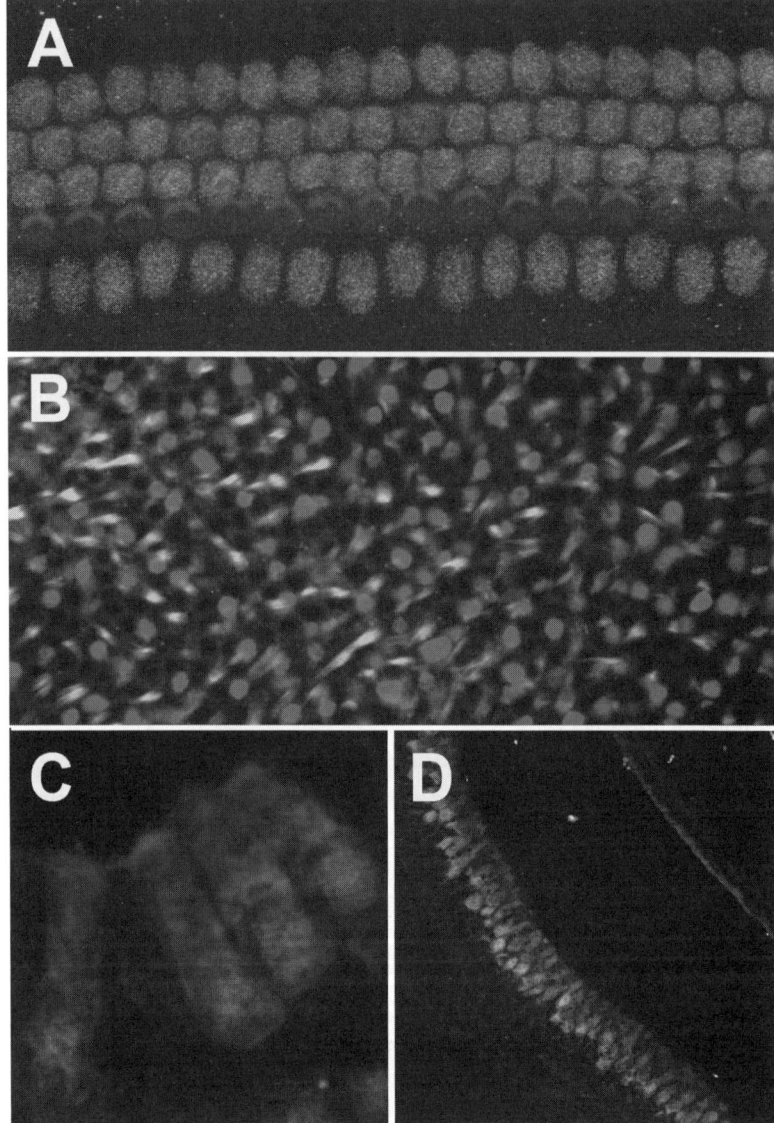

FIGURE 2.4. Expression of Pou4f3, myosin VI, and Lhx3 in the auditory and vestibular systems, demonstrated by immunohistochemistry with antibodies against each protein. (A)Whole-mount immunohistochemistry shows that Pou4f3 is expressed in the nuclei of inner and outer hair cells at E18.5 (green in online version). Actin can be visualized with phalloidin (red in online version). (Modified from Hertzano et al. 2004) (B) Whole-mount immunohistochemistry shows that the unconventional myosin VI (red in online version) is expressed in the cytoplasm of utricular hair cells at P10, while actin demonstrates the presence of stereocilia, stained with phalloidin (green in online version). (C) Cryosections shows the expression of Lhx3 (green in online version) and myosin VI (red in online version) in the cochleae and (D) the vestibular system neuroepithelial cells in the utricle from inner ears of E18.5 mice. Lhx3 is expressed in the nuclei of all hair cells. (B, C, and D provided by Amiel Dror and Karen Avraham; Hertzano et al., 2007.)

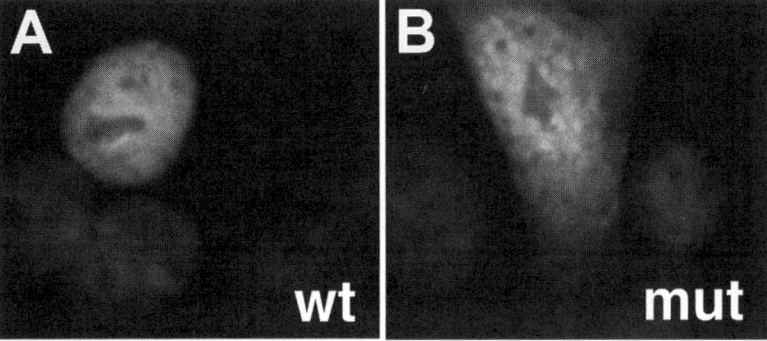

FIGURE 2.5. The dominant mutation in POU4F3 changes the subcellular localization of the protein in transfected COS-7 cells. (**A**) The wild-type form of the protein is localized to the nucleus. (**B**) The mutant form of the protein is localized to both the cytoplasm and nucleus. (Modified from Weiss et al. 2003.)

deafness causing connexin 26 mutations have been studied. For example, not only is the abnormality revealed by these experiments (Fig. 2.6), but determining whether the mutation is pathogenenic or not may be answered. For example, the M34T connexin 26 mutation has been the subject of debate for years. Though originally identified in a family with deafness (Kelsell et al. 1997), reports from other investigators revealed that this mutation exists in normal hearing individuals (Scott et al. 1998). One study suggests that the mutation is pathogenic, because

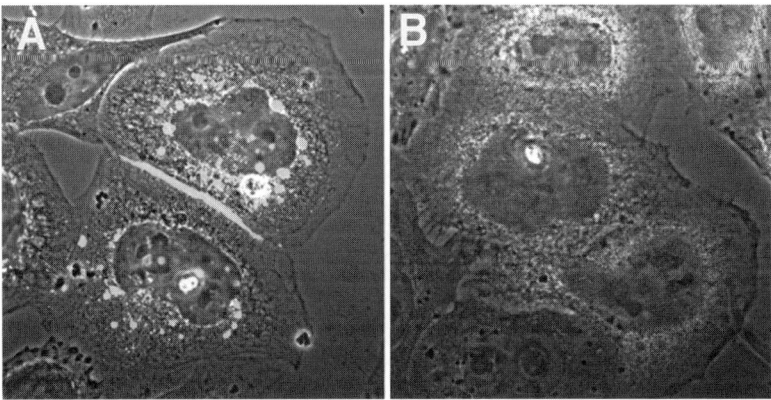

FIGURE 2.6. Connexin-GFP fused protein expressed in transfected HeLa cells. (**A**) Wild-type Cx26 is expressed in the plasma membrane, creating gap junction plaques between adjacent cells. (**B**) When expressing Cx26 carrying the deafness-causing mutation Ser139Asn in transfected cells, the protein fails to reach the plasma membrane and no gap junction plaques are formed. (Courtesy of Adi Sabag and Karen Avraham; Fleishman et al. 2006.)

although it leads to normal gap junction localization, it also leads to abnormal gating (Skerrett et al. 2004).

The espin actin-bundling proteins are involved in deafness in both *jerker* mice and DFNB36 (Zheng et al. 2000; Naz et al. 2004). A new function for this protein, the ability to assemble a large actin bundle when targeted to a specific subcellular location, was revealed by transfection into neuronal and other cells (Loomis et al. 2006). Most recently, siRNA has been developed, using the natural biological mechanism of RNA interference, where double-stranded RNA (dsRNA) induces gene silencing by targeting complementary mRNA for degradation, enabling silencing of genes in cells (Hannon and Rossi 2004). For example, RNAi reduced huntingtin mRNA and protein expression in cell culture and in the brain of a mouse model for Huntington's disease (Harper 2005). The silencing improved behavioral and neuropathological abnormalities associated with this disease. siRNA was successfully used in human HEK293 and mouse P19 cells to suppress a specific connexin 26 mutation (Maeda et al. 2005). Most important, the study demonstrated that a specific mutation could be suppressed in a transgenic mouse model, paving the way for future experiments of this sort for therapeutics.

There are numerous additional techniques available to determine gene and protein function, including microarray-based processes. DNA microarrays encompass a multifaceted approach to understanding complex interactions between genes (reviewed in Schulze and Downward 2001). Microarrays with the readily available genes from the genome can be screened to identify downstream targets for transcription factors, as was done by comparing RNA derived from Pou4f3-deficient inner ears to wild type ears (Hertzano et al. 2004). Alternatively, deafness genes can be identified by studying differential expression of genes within the cochlear using a custom mouse inner ear microarray (Morris et al. 2005). A review covering the uses of microarrays for inner ear research describes aspects of this experimentation (Chen and Corey 2002).

7. Summary: Implication of Discovery of Genes Associated with Hearing Loss

Why has the genetics of hearing loss become such a focus for researchers? First, from a biological perspective, the amount of information gained about the auditory and vestibular systems has been dramatic. Second, from a diagnostic aspect, clinicians are now able to discern the etiology of hearing loss a large number of patients by relatively simple genetic testing. Third, from a genetic counseling aspect, genetic counselors are able to predict with much greater certainty what the chances of another child being born with deafness in the family are. Fourth, from a therapeutic aspect, the discovery of genes may provide solutions for treatment and therapy for alleviating hearing loss (see Heller and Raphael, Chapter 11).

Acknowledgment. We thank Ronna Hertzano for critical review of the chapter. Funding in the Avraham laboratory is provided by the European Commission FP6 Integrated Project EuroHear LSHG-CT-20054-512063, NIH Grant R01 DC005641, The German-Israeli Foundation for Scientific Research and Development (GIF), the US-Israel Binational Science Foundation (BSF) Grant 2003335, and a gift from B. and A. Hirschfield.

References

Adato A, Vreugde S, Joensuu T, Avidan N, Hamalainen R, Belenkiy O, Olender T, Bonne-Tamir B, Ben-Asher E, Espinos C, Mill·n JM, Lehesjoki A-E, Flannery JG, Avraham KB, Pietrokovski S, Sankila E-M, Beckmann JS, Lancet D (2002) USH3A transcripts encode clarin-1, a four-transmembrane-domain protein with a possible role in sensory synapses. Eur J Hum Genet 10:339–350.

Adato A, Michel V, Kikkawa Y, Reiners J, Alagramam KN, Weil D, Yonekawa H, Wolfrum U, El-Amraoui A, Petit C (2005) Interactions in the network of Usher syndrome type 1 proteins. Hum Mol Genet 14:347–356.

Ahituv N, Avraham KB (2002) Mouse models for human deafness: current tools for new fashions. Trends Mol Med 8:447–451.

Ahmed ZM, Riazuddin S, Bernstein SL, Ahmed Z, Khan S, Griffith AJ, Morell RJ, Friedman TB, Wilcox ER (2001) Mutations of the protocadherin gene *PCDH15* cause Usher syndrome type 1F. Am J Hum Genet 69:25–34.

Ahmed ZM, Morell RJ, Riazuddin S, Gropman A, Shaukat S, Ahmad MM, Mohiddin SA, Fananapazir L, Caruso RC, Husnain T, Khan SN, Griffith AJ, Friedman TB, Wilcox ER (2003a) Mutations of *MYO6* are associated with recessive deafness, DFNB37. Am J Hum Genet 72:1315–1322.

Ahmed ZM, Riazuddin S, Ahmad J, Bernstein SL, Guo Y, Sabar MF, Sieving P, Griffith AJ, Friedman TB, Belyantseva IA, Wilcox ER (2003b) PCDH15 is expressed in the neurosensory epithelium of the eye and ear and mutant alleles are responsible for both USH1F and DFNB23. Hum Mol Genet 12:3215 3223.

Alagramam KN, Murcia CL, Kwon HY, Pawlowski KS, Wright CG, Woychik RP (2001) The mouse Ames *waltzer* hearing-loss mutant is caused by mutation of *Pcdh15*, a novel protocadherin gene. Nat Genet 27:99–102.

Anagnostopoulos AV (2002) A compendium of mouse knockouts with inner ear defect. Trends Genet 18:499.

Apaydin F, Bereketoglu M, Turan O, Hribar K, Maassen MM, Gunhan O, Zenner HP, Pfister M (2004) [Waardenburg syndrome. A heterogenic disorder with variable penetrance]. HNO 52:533–537.

Astuto LM, Bork JM, Weston MD, Askew JW, Fields RR, Orten DJ, Ohliger SJ, Riazuddin S, Morell RJ, Khan S, Kremer H, van Hauwe P, Moller CG, Cremers CW, Ayuso C, Heckenlively JR, Rohrschneider K, Spandau U, Greenberg J, Ramesar R, Reardon W, Bitoun P, Millan J, Legge R, Friedman TB, Kimberling WJ (2002) *CDH23* mutation and phenotype heterogeneity: a profile of 107 diverse families with Usher syndrome and nonsyndromic deafness. Am J Hum Genet 71:262–275.

Avraham KB, Hasson T, Steel KP, Kingsley DM, Russell LB, Mooseker MS, Copeland NG, Jenkins NA (1995) The mouse Snell's *waltzer* deafness gene encodes an unconventional myosin required for structural integrity of inner ear hair cells. Nat Genet 11:369–375.

Avraham KB, Hasson T, Sobe T, Balsara B, Testa JR, Skvorak AB, Morton CC, Copeland NG, Jenkins NA (1997) Characterization of unconventional *MYO6*, the human homologue of the gene responsible for deafness in Snell's *waltzer* mice. Hum Mol Genet 6:1225–1231.

Baldwin CT, Weiss S, Farrer LA, De Stefano AL, Adair R, Franklyn B, Kidd KK, Korostishevsky M, Bonne-Tamir B (1995) Linkage of congenital, recessive deafness (DFNB4) to chromosome 7q31 and evidence for genetic heterogeneity in the Middle Eastern Druze population. Hum Mol Genet 4:1637–1642.

Balling R (2001) ENU mutagenesis: analyzing gene function in mice. Annu Rev Genom Hum Genet 2:463–492.

Beisel KW, Nelson NC, Delimont DC, Fritzsch B (2000) Longitudinal gradients of KCNQ4 expression in spiral ganglion and cochlear hair cells correlate with progressive hearing loss in DFNA2. Brain Res Mol Brain Res 82:137–149.

Belyantseva IA, Boger ET, Friedman TB (2003) Myosin XVa localizes to the tips of inner ear sensory cell stereocilia and is essential for staircase formation of the hair bundle. Proc Natl Acad Sci USA 100:13958–13963.

Beyer LA, Odeh H, Probst FJ, Lambert EH, Dolan DF, Camper SA, Kohrman DC, Raphael Y (2000) Hair cells in the inner ear of the pirouette and shaker 2 mutant mice. J Neurocytol 29:227–240.

Bhattacharya G, Miller C, Kimberling WJ, Jablonski MM, Cosgrove D (2002) Localization and expression of usherin: a novel basement membrane protein defective in people with Usher's syndrome type IIa. Hear Res 163:1–11.

Boeda B, El-Amraoui A, Bahloul A, Goodyear R, Daviet L, Blanchard S, Perfettini I, Fath KR, Shorte S, Reiners J, Houdusse A, Legrain P, Wolfrum U, Richardson G, Petit C (2002) Myosin VIIa, harmonin and cadherin 23, three Usher I gene products that cooperate to shape the sensory hair cell bundle. EMBO J 21:6689–6699.

Bolz H, von Brederlow B, Ramirez A, Bryda EC, Kutsche K, Nothwang HG, Seeliger M, del CSCM, Vila MC, Molina OP, Gal A, Kubisch C (2001) Mutation of *CDH23*, encoding a new member of the cadherin gene family, causes Usher syndrome type 1D. Nat Genet 27:108–112.

Bom SJ, De Leenheer EM, Lemaire FX, Kemperman MH, Verhagen WI, Marres HA, Kunst HP, Ensink RJ, Bosman AJ, Van Camp G, Cremers FP, Huygen PL, Cremers CW (2001) Speech recognition scores related to age and degree of hearing impairment in DFNA2/KCNQ4 and DFNA9/COCH. Arch Otolaryngol Head Neck Surg 127: 1045–1048.

Bondurand N, Pingault V, Goerich DE, Lemort N, Sock E, Caignec CL, Wegner M, Goossens M (2000) Interaction among *SOX10*, *PAX3* and *MITF*, three genes altered in Waardenburg syndrome. Hum Mol Genet 9:1907–1917.

Bonini NM, Leiserson WM, Benzer S (1998) Multiple roles of the eyes absent gene in *Drosophila*. Dev Biol 196:42–57.

Bonne-Tamir B, DeStefano AL, Briggs CE, Adair R, Franklyn B, Weiss S, Korostishevsky M, Frydman M, Baldwin CT, Farrer LA (1996) Linkage of congenital recessive deafness (gene *DFNB10*) to chromosome 21q22.3. Am J Hum Genet 58:1254–1259.

Bork JM, Peters LM, Riazuddin S, Bernstein SL, Ahmed ZM, Ness SL, Polomeno R, Ramesh A, Schloss M, Srisailpathy CR, Wayne S, Bellman S, Desmukh D, Ahmed Z, Khan SN, Kaloustian VM, Li XC, Lalwani A, Bitner-Glindzicz M, Nance WE, Liu XZ, Wistow G, Smith RJ, Griffith AJ, Wilcox ER, Friedman TB, Morell RJ (2001) Usher syndrome 1D and nonsyndromic autosomal recessive deafness DFNB12 are caused by allelic mutations of the novel cadherin-like gene *CDH23*. Am J Hum Genet 68:26–37.

Bosman EA, Penn AC, Ambrose JC, Kettleborough R, Stemple DL, Steel KP (2005) Multiple mutations in mouse *Chd7* provide models for CHARGE syndrome. Hum Mol Genet 14:3463–3476.

Brownstein Z, Ben-Yosef T, Dagan O, Frydman M, Abeliovich D, Sagi M, Abraham FA, Taitelbaum-Swead R, Shohat M, Hildesheimer M, Friedman TB, Avraham KB (2004) The R245X mutation of *PCDH15* in Ashkenazi Jewish children diagnosed with nonsyndromic hearing loss foreshadows retinitis pigmentosa. Pediatr Res 55: 995–1000.

Brunner HG, van Bennekom A, Lambermon EM, Oei TL, Cremers WR, Wieringa B, Ropers HH (1988) The gene for X-linked progressive mixed deafness with perilymphatic gusher during stapes surgery (DFN3) is linked to PGK. Hum Genet 80:337–340.

Campbell DA, McHale DP, Brown KA, Moynihan LM, Houseman M, Karbani G, Parry G, Janjua AH, Newton V, al-Gazali L, Markham AF, Lench NJ, Mueller RF (1997) A new locus for non-syndromal, autosomal recessive, sensorineural hearing loss (DFNB16) maps to human chromosome 15q21–q22. J Med Genet 34:1015–1017.

Chen ZY, Corey DP (2002) Understanding inner ear development with gene expression profiling. J Neurobiol 53:276–285.

Cohen-Salmon M, Ott T, Michel V, Hardelin JP, Perfettini I, Eybalin M, Wu T, Marcus DC, Wangemann P, Willecke K, Petit C (2002) Targeted ablation of connexin26 in the inner ear epithelial gap junction network causes hearing impairment and cell death. Curr Biol 12:1106–1111.

Collins FS, McKusick VA (2001) Implications of the Human Genome Project for medical science. JAMA 285:540–544.

Coucke P, Van Camp G, Djoyodiharjo B, Smith SD, Frants RR, Padberg GW, Darby JK, Huizing EH, Cremers CW, Kimberling WJ, Oostra BA, Van de Heyning PH, Willems PJ. (1994) Linkage of autosomal dominant hearing loss to the short arm of chromosome 1 in two families. N Engl J Med 331:425–431.

Coucke PJ, Van Hauwe P, Kelley PM, Kunst H, Schatteman I, Van Velzen D, Meyers J, Ensink RJ, Verstreken M, Declau F, Marres H, Kastury K, Bhasin S, McGuirt WT, Smith RJ, Cremers CW, Van de Heyning P, Willems PJ, Smith SD, Van Camp G (1999) Mutations in the *KCNQ4* gene are responsible for autosomal dominant deafness in four DFNA2 families. Hum Mol Genet 8:1321–1328.

del Castillo I, Villamar M, Moreno-Pelayo MA, del Castillo FJ, Alvarez A, Telleria D, Menendez I, Moreno F (2002) A deletion involving the connexin 30 gene in nonsyndromic hearing impairment. N Engl J Med 346:243–249.

Delprat B, Michel V, Goodyear R, Yamasaki Y, Michalski N, El-Amraoui A, Perfettini I, Legrain P, Richardson G, Hardelin JP, Petit C (2005) Myosin XVa and whirlin, two deafness gene products required for hair bundle growth, are located at the stereocilia tips and interact directly. Hum Mol Genet 14:401–410.

Denoyelle F, Weil , Maw MA, Wilcox SA, Lench NJ, Allen-Powell DR, Osborn AH, Dahl HH, Middleton A, Houseman MJ, Dode C, Marlin S, Boulila-ElGaied A, Grati M, Ayadi H, BenArab S, Bitoun P, Lina-Granade G, Godet J, Mustapha M, Loiselet J, El-Zir E, Aubois A, Joannard A, Levilliers J, Garabédian E-N, Mueller RF, McKinlay Gardner RJ, Petit C (1997) Prelingual deafness: high prevalence of a 30delG mutation in the connexin 26 gene. Hum Mol Genet 6:2173–2177.

Deol MS, Green MC (1966) Snell's *waltzer*, a new mutation affecting behaviour and the inner ear in the mouse. Genet Res 8:339–345.

Di Palma F, Holme RH, Bryda EC, Belyantseva IA, Pellegrino R, Kachar B, Steel KP, Noben-Trauth K (2001) Mutations in *Cdh23*, encoding a new type of cadherin, cause stereocilia disorganization in *waltzer*, the mouse model for Usher syndrome type 1D. Nat Genet 27:103–107.

Doyle DA, Morais Cabral J, Pfuetzner RA, Kuo A, Gulbis JM, Cohen SL, Chait BT, MacKinnon R (1998) The structure of the potassium channel: molecular basis of K+ conduction and selectivity. Science 280:69–77.

Endreffy E, Ondrik Z, Kemeny E, Vas Z, Maroti Z, Lencse G, Bereczki C, Haszon I, Turi S, Ivanyi B (2005) [Collagen type IV nephropathy: from thin basement membrane nephropathy to Alport syndrome]. Orv Hetil 146:2647–2653.

Everett LA, Glaser B, Beck JC, Idol JR, Buchs A, Heyman M, Adawi F, Hazani E, Nassir E, Baxevanis AD, Sheffield VC, Green ED (1997) Pendred syndrome is caused by mutations in a putative sulphate transporter gene (PDS). Nat Genet 17:411–422.

Fleishman SJ, Sabag AD, Ophir E, Avraham KB, Ben-Tal N. (2006) The structural context of disease-causing mutations in gap junctions. J Biol Chem 281:28958–28963.

Forge A, Becker D, Casalotti S, Edwards J, Evans WH, Lench N, Souter M (1999) Gap junctions and connexin expression in the inner ear. Novartis Found Symp 219:134–150.

Frank DJ, Noguchi T, Miller KG (2004) Myosin VI: a structural role in actin organization important for protein and organelle localization and trafficking. Curr Opin Cell Biol 16:189–194.

Fraser FC, Ling D, Clogg D, Nogrady B (1978) Genetic aspects of the BOR syndrome— branchial fistulas, ear pits, hearing loss, and renal anomalies. Am J Med Genet 2:241–252.

Friedman TB, Liang Y, Weber JL, Hinnant JT, Barber TD, Winata S, Arhya IN, Asher JH, Jr (1995) A gene for congenital, recessive deafness DFNB3 maps to the pericentromeric region of chromosome 17. Nat Genet 9:86–91.

Friedman TB, Schultz JM, Ben-Yosef T, Pryor SP, Lagziel A, Fisher RA, Wilcox ER, Riazuddin S, Ahmed ZM, Belyantseva I, Griffith AJ (2003) Recent advances in the understanding of syndromic forms of hearing loss. Ear Hear 24:289–302.

Gabriel HD, Jung D, Butzler C, Temme A, Traub O, Winterhager E, Willecke K (1998) Transplacental uptake of glucose is decreased in embryonic lethal connexin26-deficient mice. J Cell Biol 140:1453–1461.

Gerido DA, White TW (2004) Connexin disorders of the ear, skin, and lens. Biochim Biophys Acta 1662:159–170.

Gibson F, Walsh J, Mburu P, Varela A, Brown KA, Antonio M, Beisel KW, Steel KP, Brown SD (1995) A type VII myosin encoded by the mouse deafness gene shaker-1. Nature 374:62–64.

Gorlin RJ, Toriello HV, Cohen MM (1995) Hereditary Hearing Loss and Its Syndromes. New York: Oxford University Press.

Gratton MA, Rao VH, Meehan DT, Askew C, Cosgrove D (2005) Matrix metalloproteinase dysregulation in the stria vascularis of mice with Alport syndrome: implications for capillary basement membrane pathology. Am J Pathol 166:1465–1474.

Greinwald JH Jr, Hartnick CJ (2002) The evaluation of children with sensorineural hearing loss. Arch Otolaryngol Head Neck Surg 128:84–87.

Greinwald JH Jr, Wayne S, Chen AH, Scott DA, Zbar RI, Kraft ML, Prasad S, Ramesh A, Coucke P, Srisailapathy CR, Lovett M, Van Camp G, Smith RJ (1998) Localization of a novel gene for nonsyndromic hearing loss (DFNB17) to chromosome region 7q31. Am J Med Genet 78:107–113.

Guilford P, Ben Arab S, Blanchard S, Levilliers J, Weissenbach J, Belkahia A, Petit C (1994) A non-syndrome form of neurosensory, recessive deafness maps to the pericentromeric region of chromosome 13q. Nat Genet 6:24–28.

Gumbiner BM (1996) Cell adhesion: the molecular basis of tissue architecture and morphogenesis. Cell 84:345–357.

Hannon GJ, Rossi JJ (2004) Unlocking the potential of the human genome with RNA interference. Nature 431:371–378.

Harper B (2005) Huntington disease. J R Soc Med 98:550.

Hasson T, Gillespie PG, Garcia JA, MacDonald RB, Zhao Y, Yee AG, Mooseker MS, Corey DP (1997) Unconventional myosins in inner-ear sensory epithelia. J Cell Biol 137:1287–1307.

Hertzano R, Avraham KB (2005) Developmental genes associated with human hearing loss. In: Kelley M, Wu D, Popper AN, Fay RR (eds) Development of the Inner Ear. New York: Springer, pp. 204–232.

Hertzano R, Dror AA, Montcouquiol M, Ahmed Z, Ellsworth B, Camper S, Friedman TB, Kelley MW, Avraham KB (2007) Lhx3, a LIM domain transcription factor, is regulated by Pou4f3 in the auditory, but not in the vestibular system. Eur J Neurosci 25:999–1005.

Hertzano R, Montcouquiol M, Rashi-Elkeles S, Elkon R, Yucel R, Frankel WN, Rechavi G, Moroy T, Friedman TB, Kelley MW, Avraham KB (2004) Transcription profiling of inner ears from $Pou4f3^{(ddl/ddl)}$ identifies Gfi1 as a target of the Pou4f3 deafness gene. Hum Mol Genet 13:2143–2153.

Hone SW, Smith RJ (2003) Genetic screening for hearing loss. Clin Otolaryngol Allied Sci 28:285–290.

Hudspeth AJ, Corey DP (1977) Sensitivity, polarity, and conductance change in the response of vertebrate hair cells to controlled mechanical stimuli. Proc Natl Acad Sci USA 74:2407–2411.

Karjalainen S, Terasvirta M, Karja J, Kaariainen H (1983) An unusual otological manifestation of Usher's syndrome in four siblings. Clin Genet 24:273–279.

Kelley PM, Harris DJ, Comer BC, Askew JW, Fowler T, Smith SD, Kimberling WJ (1998) Novel mutations in the connexin 26 gene (GJB2) that cause autosomal recessive (DFNB1) hearing loss. Am J Hum Genet 62:792–799.

Kelsell DP, Dunlop J, Stevens HP, Lench NJ, Liang JN, Parry G, Mueller RF, Leigh IM (1997) Connexin 26 mutations in hereditary non-syndromic sensorineural deafness. Nature 387:80–83.

Kharkovets T, Hardelin JP, Safieddine S, Schweizer M, El-Amraoui A, Petit C, Jentsch TJ (2000) KCNQ4, a K^+ channel mutated in a form of dominant deafness, is expressed in the inner ear and the central auditory pathway. Proc Natl Acad Sci USA 97:4333–4338.

Kharkovets T, Dedek K, Maier H, Schweizer M, Khimich D, Nouvian R, Vardanyan V, Leuwer R, Moser T, Jentsch TJ (2006) Mice with altered KCNQ4 K^+ channels implicate sensory outer hair cells in human progressive deafness. EMBO J 25:642–652.

Kikkawa Y, Shitara H, Wakana S, Kohara Y, Takada T, Okamoto M, Taya C, Kamiya K, Yoshikawa Y, Tokano H, Kitamura K, Shimizu K, Wakabayashi Y, Shiroishi T, Kominami R, Yonekawa H (2003) Mutations in a new scaffold protein Sans cause deafness in Jackson shaker mice. Hum Mol Genet 12:453–461.

Kikuchi T, Kimura RS, Paul DL, Adams JC (1995) Gap junctions in the rat cochlea: immunohistochemical and ultrastructural analysis. Anat Embryol (Berl) 191:101–118.

Kubisch C, Schroeder BC, Friedrich T, Lutjohann B, El-Amraoui A, Marlin S, Petit C, Jentsch TJ (1999) KCNQ4, a novel potassium channel expressed in sensory outer hair cells, is mutated in dominant deafness. Cell 96:437–446.

Kurima K, Peters LM, Yang Y, Riazuddin S, Ahmed ZM, Naz S, Arnaud D, Drury S, Mo J, Makishima T, Ghosh M, Menon PS, Deshmukh D, Oddoux C, Ostrer H, Khan S, Deininger PL, Hampton LL, Sullivan SL, Battey JF, Jr, Keats BJ, Wilcox ER,

Friedman TB, Griffith AJ (2002) Dominant and recessive deafness caused by mutations of a novel gene, TMC1, required for cochlear hair-cell function. Nat Genet 30:277–284.

Lautermann J, ten Cate WJ, Altenhoff P, Grummer R, Traub O, Frank H, Jahnke K, Winterhager E (1998) Expression of the gap-junction connexins 26 and 30 in the rat cochlea. Cell Tissue Res 294:415–420.

Leon PE, Raventos H, Lynch E, Morrow J, King MC (1992) The gene for an inherited form of deafness maps to chromosome 5q31.Proc Natl Acad Sci USA 89:5181–5184.

Lerer I, Sagi M, Ben-Neriah Z, Wang T, Levi H, Abeliovich D (2001) A deletion mutation in *GJB6* cooperating with a *GJB2* mutation in trans in non-syndromic deafness: a novel founder mutation in Ashkenazi Jews. Hum Mutat 18:460.

Liang Y, Wang A, Probst FJ, Arhya IN, Barber TD, Chen KS, Deshmukh D, Dolan DF, Hinnant JT, Carter LE, Jain PK, Lalwani AK, Li XC, Lupski JR, Moeljopawiro S, Morell R, Negrini C, Wilcox ER, Winata S, Camper SA, Friedman TB (1998) Genetic mapping refines DFNB3 to 17p11.2, suggests multiple alleles of DFNB3, and supports homology to the mouse model shaker-2. Am J Hum Genet 62:904–915.

Libby RT, Steel KP (2001) Electroretinographic anomalies in mice with mutations in *Myo7a*, the gene involved in human Usher syndrome type 1B. Invest Ophthalmol Vis Sci 42:770–778.

Liburd N, Ghosh M, Riazuddin S, Naz S, Khan S, Ahmed Z, Liang Y, Menon PS, Smith T, Smith AC, Chen KS, Lupski JR, Wilcox ER, Potocki L, Friedman TB (2001) Novel mutations of *MYO15A* associated with profound deafness in consanguineous families and moderately severe hearing loss in a patient with Smith-Magenis syndrome. Hum Genet 109:535–541.

Liu XZ, Walsh J, Mburu P, Kendrick-Jones J, Cope MJ, Steel KP, Brown SD (1997) Mutations in the myosin VIIA gene cause non-syndromic recessive deafness. Nat Genet 16:188–190.

Lloyd RV, Vidal S, Jin L, Zhang S, Kovacs K, Horvath E, Scheithauer BW, Boger ET, Fridell RA, Friedman TB (2001) Myosin XVA expression in the pituitary and in other neuroendocrine tissues and tumors. Am J Pathol 159:1375–1382.

Loomis PA, Kelly AE, Zheng L, Changyaleket B, Sekerkova G, Mugnaini E, Ferreira A, Mullins RD, Bartles JR (2006) Targeted wild-type and jerker espins reveal a novel, WH2-domain-dependent way to make actin bundles in cells. J Cell Sci 119:1655–1665.

Lynch ED, Lee MK, Morrow JE, Welcsh PL, Leon PE, King MC (1997) Nonsyndromic deafness DFNA1 associated with mutation of a human homolog of the *Drosophila* gene diaphanous. Science 278:1315–1318.

Maeda Y, Fukushima K, Nishizaki K, Smith RJ (2005) In vitro and in vivo suppression of *GJB2* expression by RNA interference. Hum Mol Genet 14:1641–1650.

Marazita ML, Ploughman LM, Rawlings B, Remington E, Arnos KS, Nance WE (1993) Genetic epidemiological studies of early-onset deafness in the U.S. school-age population. Am J Med Genet 46:486–491.

Maubaret C, Griffoin JM, Arnaud B, Hamel C (2005) Novel mutations in *MYO7A* and *USH2A* in Usher syndrome. Ophthalmic Genet 26:25–29.

Maw MA, Allen-Powell DR, Goodey RJ, Stewart IA, Nancarrow DJ, Hayward NK, Gardner RJ (1995) The contribution of the DFNB1 locus to neurosensory deafness in a Caucasian population. Am J Hum Genet 57:629–635.

Mburu P, Mustapha M, Varela A, Weil D, El-Amraoui A, Holme RH, Rump A, Hardisty RE, Blanchard S, Coimbra RS, Perfettini I, Parkinson N, Mallon AM, Glenister P, Rogers MJ, Paige AJ, Moir L, Clay J, Rosenthal A, Liu XZ, Blanco G, Steel KP, Petit C, Brown SD (2003) Defects in whirlin, a PDZ domain molecule involved in stereocilia elongation, cause deafness in the whirler mouse and families with DFNB31. Nat Genet 2003 34:421–428.

Medlej-Hashim M, Mustapha M, Chouery E, Weil D, Parronaud J, Salem N, Delague V, Loiselet J, Lathrop M, Petit C, Megarbane A (2002) Non-syndromic recessive deafness in Jordan: mapping of a new locus to chromosome 9q34.3 and prevalence of DFNB1 mutations. Eur J Hum Genet 10:391–394.

Melchionda S, Ahituv N, Bisceglia L, Sobe T, Glaser F, Rabionet R, Arbones ML, Notarangelo A, Di Iorio E, Carella M, Zelante L, Estivill X, Avraham KB, Gasparini P (2001) *MYO6*, the human homologue of the gene responsible for deafness in Snell's *waltzer* mice, is mutated in autosomal dominant nonsyndromic hearing loss. Am J Hum Genet 69:635–640.

Melnick M, Bixler D, Nance WE, Silk K, Yune H (1976) Familial branchio-oto-renal dysplasia: a new addition to the branchial arch syndromes. Clin Genet 9:25–34.

Mermall V, Post PL, Mooseker MS (1998) Unconventional myosins in cell movement, membrane traffic, and signal transduction. Science 279:527–533.

Mohiddin SA, Ahmed ZM, Griffith AJ, Tripodi D, Friedman TB, Fananapazir L, Morell RJ (2004) Novel association of hypertrophic cardiomyopathy, sensorineural deafness, and a mutation in unconventional myosin VI (*MYO6*). J Med Genet 41: 309–314.

Morris KA, Snir E, Pompeia C, Koroleva IV, Kachar B, Hayashizaki Y, Carninci P, Soares MB, Beisel KW (2005) Differential expression of genes within the cochlea as defined by a custom mouse inner ear microarray. J Assoc Res Otolaryngol 6:75–89.

Mustapha M, Chouery E, Torchard-Pagnez D, Nouaille S, Khrais A, Sayegh FN, Megarbane A, Loiselet J, Lathrop M, Petit C, Weil D (2002) A novel locus for Usher syndrome type I, USH1G, maps to chromosome 17q24–25. Hum Genet 110:348–350.

Nadol JB, Jr., Merchant SN (2001) Histopathology and molecular genetics of hearing loss in the human. Int J Pediatr Otorhinolaryngol 61:1–15.

Nance WE (2003) The genetics of deafness. Ment Retard Dev Disabil Res Rev 9:109–119.

Naz S, Griffith AJ, Riazuddin S, Hampton LL, Battey JF, Jr., Khan SN, Wilcox ER, Friedman TB (2004) Mutations of ESPN cause autosomal recessive deafness and vestibular dysfunction. J Med Genet 41:591–595.

Newby HA (1992) Audiology. Englewood Cliffs, NJ: Prentice-Hall.

Nolan PM, Hugill A, Cox RD (2002) ENU mutagenesis in the mouse: application to human genetic disease. Brief Funct Genomic Proteomic 1:278–289.

Nollet F, Kools P, van Roy F (2000) Phylogenetic analysis of the cadherin superfamily allows identification of six major subfamilies besides several solitary members. J Mol Biol 299:551–572.

Pakarinen L, Tuppurainen K, Laippala P, Mantyjarvi M, Puhakka H (1995) The ophthalmological course of Usher syndrome type III. Int Ophthalmol 19:307–311.

Park HJ, Shaukat S, Liu XZ, Hahn SH, Naz S, Ghosh M, Kim HN, Moon SK, Abe S, Tukamoto K, Riazuddin S, Kabra M, Erdenetungalag R, Radnaabazar J, Khan S, Pandya A, Usami SI, Nance WE, Wilcox ER, Griffith AJ (2003) Origins and frequencies of *SLC26A4* (PDS) mutations in east and south Asians: global implications for the epidemiology of deafness. J Med Genet 40:242–248.

Petit C, Levilliers J, Hardelin JP (2001) Molecular genetics of hearing loss. Annu Rev Genet 35:589–646.

Pickles JO, Comis SD, Osborne MP (1984) Cross-links between stereocilia in the guinea pig organ of Corti, and their possible relation to sensory transduction. Hear Res 15: 103–112.

Popp RA, Bailiff EG, Skow LC, Johnson FM, Lewis SE (1983) Analysis of a mouse alpha-globin gene mutation induced by ethylnitreosourea. Genetics 105:157–167.

Probst FJ, Fridell RA, Raphael Y, Saunders TL, Wang A, Liang Y, Morell RJ, Touchman JW, Lyons RH, Noben-Trauth K, Friedman TB, Camper SA (1998) Correction of deafness in shaker-2 mice by an unconventional myosin in a BAC transgene. Science 280:1444–1447.

Reiners J, Nagel-Wolfrum K, Jurgens K, Marker T, Wolfrum U (2006) Molecular basis of human Usher syndrome: deciphering the meshes of the Usher protein network provides insights into the pathomechanisms of the Usher disease. Exp Eye Res 83:97–119.

Rhodes CR, Hertzano R, Fuchs H, Bell RE, de Angelis MH, Steel KP, Avraham KB (2004) A *Myo7a* mutation cosegregates with stereocilia defects and low-frequency hearing impairment. Mamm Genome 15:686–697.

Riazuddin S, Khan SN, Ahmed ZM, Ghosh M, Caution K, Nazli S, Kabra M, Zafar AU, Chen K, Naz S, Antonellis A, Pavan WJ, Green ED, Wilcox ER, Friedman PL, Morell RJ, Friedman TB (2006) Mutations in *TRIOBP*, which encodes a putative cytoskeletal-organizing protein, are associated with nonsyndromic recessive deafness. Am J Hum Genet 78:137–143.

Royaux IE, Belyantseva IA, Wu T, Kachar B, Everett LA, Marcus DC, Green ED (2003) Localization and functional studies of pendrin in the mouse inner ear provide insight about the etiology of deafness in pendred syndrome. J Assoc Res Otolaryngol 4: 394–404.

Ruf RG, Xu PX, Silvius D, Otto EA, Beekmann F, Muerb UT, Kumar S, Neuhaus TJ, Kemper MJ, Raymond RM, Jr., Brophy PD, Berkman J, Gattas M, Hyland V, Ruf EM, Schwartz C, Chang EH, Smith RJ, Stratakis CA, Weil D, Petit C, Hildebrandt F (2004) SIX1 mutations cause branchio-oto-renal syndrome by disruption of EYA1-SIX1-DNA complexes. Proc Natl Acad Sci USA 101:8090–8095.

Sabag AD, Dagan O, Avraham KB (2005) Connexins in hearing loss: a comprehensive overview. J Basic Clin Physiol Pharmacol 16:101–116.

Sambrook J, Russell D (2001) Molecular Cloning: A Laboratory Manual. Cold Spring Harbor, NY: Cold Spring Harbor Laboratory Press.

Sanchez-Martin M, Rodriguez-Garcia A, Perez-Losada J, Sagrera A, Read AP, Sanchez-Garcia I (2002) SLUG (SNAI2) deletions in patients with Waardenburg disease. Hum Mol Genet 11:3231–3236.

Schulze A, Downward J (2001) Navigating gene expression using microarrays-a technology review. Nat Cell Biol 3:E190–195.

Scott DA, Kraft ML, Stone EM, Sheffield VC, Smith RJ (1998) Connexin mutations and hearing loss. Nature 391:32.

Scott HS, Kudoh J, Wattenhofer M, Shibuya K, Berry A, Chrast R, Guipponi M, Wang J, Kawasaki K, Asakawa S, Minoshima S, Younus F, Mehdi SQ, Radhakrishna U, Papasavvas MP, Gehrig C, Rossier C, Korostishevsky M, Gal A, Shimizu N, Bonne-Tamir B, Antonarakis SE (2001) Insertion of beta-satellite repeats identifies a transmembrane protease causing both congenital and childhood onset autosomal recessive deafness. Nat Genet 27:59–63.

Sellers JR (2000) Myosins: a diverse superfamily. Biochim Biophys Acta 1496:3–22.

Senften M, Schwander M, Kazmierczak P, Lillo C, Shin JB, Hasson T, Geleoc GS, Gillespie PG, Williams D, Holt JR, Muller U (2006) Physical and functional interaction between protocadherin 15 and myosin VIIa in mechanosensory hair cells. J Neurosci 26:2060–2071.

Shahin H, Walsh T, Sobe T, Lynch E, King MC, Avraham KB, Kanaan M (2002) Genetics of congenital deafness in the Palestinian population: multiple connexin 26 alleles with shared origins in the Middle East. Hum Genet 110:284–289.

Shahin H, Walsh T, Sobe T, Abu Sa'ed J, Abu Rayan A, Lynch ED, Lee MK, Avraham KB, King MC, Kanaan M (2006) Mutations in a novel isoform of TRIOBP that encodes a filamentous-actin binding protein are responsible for DFNB28 recessive nonsyndromic hearing loss. Am J Hum Genet 78:144–152.

Siemens J, Lillo C, Dumont RA, Reynolds A, Williams DS, Gillespie PG. Muller U (2004) Cadherin 23 is a component of the tip link in hair-cell stereocilia. Nature 428:950–955.

Skerrett IM, Di WL, Kasperek EM, Kelsell DP, Nicholson BJ (2004) Aberrant gating, but a normal expression pattern, underlies the recessive phenotype of the deafness mutant Connexin26 M34T. Faseb J 18:860–862.

Skvorak Giersch AB, Morton CC (1999) Genetic causes of nonsyndromic hearing loss. Curr Opin Pediatr 11:551–557.

Snoeckx RL, Huygen PL, Feldmann D, Marlin S, Denoyelle F, Waligora J, Mueller-Malesinska M, Pollak A, Ploski R, Murgia A, Orzan E, Castorina P, Ambrosetti U, Nowakowska-Szyrwinska E, Bal J, Wiszniewski W, Janecke AR, Nekahm-Heis D, Seeman P, Bendova O, Kenna MA, Frangulov A, Rehm HL, Tekin M, Incesulu A, Dahl HH, du Sart D, Jenkins L, Lucas D, Bitner-Glindzicz M, Avraham KB, Brownstein Z, del Castillo I, Moreno F, Blin N, Pfister M, Sziklai I, Toth T, Kelley PM, Cohn ES, Van Maldergem L, Hilbert P, Roux AF, Mondain M, Hoefsloot LH, Cremers CW, Lopponen T, Lopponen H, Parving A, Gronskov K, Schrijver I, Roberson J, Gualandi F, Martini A, Lina-Granade G, Pallares-Ruiz N, Correia C, Fialho G, Cryns K, Hilgert N, Van de Heyning P, Nishimura CJ, Smith RJ, Van Camp G (2005) GJB2 mutations and degree of hearing loss: a multicenter study. Am J Hum Genet 77:945–957.

Sobe T, Erlich P, Berry A, Korostichevsky M, Vreugde S, Avraham KB, Bonne-Tamir B, Shohat M (1999) High frequency of the deafness-associated 167delT mutation in the connexin 26 (GJB2) gene in Israeli Ashkenazim. Am J Med Genet 86:499–500.

Steel KP, Kros CJ (2001) A genetic approach to understanding auditory function. Nat Genet 27:143–149.

Street VA, Kallman JC, Kiemele KL (2004) Modifier controls severity of a novel dominant low-frequency myosinVIIA (MYO7A) auditory mutation. J Med Genet 41:e62.

Suzuki ST (1996) Structural and functional diversity of cadherin superfamily: are new members of cadherin superfamily involved in signal transduction pathway? J Cell Biochem 61:531–542.

Tachibana M, Takeda K, Nobukuni Y, Urabe K, Long JE, Meyers KA, Aaronson SA, Miki T (1996) Ectopic expression of MITF, a gene for Waardenburg syndrome type 2, converts fibroblasts to cells with melanocyte characteristics. Nat Genet 14:50–54.

Tekin M, Arnos KS, Pandya A (2001) Advances in hereditary deafness. Lancet 358:1082–1090.

Verpy E, Leibovici M, Zwaenepoel I, Liu XZ, Gal A, Salem N, Mansour A, Blanchard S, Kobayashi I, Keats BJ, Slim R, Petit C (2000) A defect in harmonin, a PDZ domain-containing protein expressed in the inner ear sensory hair cells, underlies Usher syndrome type 1C. Nat Genet 26:51–55.

Verpy E, Masmoudi S, Zwaenepoel I, Leibovici M, Hutchin TP, Del Castillo I, Nouaille S, Blanchard S, Laine S, Popot JL, Moreno F, Mueller RF, Petit C (2001) Mutations in a new gene encoding a protein of the hair bundle cause non-syndromic deafness at the DFNB16 locus. Nat Genet 29:345–349.

Vreugde S, Erven A, Kros CJ, Marcotti W, Fuchs H, Kurima K, Wilcox ER, Friedman TB, Griffith AJ, Balling R, Hrabe De Angelis M, Avraham KB, Steel KP (2002) Beethoven, a mouse model for dominant, progressive hearing loss DFNA36. Nat Genet 30:257–258.

Waardenburg PJ (1951) A new syndrome combining developmental anomalies of the eyelids, eyebrows and nose root with pigmentary defects of the iris and head hair and with congenital deafness. Am J Hum Genet 3:195–253.

Walsh T, Walsh V, Vreudge S, Hertzano R, Shahin H, Haika S, Lee MK, Kanaan M, King M-C, Avraham KB (2002) From flies' eyes to our ears: mutations in a human class III myosin cause progressive nonsyndromic hearing loss DFNB30. Proc Natl Acad Sci USA 99:7518–7523.

Walsh T, Abu Rayan A, Abu Sa'ed J, Shahin H, Shepshelovich J, Lee MK, Hirschberg K, Tekin M, Salhab W, Avraham KB, King MC, Kanaan M (2006) Genomic analysis of a heterogeneous Mendelian phenotype: multiple novel alleles for inherited hearing loss in the Palestinian population. Hum Genomics 2:203–211.

Wang A, Liang Y, Fridell RA, Probst FJ, Wilcox ER, Touchman JW, Morton CC, Morell RJ, Noben-Trauth K, Camper SA, Friedman TB (1998) Association of unconventional myosin *MYO15* mutations with human nonsyndromic deafness DFNB3. Science 280:1447–1451.

Watanabe A, Takeda K, Ploplis B, Tachibana M (1998) Epistatic relationship between Waardenburg syndrome genes *MITF* and *PAX3*. Nat Genet 18:283–286.

Weil D, Blanchard S, Kaplan J, Guilford P, Gibson F, Walsh J, Mburu P, Varela A, Levilliers J, Weston MD, Kelley PM, Kimberling WJ, Wagenaar M, Levi-Acobas F, Larget-Piet D, Munnich A, Steel KP, Brown SDM, Petit C (1995) Defective myosin VIIA gene responsible for Usher syndrome type 1B. Nature 374:60–61.

Weil D, Levy G, Sahly I, Levi-Acobas F, Blanchard S, El-Amraoui A, Crozet F, Philippe H, Abitbol M, Petit C (1996) Human myosin VIIA responsible for the Usher 1B syndrome: a predicted membrane-associated motor protein expressed in developing sensory epithelia. Proc Natl Acad Sci USA 93:3232–3237.

Weil D, Kussel P, Blanchard S, Levy G, Levi-Acobas F, Drira M, Ayadi H, Petit C (1997) The autosomal recessive isolated deafness, DFNB2, and the Usher 1B syndrome are allelic defects of the myosin-VIIA gene. Nat Genet 16:191–193.

Weiss S, Gottfried I, Mayrose I, Khare SL, Xiang M, Dawson SJ, Avraham KB (2003) The DFNA15 deafness mutation affects POU4F3 protein stability, localization, and transcriptional activity. Mol Cell Biol 23:7957–7964.

Wells AL, Lin AW, Chen LQ, Safer D, Cain SM, Hasson T, Carragher BO, Milligan RA, Sweeney HL (1999) Myosin VI is an actin-based motor that moves backwards. Nature 401:505–508.

Zehnder AF, Adams JC, Santi PA, Kristiansen AG, Wacharasindhu C, Mann S, Kalluri R, Gregory MC, Kashtan CE, Merchant SN (2005) Distribution of type IV collagen in the cochlea in Alport syndrome. Arch Otolaryngol Head Neck Surg 131:1007–1013.

Zelante L, Gasparini P, Estivill X, Melchionda S, D'Agruma L, Govea N, Mila M, Monica MD, Lutfi J, Shohat M, Mansfield E, Delgrosso K, Rappaport E, Surrey, Fortina P (1997) Connexin26 mutations associated with the most common form of non-syndromic neurosensory autosomal recessive deafness (DFNB1) in Mediterraneans. Hum Mol Genet 6:1605–1609.

Zheng L, Sekerkova G, Vranich K, Tilney LG, Mugnaini E, Bartles JR (2000) The deaf jerker mouse has a mutation in the gene encoding the espin actin-bundling proteins of hair cell stereocilia and lacks espins. Cell 102:377–385.

Zwaenepoel I, Mustapha M, Leibovici M, Verpy E, Goodyear R, Liu XZ, Nouaille S, Nance WE, Kanaan M, Avraham KB, Tekaia F, Loiselet J, Lathrop M, Richardson G, Petit C (2002) Otoancorin, an inner ear protein restricted to the interface between the apical surface of sensory epithelia and their overlying acellular gels, is defective in autosomal recessive deafness DFNB22. Proc Natl Acad Sci USA 99:6240–6245.

3
Cochlear Homeostasis and Homeostatic Disorders

PHILINE WANGEMANN

1. Principles of Homeostasis

The concept of homeostasis goes back to the French physiologist Claude Bernard (1813–1878), who stressed that two environments are most important in multicellular organisms, a "milieu extérieur" that surrounds the organism and a "milieu intérieur" as an extracellular fluid space in which the cells of the organism live (Bernard 1878). Bernard recognized the "fixité du milieu intérieur," which means that multicellular organisms strive to compensate and equilibrate the extracellular fluid environment against changes in the external environment. This concept was later termed *homeostasis* (Cannon 1929). Homeostasis is the nearly all-encompassing control of vital parameters and is maintained on more than one level, for example, on the level of the entire body as well as on the level of individual cells. Countless parameters are kept by the body within tight tolerances. Examples include core temperature, levels of O_2, CO_2, plasma pH, glucose, osmolarity, and plasma K^+ concentration. Cells also maintain many parameters within tight tolerances including cytosolic ion concentrations of K^+, Na^+, Cl^-, and Ca^{2+}, as well as the cytosolic pH, osmolarity, glucose, and ATP. These homeostatic efforts, which consume a good portion of most cells' energy, is to support life, which at the most molecular level consists of physical interactions and chemical reactions between proteins, lipids, carbohydrates, and nucleic acids.

2. Intracellular Homeostasis

Cochlear homeostasis encompasses all aspects of cochlear physiology except the sensory transduction. Major aspects include cellular energy production, maintenance of cation and anion concentrations and of cell volume, pH, and Ca^{2+}.

2.1 Cellular Energy Production and Redox Homeostasis

Glucose is the major energy substrate of the cochlea although other substrates can support cochlear function (Kambayashi et al. 1982). Glucose is supplied via the bloodstream, enters perilymph by facilitated diffusion, and is taken up into cells independent of insulin (Ferrary et al. 1987; Wang and Schacht 1990). Several glucose transporters have been identified: SLC2A5[1] (GLUT5) has been localized to outer hair cells (Nakazawa et al. 1995) and SLC2A1 (GLUT1) appears to be the main uptake mechanism in stria vascularis (Ito et al. 1993; Nakazawa et al. 1995; Takeuchi and Ando 1997). As an additional fuel, marginal cells of stria vascularis can take up pyruvate and lactate via the monocarboxylate transporters SLC16A1 (MCT1) and convert lactate to pyruvate (Shimozono et al. 1997; Okamura et al. 2001). Glucose and pyruvate are metabolized to CO_2 to yield energy mainly in the form of ATP (Fig. 3.1).

Several key enzymes and intermediates of glycolysis and the citric acid cycle had been analyzed in some of the earliest biochemical assays of microdissected structures of the cochlea (Thalmann et al. 1970). Oxygen consumption and metabolic activity are higher in stria vascularis than in the organ of Corti, which supported the concept that the cochlear transducer is powered by stria vascularis and that stria vascularis is the structure that generates the endocochlear potential (Thalmann et al. 1972; Marcus et al. 1978b; Ryan et al. 1982). In addition to glycolysis and the citric acid cycle, stria vascularis uses the pentose phosphate pathway (Marcus et al. 1978a) to generate large quantities of NADPH (Fig. 3.1) required in defense of free radical stress associated with high metabolic rates.

Free radicals, generated in controlled amounts, can serve as signaling molecules and are part of the cellular redox homeostasis. Unalleviated free radical stress, however, leads to redox imbalance, uncontrolled oxidation of proteins and lipids, and causes cell damage and unwanted cell death. Free radicals are a byproduct of an inefficient "leaky" electron transfer in the mitochondrial electron transport chain and of the mitochondrial cytochrome P450 systems. Incomplete reduction of O_2 generates the free radical superoxide anion $\cdot O_2^-$. In addition, $\cdot O_2^-$ can be generated by mitochondrial xanthine oxidases, by NADPH oxidases (Banfi et al. 2004), and by the cyclooxygenase pathway that is part of the arachidonic acid metabolism (Ziegler et al. 2004). Superoxide anion $\cdot O_2^-$ dismutates to hydrogen peroxide H_2O_2 (see reaction 3.1), which in the presence of Fe^{2+} or $\cdot O_2^-$ gives rise to the formation of the extremely aggressive hydroxyl radial $\cdot OH^-$ (Fenton reaction, see reactions 3.2 and Haber-Weiss reaction, see reaction 3.3). Alternatively, $\cdot O_2^-$ can react with nitric oxide radical $\cdot NO$ to form peroxynitrate $ONOO^-$ (reaction 3.4) that under acidic conditions (reactions 3.5 and 3.6) or in the presence of CO_2 (reactions 3.7 and 3.8) causes nitration of proteins, lipids

---

[1]Throughout this chapter, proteins are identified by their human gene names according the HUGO human genome nomenclature committee (http://www.gene.ucl.ac.uk/ nomenclature/). Commonly used alternative names are given in brackets.

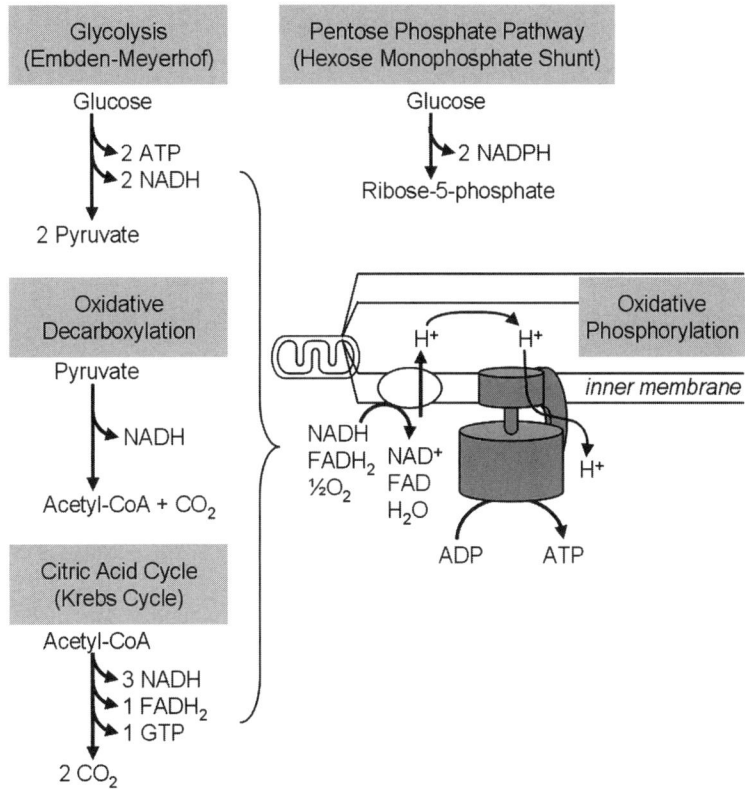

FIGURE 3.1. Metabolic energy production. Metabolism of glucose to CO_2 via glycolysis, oxidative decarboxylation, the citric acid cycle, and oxidative phosphorylation yields energy mainly in the form of ATP. Alternatively, glucose is metabolized by the pentose phosphate pathway to yield the reducing equivalent NADPH.

and nucleic acids (represented as R) via formation of the very aggressive nitrate radical $\bullet NO_2$ (Beckman et al. 1990; Lymar et al. 1996).

$$2\bullet O_2^- + 2H^+ \rightarrow O_2 + H_2O_2 \tag{3.1}$$

$$H_2O_2 + Fe^{2+} \rightarrow Fe^{3+} + OH^- + \bullet OH \tag{3.2}$$

$$H_2O_2 + \bullet O_2^- \rightarrow OH^- + \bullet OH + O_2 \tag{3.3}$$

$$\bullet O_2^- + \bullet NO \rightarrow ONOO^- \tag{3.4}$$

$$ONOO^- + H^+ \rightarrow [HOONO] \rightarrow \bullet OH^- + \bullet NO_2 \tag{3.5}$$

$$2 \bullet NO_2 + R^- \rightarrow R + NO_2^- + NO - R \tag{3.6}$$

$$ONOO^- + CO_2 \rightarrow [ONO_2CO_2^-] \tag{3.7}$$

$$[ONO_2CO_2^-] + R + H^+ \rightarrow HCO_3^- + NO - R \tag{3.8}$$

Maintenance of redox homeostasis is accomplished by cellular defenses against free radical stress, which include the prevention of the formation of free radicals and detoxification of free radicals. Given that iron is a catalyst for free radicals, the availability of free iron is carefully controlled. Iron is chelated by transferrin, a protein dimer that binds 2 iron atoms. Iron bound to transferrin is taken up into cells via transferrin receptor. In the cytosol, iron is either bound to transferrin or confined by ferritin, a cannonball-like protein multimer that houses up to 4500 iron atoms (Eisenstein 2000). The role of iron homeostasis for inner ear function is evident from the findings that ototoxicity is enhanced by iron and alleviated by the administration of iron chelators (Song et al. 1998; Conlon and Smith 1998). Control of tissue pH, CO_2, and $\bullet NO$ are additional homeostatic mechanisms intertwined with the defense against free radicals. Acidification aggravates and alkalinization alleviates ototoxicity (Tanaka et al. 2004), and inhibition of nitric oxide synthase alleviates ischemia reperfusion injury to the inner ear (Tabuchi et al. 2001).

Direct defense mechanisms against free radicals aim to scavenge radicals or convert aggressive radicals to less toxic radicals. The first line of defense is superoxide dismutase that converts superoxide anion $\bullet O_2^-$ to hydrogen peroxide H_2O_2 (see reaction 3.1). Although H_2O_2 is potentially harmful, this reaction is of benefit, as long as H_2O_2 is rapidly detoxified before it can give rise to the formation of the more harmful $\bullet OH$ radicals.

$$\text{Superoxide dismutase}: 2 \bullet O_2^- + 2H^+ \rightarrow O_2 + H_2O_2 \tag{3.9}$$

$$\text{Catalase}: 2H_2O_2 \rightarrow O_2 + 2H_2O \tag{3.10}$$

$$\text{Glutathione peroxidase}: H_2O_2 + 2GSH \rightarrow 2H_2O + GSSG \tag{3.11}$$

$$\text{Glutathione reductase}: GSSG + NADPH + H^+ \rightarrow GSH + NADP^+ \tag{3.12}$$

Catalase and glutathione peroxidase are the second line of defense catalyzing the conversion of hydrogen peroxide to O_2 and H_2O (see reactions 3.10 and 3.11). Glutathione peroxidase oxidizes glutathione (GSH) to glutathione disulfide (GSSG) in this process. GSH is a tripeptide composed of glutamate, cysteine, and glycine that is present in the cytosol in millimolar concentration and can convert hydrogen peroxide to O_2 and H_2O even in the absence of glutathione peroxidase, albeit more slowly. GSH is restored by reduction of GSSG into 2 moles of GSH in a reaction catalyzed by glutathione reductase, an enzyme that uses NADPH as a cofactor (see reaction 3.12).

Stria vascularis and organ of Corti contain high activities of superoxide dismutase, catalase, and glutathione peroxidase (Spector and Carr 1979; Pierson and Gray 1982). In the lateral wall, GSH immunoreactivity is preferentially distributed in basal cells and intermediate cells of stria vascularis, as well as in fibrocytes and capillary endothelial cells of the spiral ligament (Usami et al. 1996). Glutathione does not only play a role in free radical defense but also in detoxification reactions in which denatured proteins or xenobiotics are conjugated to glutathione in a reaction catalyzed by glutathione S-transferases (GSTs). Several isoforms of GST have been found in intermediate cells and the basal cells of the stria vascularis and various types of fibrocytes in the spiral ligament that may be serve the need to detoxify metabolic waste products (El Barbary et al. 1993; Takumi et al. 2001.

Defenses against free radical stress are not static but adaptable. Moderate noise stimulation, which may lead to low levels of free radicals, has been shown to enhance defense mechanisms and cause protection against noise induced hearing loss (Jacono et al. 1998). Similarly, preadministration of ototoxic drugs at nontoxic concentrations provides protection (Oliveira et al. 2004). Defense mechanisms, however, can be overwhelmed by ischemia reperfusion, intense noise stimulation, in homeostatic disorders, and by ototoxic drugs including aminoglycoside antibiotics and platinum-based chemotherapeutic agents (see Henderson et al., Chapter 7; Rybak et al., Chapter 8).

2.2 Cellular Na^+, K^+, and Cl^- Homeostasis

Cells use much of the energy available in form of ATP for the maintenance of steep Na^+ and K^+ gradients across the plasma membrane. Typical cytosolic concentrations of 5–15 mM Na^+ and 100–120 mM K^+ are 10- to 40-fold different from typical extracellular concentrations of 150 mM Na^+ and 3–4 mM K^+. These steep Na^+ and K^+ gradients that are established by Na^+/K^+-ATPases in virtually all cells and serve as sources of energy for cellular homeostasis and for cell signaling. Na^+-coupled transporters, which use the steep Na^+ gradient as energy source, include cotransporters and exchangers (Fig. 3.2). Some of these transporters are electrogenic and contribute to the membrane potential or use the membrane potential as an additional source of energy. The steep K^+ gradient in conjunction with K^+-selective channels defines the resting membrane potential in most cells. A notable exception are strial marginal cells and vestibular dark cells that maintain their membrane potential predominantly by a basolateral Cl^- conductance (Wangemann and Marcus 1992; Takeuchi and Irimajiri 1994).

Cytosolic concentrations for Cl^- vary greatly between cell types ranging between 5 and 50 mM. Uptake of Cl^- can be mediated in exchange for metabolically

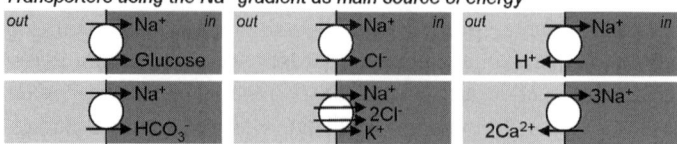

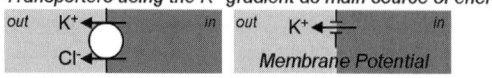

FIGURE 3.2. Establishment and use of Na^+ and K^+ gradients. The Na^+/K^+-ATPases establishes steep Na^+ and K^+ gradients that are used by cotransporters and exchangers as energy sources. The K^+ gradient in conjunction with K^+ channels defines in most cells the cytosolic side negative membrane potential.

generated HCO_3^- or driven by the Na^+ gradient via a $Na^+/2Cl^-/K^+$ cotransporter or a Na^+/Cl^- cotransporter (Fig. 3.2). Elimination of Cl^- can be driven by the membrane potential via Cl^- channels or by the K^+ gradient via KCl cotransporters. Homeostasis of the cytosolic salinity, which is defined mainly by the concentrations of K^+, Cl^-, and HCO_3^- ensures the maintenance of a suitable environment for proteins, lipids, and nucleic acids.

2.3 Cell Volume Regulation

Cellular function depends on a normal cell volume that ensures appropriate proximities between molecules in an environment of normal ionic strength and osmolarity, and on factors that affect protein configuration, protein–protein interactions, and enzyme activity (Lang et al. 1998). Challenges to the constancy of cell volume can originate with changes in the osmolarity or electrolyte composition of the extracellular environment, with mismatches between uptake and release of osmolytes such as salts and sugars, and with catabolic formation of osmolytes from osmotically inactive macromolecules or the anabolic fixation of osmolytes into macromolecules. In response to excessive cell swelling, cells respond with a regulatory volume decrease that may consist of the release of osmolytes. Most commonly, cells release KCl via K^+ or nonselective cation channels and Cl^- channels or via KCl cotransporters (Fig. 3.3). Conversely, in response to excessive cell shrinking, cells respond with a regulatory volume increase that commonly consists of the uptake of NaCl via activation of $Na^+/2Cl^-/K^+$ cotransporters or parallel activation of Na^+/H^+ and Cl^-/HCO_3^- exchangers.

Cell volume, however, is not just a parameter that is kept constant. Mechanisms of cell volume regulation have been incorporated into normal cell function,

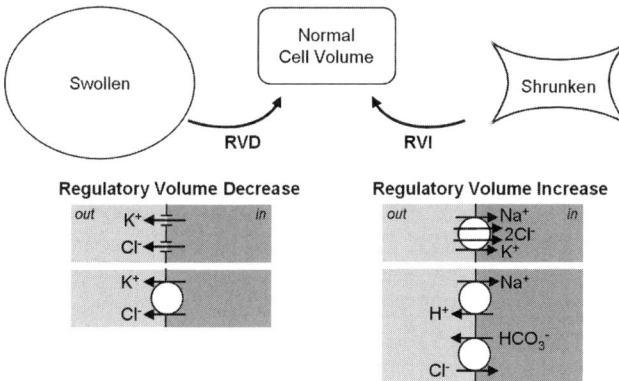

FIGURE 3.3. Cell volume regulation. Cell swelling is corrected by a regulatory volume decrease (RVD) and cell shrinking by a regulatory volume increases (RVI). Mechanisms of RVD include release of osmolytes via K^+ and Cl^- channels or KCl cotransporters. Mechanisms of RVI include uptake of osmolytes via the $Na^+/2Cl^-/K^+$ cotransporter.

where cell volume plays the role of a cellular signaling mechanisms much like a second messenger. Cell volume can communicate changes in ion transport across the basolateral membrane to the apical membrane and vice versa. In vestibular dark cells, for example, cell volume communicates changes in the rate of transport across the basolateral membrane to the apical membrane. The apical K^+ channel in vestibular dark cells is activated by cell swelling and the basolateral $Na^+/2Cl^-/K^+$ cotransporter by cell shrinkage (Wangemann and Shiga 1994; Wangemann et al. 1995b). Cell volume couples the two cell membranes; increased uptake of ions across the basolateral membrane causes cell swelling, which provides the signal to activate K^+ secretion across the apical membrane.

2.4 Cellular pH Regulation

The cytosolic pH is another critical variable in the cytosolic environment that affects many fundamental homeostatic mechanisms. For example, cytosolic acidification inhibits protein synthesis and favors the generation of toxic free radicals. Cells are under a constant thread of acidification since energy metabolism generates CO_2 as an end product (Fig. 3.1), which causes cellular acidification through the generation of H^+ (See reaction 3.13).

$$CO_2 + H_2O \leftrightarrow HCO_3^- + H^+ \tag{3.13}$$

The spontaneous conversion of CO_2 to HCO_3^- and H^+ occurs very slowly but can be facilitated by carbonic anhydrases. Metabolically active tissues as well as epithelia engaged in acid and base transport express large quantities of carbonic anhydrases to maintain the equilibrium between CO_2, HCO_3^-, and

H^+. The expression of carbonic anhydrase has therefore long been associated with high rates of ion transport. The dissolved gas CO_2 crosses most cell membranes relatively freely, in contrast to the solutes HCO_3^- and H^+, which require specialized transporters. Virtually all cells contain one or more acid removal mechanisms that either export H^+ or take up HCO_3^- to maintain the usual cytosolic pH of 7.2 (Fig. 3.4). Uptake of HCO_3^- is a mechanism of acid removal equivalent to H^+ extrusion because HCO_3^- traps H^+ and forms the freely diffusible CO_2 that leaves the cell by diffusion (Romero et al. 2004; Mount and Romero 2004). Carbonic anhydrases associate with HCO_3^- transporters to generate highly effective transport metabolons. The first metabolon described consisted of a Cl^-/HCO_3^- exchanger and two carbonic anhydrases, one with an extracellular and one with an intracellular located catalytic domain (Vince and Reithmeier 1998; Sterling et al. 2001). Analogous metabolons include a Na^+/HCO_3^- cotransporter.

Cellular pH regulation is not limited to the homeostasis of the cytosol. Secretory vesicles including vesicles storing neurotransmitters, lysosomes and internalized vesicles require for normal function a luminal pH of 4.5–6.5, which is much more acidic than the cytosol (Moriyama et al. 1992). Acid secretion into these intracellular compartments is mediated by vacuolar H^+-ATPases, which are large protein complexes consisting of at least 10 subunits (Wagner et al. 2004).

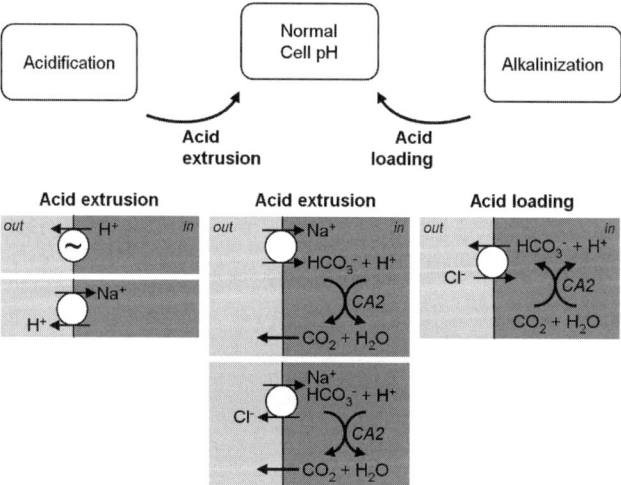

FIGURE 3.4. Cellular pH regulation. Cellular pH regulation consists of a balance between acid production, pH buffering, acid extrusion, and acid loading. Acid extrusion can be mediated by H^+-ATPases or driven by the Na^+ gradient and mediated by Na^+/H^+ exchangers. Uptake of HCO_3^- is an alternative mechanism of acid extrusion because HCO_3^- traps H^+ and forms the freely diffusible CO_2 that leaves the cell by diffusion. Uptake of HCO_3^- can be driven by the Na^+ gradient via the Na^+/HCO_3^- cotransporters or by Na^+-coupled Cl^-/HCO_3^- exchangers. Acid loading in response to cytosolic alkalinization can be driven by Cl/HCO_3^- exchangers.

The steep H^+ gradient across the vesicular membrane is used by neurotransmitter containing vesicles to drive the uptake of neurotransmitters. Lysosomes use the low luminal pH for the activation of enzymes and vesicles containing internalized receptors require the low luminal pH for the dissociation of ligand receptor complexes. The expression of vacuolar H^+-ATPases is not limited to cellular vesicles and organelles but occurs also in the plasma membrane where vacuolar H^+-ATPases mediate cytosolic acid extrusion and participate in the vectorial transport of acids or bases.

In the cochlea prominent acid extrusion mechanisms exist in the stria vascularis and spiral ligament, in the outer sulcus, in outer hair cells, and in interdental cells of the spiral limbus. Strial marginal cells contain vacuolar H^+-ATPase in their apical membrane, which was identified by the subunit ATP6V1E, and K^+/H^+-ATPase in their basolateral membrane (Stankovic et al. 1997; Shibata et al. 2006). Notably, marginal cells lack the nonessential H^+-ATPase subunit ATP6V1B1 (Karet et al. 1999b). Functional evidence for H^+-ATPases in marginal cells has not yet been obtained. Further, marginal cells contain the Na^+/H^+ exchanger SLC9A1 (NHE1) in their basolateral membrane, which likely plays a major role in acid extrusion (Wangemann et al. 1996a; Bond et al. 1998). Interestingly, although marginal cells are metabolically highly active and thus can be expected to generate notable amounts of CO_2, they appear to lack carbonic anhydrase (Lim et al. 1983; Yamashita et al. 1992; Okamura et al. 1996). Several cells adjacent cells, however, contain this enzyme including erythrocytes in the bloodstream, strial intermediate, and basal cells as well as fibrocytes of the spiral ligament (Lim et al. 1983; Okamura et al. 1996). In fact, erythrocytes as well as type I and III fibrocytes bordering stria vascularis and bone contain the highly active carbonic anhydrase isoform CA2 (Spicer and Schulte 1991). In addition, several HCO_3^- transporters have been identified in fibrocytes of the spiral ligament as well as in outer sulcus and spiral prominence epithelial cells including the Cl^-/HCO_3^- exchangers SLC4A2 (AE2) and SLC26A4 (pendrin) and the Na^+/HCO_3^- cotransporter SLC4A7 (NBC3) (Stankovic et al. 1997; Bok et al. 2003; Wangemann et al. 2004). It is conceivable that intermediate cells and basal cells of stria vascularis in conjunction with spiral ligament fibrocytes provide a buffer system between sources and sinks for CO_2. Metabolically active strial marginal cells as well as certain fibrocytes may be seen as sources of CO_2 whereas plasma, endolymph, and perilymph may be seen as sinks.

Outer hair cells contain in their basolateral membrane the Na^+/H^+ exchanger SLC9A1 (NHE1), which most likely functions as an acid extrusion mechanism (Ikeda et al. 1992a; Bond et al. 1998). Carbonic anhydrase activity has been found to be limited in outer hair cells to the area of the cuticular plate (Okamura et al. 1996). In addition, outer hair cells contain the Cl^-/HCO_3^- exchanger SLC4A2 (Zimmermann et al. 2000). It is unclear whether SLC4A2 functions as an acid extrusion mechanism because it is unclear whether outer hair cells maintain an outwardly directed Cl^- gradient that could drive the uptake of HCO_3^-. It may be more likely that SLC4A2 participates in the maintenance of cell volume (Cecola and Bobbin 1992).

Interdental cells of the spiral limbus express in their apical membrane vacuolar H^+-ATPase, which was identified by two subunits, ATP6V1E and ATP6V1B1 (Stankovic et al. 1997; Karet et al. 1999b). Further, interdental cells express the cytosolic carbonic anhydrases CA1 and CA3 and the Cl^-/HCO_3^- exchangers SLC4A2 in their basolateral membrane (Yamashita et al. 1992; Stankovic et al. 1997). In the absence of functional data, the significance of these transporters is currently unclear although it is conceivable that interdental cells are engaged in H^+ secretion and HCO_3^- reabsorption and the maintenance of pH in endolymph.

2.5 Cellular Ca^{2+} Regulation

Ca^{2+} is both a key second messenger involved in cell signaling as well as a cytotoxin. The cytosolic Ca^{2+} concentration under resting conditions of approximately 100 nM is approximately 10,000-fold lower than the interstitial Ca^{2+} concentration of 1 mM. This enormous gradient is carefully controlled to prevent ambiguity in cell signaling and cell death due to Ca^{2+} overload (Fig. 3.5). Increases in the cytosolic Ca^{2+} concentration are used to translate mechanical signals such as cellular deformation and chemical signals such as hormones, neurotransmitters, and growth factors into a variety of cellular actions such as regulation of enzyme activities, neurotransmitter release, salt and water secretion, contraction, proliferation, and cell death. Increases in the cytosolic

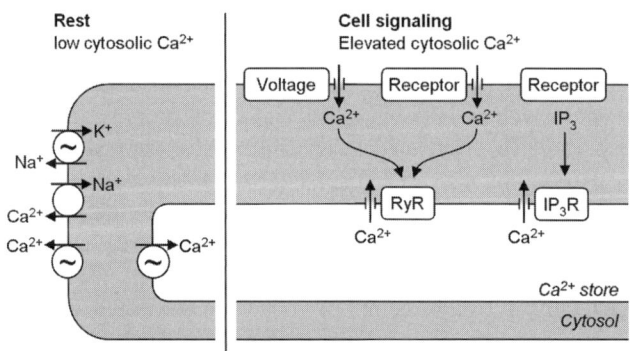

FIGURE 3.5. Cellular Ca^{2+} regulation. Cellular Ca^{2+} regulation consists mainly of Ca^{2+} export, sequestration in Ca^{2+} stores and Ca^{2+} buffering. Ca^{2+} extrusion can be mediated by PMCA Ca^{2+}-ATPases that pump Ca^{2+} out of the cell or by SERCA Ca^{2+}-ATPases that pump Ca^{2+} into cytosolic stores. Alternatively, Ca^{2+} extrusion can be driven by the Na^+ gradient established by the Na^+/K^+-ATPase and mediated by Na^+/Ca^{2+} exchangers. Ca^{2+} mediated cell signaling entails increases in the cytosolic Ca^{2+} concentration. Ca^{2+} increases can be mediated by voltage- or receptor-gated Ca^{2+} channels in the plasma membrane or by plasma membrane receptors that cause release of Ca^{2+} from cytosolic stores via IP_3 receptors (IP_3R). Increases in the cytosolic Ca^{2+} concentration can be enhanced by Ca^{2+}-induced Ca^{2+} release via ryanodine receptors (RyR).

Ca^{2+} concentration are mostly brief and limited in amplitude. A well defined spatial and temporal organization of signals makes it possible for cells to use Ca^{2+} as a second messenger for a variety of actions. Ca^{2+} stores mainly consisting of the endoplasmic reticulum provide a cytosolic Ca^{2+} store from which Ca^{2+} can be released via inositol-3-phosphate receptors or ryanodine receptors. Ca^{2+} binding proteins in the cytosol contribute to the buffering of Ca^{2+} necessary to keep increases in the cytosolic Ca^{2+} concentration local.

Several mechanisms are employed to maintain the resting cytosolic Ca^{2+} concentration or to terminate Ca^{2+}-mediated cell signaling. Mechanisms that directly use ATP as source of energy to move Ca^{2+} out of the cytosol include PMCA Ca^{2+}-ATPases located in the plasma membrane that pump Ca^{2+} into the extracellular space and SERCA Ca^{2+}-ATPase located in the sarco- or endoplasmic reticulum that pump Ca^{2+} out of the cytoplasm into the lumen of the sarco- or endoplasmic reticulum, which functions as a Ca^{2+} store. Other mechanisms rely on the Na^{+} gradient as energy source. Na^{+}/Ca^{2+} exchangers in the plasma membrane are driven by the Na^{+} gradient, which is established by the Na^{+}/K^{+}-ATPase, and under normal circumstances pump Ca^{2+} from the cytosol into the extracellular space.

Prominent expression of Ca^{2+} control mechanisms have been observed in the cochlea in inner and outer hair cells, in stria vascularis, Reissner's membrane, and in interdental cells. Inner and outer hair cells sequester Ca^{2+} into cytosolic stores. Ca^{2+} stores in inner and outer hair cells express ryanodine receptors and inositol-1,4,5-trisphosphate receptors, which permit highly localized releases of Ca^{2+} into the cytosol and thereby can function as amplification mechanisms for Ca^{2+} mediated cell signaling (Mammano et al. 1999; Grant et al. 2006). In addition, inner and outer hair cells buffer Ca^{2+} with Ca^{2+}-binding proteins such as calmodulin and calbindin-D28k (Pack and Slepecky 1995; Imamura and Adams 1996). High concentrations of Ca^{2+} buffer have been found particularly in outer hair cells (Hackney et al. 2005). Some Ca^{2+}-binding proteins are preferentially expressed in outer hair cells including oncomodulin (β-parvalbumin) (Sakaguchi et al. 1998) and others are preferentially expressed in inner hair cells including calretinin and α-parvalbumin (Dechesne et al. 1991; Pack and Slepecky 1995). The physiological significance of the high concentration of Ca^{2+}-binding proteins in outer hair cells is largely unclear. The lower concentration of Ca^{2+}-binding proteins in inner hair cells, however, may be related to synaptic transmission and the need for fast and unambiguous Ca^{2+} spikes.

Inner and outer hair cells also differ in their complement of mechanisms for Ca^{2+} extrusion across the basolateral membrane. Outer hair cells have been shown to contain the Na^{+}/Ca^{2+} exchanger SLC8A1 (NCX1) but lack prominent expression of PMCA Ca^{2+}-ATPases (Ikeda et al. 1992b; Oshima et al. 1997; Furuta et al. 1998; Wood, et al. 2004b). In contrast, inner hair cells express the PMCA Ca^{2+}-ATPase ATP2B1 (PMCA1b) in the basolateral membrane (Ikeda et al. 1992b; Oshima et al. 1997; Furuta et al. 1998). Both inner and outer hair cells express in the hair bundles the PMCA Ca^{2+}-ATPase ATP2B2 (PMCA2) that exports Ca^{2+} that enters the hair bundle via the transduction channel (Furuta et al. 1998; Street et al. 1998; Agrup et al. 1999).

Ca^{2+}-binding proteins and PMCA Ca^{2+}-ATPases are expressed not only in hair cells but also in other parts of the cochlea. Stria vascularis and interdental cells express Ca^{2+}-ATPase ATP2B1 (PMCA1b) and Reissner's membrane expresses Ca^{2+}-ATPase ATP2B2 (Furuta et al. 1998; Wood, et al. 2004b). Ca^{2+}-binding proteins are also prominently expressed in stria vascularis and spiral ligament fibrocytes (Ichimiya et al. 1994; Foster et al. 1994; Imamura and Adams 1996; Nakazawa 2001).

3. Control of the Pericellular Environment

Homeostasis is maintained not on the level of the cytosol but also on the level of the pericellular environment. Maintenance of the pericellular environment encompasses virtually the entire field of physiology including salt and fluid regulation, pH and Ca^{2+} regulation. Two aspects in the control of the pericellular environment, K^+ and glutamate buffering, are prominent in the cochlea and are thought to relate to homeostatic disorders.

3.1 K^+ Buffering

Most cells maintain their resting membrane potential via K^+ channels in conjunction with a high cytosolic and low extracellular K^+ concentration. A notable exception to this general rule are strial marginal cells and vestibular dark cells of the inner ear that maintain their membrane potential mainly by a Cl^- conductance (Wangemann and Marcus 1992; Takeuchi et al. 1995). Stimulation of cells that maintain their membrane potential by K^+ channels can lead to an efflux of K^+ that increases the K^+ concentration in the extracellular space and affects cellular responsiveness. Several mechanisms have been recognized to limit the amplitude of K^+ concentration changes in the extracellular environment. The premier mechanism is diffusion into unobstructed open fluid spaces. In the absence of such open spaces, K^+ can be buffered or siphoned away by adjacent cells or cellular networks (Orkand et al. 1966; Kuffler et al. 1966; Newman et al. 1984). For example, increases in the extracellular K^+ concentration have been observed during stimulation near sensory hair cells in the cochlea and the vestibular system (Johnstone et al. 1989; Valli et al. 1990). More general, neurons lose K^+ during action potential repolarization. The resulting accumulation of K^+ in the pericellular environment is buffered by adjacent glia cells (Orkand et al. 1966; Kuffler et al. 1966). Cellular buffering of extracellular K^+ depends on the uptake of K^+ at the site of extracellular accumulation and release of K^+ at remote and less critical sites. Several transport mechanisms may serve to take up K^+ including inward-rectifying K^+ channels, $Na^+/2Cl^-/K^+$ cotransporters, Na^+/K^+-ATPases and, at least under some circumstances, KCl contransporters.

Strongly inward-rectifying K^+ channels are well suited for K^+ buffering since they conduct K^+ influx more efficiently than K^+ efflux. A local increase in the extracellular K^+ concentration can set the local K^+ equilibrium potential below the membrane potential, which promotes K^+ influx into the buffering cell. The ensuing elevation of the cytosolic K^+ concentration sets the K^+ equilibrium potential above the membrane potential and promotes K^+ efflux preferentially through non- or less inward-rectifying K^+ channels. Strategic localization of more and less inward-rectifying K^+ channels can result in a siphoning of K^+ away from the site of extracellular accumulation. Such a mechanism has been observed in Mueller glia of the retina, which contains highly rectifying KCNJ2 (Kir2.1) K^+ channels in parts of the cell that contact retinal neurons and weakly rectifying homomeric (KCNJ10) Kir4.1 K^+ channel in the endfeet, which face the vitreous humor and blood vessels both of which may function as a large open space for the release of K^+ (Kofuji et al. 2002).

KCl cotransporters can serve in the uptake of K^+ and K^+ buffering under special circumstances, when the cytosolic Cl^- concentration is unusually low. Buffering the extracellular K^+ that varies around 5 mM can be accomplished by KCl cotransporters in cells that have a normal intracellular K^+ concentration of approximately 100 mM and a low intracellular Cl^- concentration of about 7 mM (Payne 1997). Such extremely low intracellular Cl^- concentrations, however, are uncommon.

Na^+/K^+-ATPases and $Na^+/2Cl^-/K^+$ cotransporters are well suited for K^+ buffering because they can mediate K^+ uptake under less stringent conditions than KCl cotransporters. Uptake of K^+ via a Na^+/K^+-ATPases is driven by hydrolysis of ATP and occurs with a high affinity for K^+ of approximately 0.9 mM (Kuijpers and Bonting 1969; Sweadner 1985). Uptake of K^+ via the $Na^+/2Cl^-/K^+$ cotransporters is driven by the Na^+ gradient that is established by an Na^+/K^+-ATPase and occurs with a slightly lower affinity for K^+ of 2.7 mM. $Na^+/2Cl^-/K^+$ cotransporters in conjunction with Na^+/K^+-ATPases can maintain very low extracellular K^+ concentrations and respond to minute changes in the extracellular K^+ concentration (Payne et al. 1995). For example, $Na^+/2Cl^-/K^+$ cotransporters and Na^+/K^+-ATPases expressed in strial marginal cells maintain a very low extracellular K^+ concentrations of estimated 1–2 mM in the intrastrial fluid space that is required for the generation of the endocochlear potential (Takeuchi et al. 2000). Consistent with the K^+ buffering activity, strial marginal cells and vestibular dark cells respond to increases in the extracellular K^+ concentration as small as 1 mM (Wangemann et al. 1996b). In contrast, K^+ buffering in other tissues, for example in the vicinity of neurons, requires the maintenance of higher K^+ concentrations in the range of 3–5 mM. Na^+/K^+-ATPases serve this need of K^+ buffering in conjunction with inward-rectifying K^+ channels that provide a back leak and prevent extracellular K^+ to drop to too low concentrations (D'Ambrosio et al. 2002).

3.2 Glutamate Buffering

Glutamate is generally accepted to be the main afferent neurotransmitter in the cochlea and the vestibular labyrinth. Buffering of glutamate is necessary since extracellular glutamate is cytotoxic. Cytotoxicity is mainly mediated by ionotropic glutamate receptors and an unduly elevation of the cytosolic Ca^{2+} concentration. Further, an elevated extracellular glutamate concentration can reverse the transport direction of cystine/glutamate antiporters that results in cytosolic cystine depletion followed by glutathione depletion and loss of protection from oxidative stress leading to cell death (Rimaniol et al. 2001). The glutamate transporter SLC1A3 (GLAST) is localized in supporting cells surrounding cochlear inner hair cells and vestibular hair cells (Li et al. 1994; Furness and Lehre 1997). No evidence for other glutamate transporters has been found in the cochlear or vestibular labyrinth (Takumi et al. 1997).

Glutamate metabolism is best understood in the brain, where glutamate released from neurons is taken up via glutamate transporters into glia cells. Glia cells feed glutamate into two metabolic pathways, the glutamate–glutamine shuttle that serves to return glutamate to the neuron and the glutamate–lactate conversion that enters glutamate into the general metabolism (Danbolt 2001). Glutamate is converted to glutamine for the glutamate–glutamine shuttle in a reaction (glutamate + NH_4^+ + ATP $\rightarrow$ glutamine + ADP) that consumes ATP, requires NH_4^+, and is catalyzed by glutamine synthase. Glutamine is then exported and returned to neurons via glutamine transporters. Neurons convert glutamine to glutamate in a reaction that is catalyzed by phosphate-activated glutaminase, which completes the cycle. The alternative to the glutamate–glutamine shuttle is that glutamate is converted to lactate and entered as fuel into the general energy metabolism. Glutamate can also be converted to α-ketoglutarate, fed into the Krebs cycle, and metabolized to maleate that is decarboxylated to pyruvate and may be reduced to lactate (Danbolt 2001). Pyruvate or lactate is then shuttled back to neurons. Neurons recarboxylate pyruvate to maleate in order to replenish the Krebs cycle for the loss of α-ketoglutarate that is used for glutamate production. Glutamate synthesis may either depend on a transamination reaction or obtained in a reaction of α-ketoglutarate and NH_4^+ that is catalyzed by glutamate dehydrogenase.

Convincing evidence for a glutamate–glutamine shuttle has been obtained in the vestibular labyrinth, where SLC1A3 and glutamine synthase were found colocalized in supporting cells and where phosphate-activated glutaminase was found in hair cells (Takumi et al. 1997, 1999; Ottersen et al. 1998). In the cochlea, however, the issue is less clear. SLC1A3 and glutamine synthase are not clearly colocalized. Expression of SLC1A3 in the organ of Corti is limited to the inner hair cells' supporting cells, whereas glutamine synthase is expressed in the spiral limbus and in outer sulcus cells (Eybalin et al. 1996; Ottersen et al. 1998). It is conceivable that transcellular metabolism of glutamate contributes to glutamate buffering in the organ of Corti. Transcellular metabolism may consist of the uptake of glutamate by the inner hair cells' supporting cells, diffusion via gap junctions to neighboring supporting cells that assist in the metabolism of glutamate. Whether supporting cells of the cochlea and the vestibular labyrinth contain mechanisms for NH_4^+ uptake as seen in retinal glia cells is currently unclear (Marcaggi and Coles 2001).

Two lines of evidence support the importance of glutamate buffering in the cochlea. Acoustic overstimulation of mice deficient in the glutamate transporter SLC1A3 causes accumulation of glutamate in perilymph of scala tympani and exacerbation of noise-induced hearing loss (Hakuba et al. 2000). Further, selective ablation of the gap junction GJB2 (Cx26) from the organ of Corti, which can be expected to interrupt transcellular metabolism of glutamate, results in the apoptotic cells death of supporting cells of the organ of Corti that begins with the inner hair cells' supporting cells which express SLC1A3 (Cohen-Salmon et al. 2002).

4. Cochlear Blood Flow Regulation

Blood flow serves to provide the delivery of O_2 and metabolic substrates and the removal of CO_2 and metabolic waste products. Rates of blood flow in essential structures such as the brain are held constant in the presence of variations in systemic blood pressure. A similar autoregulation has been found in the cochlea (Quirk et al. 1989; Brown and Nuttall 1994). Blood flow and metabolic rates within the cochlea, however, are not uniform and thus provide different challenges for the regulatory processes. Among the different tissues of the cochlea, stria vascularis and spiral ligament have the highest capillary densities, metabolic rates, and acute sensitivity to hypoxia (Konishi et al. 1961; Thalmann et al. 1972; Ryan et al. 1982). Auditory stimulation imposes another variable need for metabolically driven increases in blood flow. Large increases in the metabolic rate in response to acoustic stimulation have been observed in the organ of Corti, in spiral ganglion cells, and in the eighth nerve. Smaller increases in metabolism were seen in stria vascularis and spiral ligament that have very high basal rates (Ryan et al 1982). These increases in metabolic rate apparently increase cochlear blood flow as evident from an increase in O_2 levels in perilymph and an increase in blood cell velocity in the lateral wall (Quirk et al. 1992; Scheibe et al. 1992). The causal links between blood flow and metabolic need in spiral ganglia, stria vascularis, or spiral ligament are largely unknown. Astrocytes in the central nervous system have been implicated as links between neuronal activity and cerebrovascular blood flow regulation. Evidence supports a role for cyclooxygenase-2 metabolites, epoxyeicosatrienoic acids, adenosine, and neuronally derived nitric oxide in the coupling of blood flow to neuronal activity.

Part of the regulation of cochlear blood flow occurs in the spiral modiolar artery, which provides the main blood supply to the cochlea. Whereas flow through the lateral wall needs to be maintained at high rates, flow into branches supplying the spiral ganglion may be regulated and adaptable to auditory stimulation. The finding that smooth muscle cells at branch points of the spiral modiolar artery are exclusively endowed with α_{1A} adrenergic receptors supports the hypothesis that constriction of these cells alters the angle and thereby the flow distribution into the branch (Gruber et al. 1998; Wangemann and Wonneberger 2005).

Another level of cochlear blood flow regulation may occur in the lateral wall. Capillaries of the spiral ligament comprise a vascular bed parallel to the capillary bed of stria vascularis. Capillaries of the spiral ligament have been shown to be contractile and responsive to the vasocontractor endothelin and the vasodilator nitric oxide (Sadanaga et al. 1997). In addition, prostacyclin synthase, an enzyme necessary for the synthesis of the vasodilator prostacyclin, and nitric oxide synthase have been found in capillary endothelial cells of the spiral ligament, which supports the concept that blood flow in the spiral ligament can be regulated (Konishi et al. 1998).

In general, blood flow in a vessel is determined by the vascular diameter, the blood pressure differential between the inflow and the outflow site, the length of the vessel and the viscosity of the blood. Control of the vascular diameter is the most effective means of controlling blood flow since, according to the law of Hagen–Poiseuille, a change in flow is related to the fourth power of the change in the diameter. The vascular diameter is determined by the degree of constriction or relaxation of the smooth muscle cells in the vascular wall. A constriction of the smooth muscle cells reduces the diameter of the vascular lumen and thereby decreases blood flow, whereas a relaxation increases blood flow. Constriction of vascular smooth muscle cells can be mediated by an increase in intracellular Ca^{2+} concentration and by an increase in the Ca^{2+} sensitivity of the contractile apparatus (Fig. 3.6; Somlyo and Somlyo 2003). The Ca^{2+} sensitivity of the contractile apparatus is controlled by the phosphorylation state of myosin light chain phosphatase, which is a substrate of Rho-kinase (Kimura et al. 1996). Rho-kinase is a major regulator of smooth muscle contractility and a key target for treating cardiac and cerebral vasospasms (Sato et al. 2000; Shimokawa and Takeshita 2005; Yada et al. 2006).

Some textbooks imply that vascular function is quite uniform and that mechanisms are more or less present in all vessels (Vanhoutte 1978). This view, however, is not valid anymore. Recent advances in the field of vascular smooth muscle cell biology revealed that smooth muscle cells from different arterioles differ widely in their endowment with mechanisms that regulate the degree of contraction and relaxation. Investigations of the spiral modiolar artery at the cellular and molecular level are aided by recently developed in vitro preparations of cochlear arterials (Lamm et al. 1994; Wangemann and Gruber 1998; Wangemann et al. 1998). These preparations in concert with video-microscopy; microfluorometry; confocal microscopy; and electrophysiological, biochemical, and molecular biological techniques promise to lead to significant advances in the understanding of cochlear blood flow regulation.

Inappropriate vasoconstriction is a likely pathobiological mechanism that may lead to inner ear disorders that result in sudden sensorineural hearing loss. Increased levels of endothelin-1 and sphingosine-1-phosphate play an important role in some forms of cerebral and coronary vasospasms (Rubanyi and Polokoff 1994; Yatomi 2006). Endothelin-1 is synthesized by endothelial

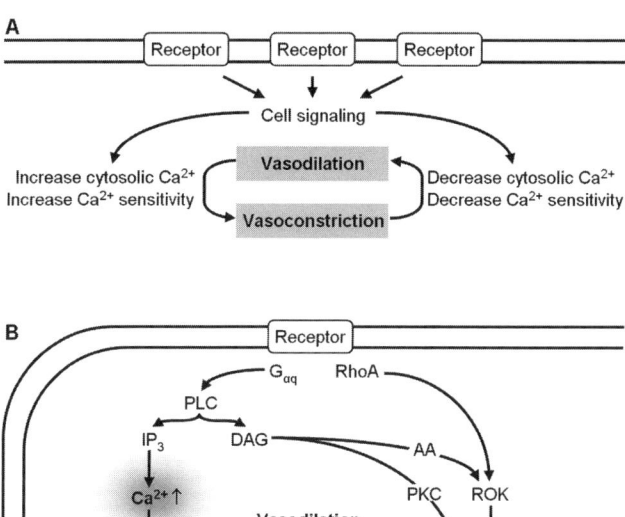

FIGURE 3.6. Cellular blood flow regulation. (**A**) The state of contractility of vascular smooth muscle cells is determined by the cytosolic Ca^{2+} concentration and the Ca^{2+} sensitivity of the myofilaments. Receptors that increase the cytosolic Ca^{2+} concentration and/or increase the Ca^{2+} sensitivity mediate vasoconstriction and receptors that decrease the cytosolic Ca^{2+} concentration and/or decrease the Ca^{2+} sensitivity mediate vasodilation. (**B**) Vasoconstriction is mediated by Ca^{2+} calmodulin (Ca^{2+} CaM) dependent activation of myosin light chain kinase (MLCK) and phosphorylation of myosin light chain (LC). Dephosphorylation of myosin light chain by myosin light chain phosphatase (MLCP) mediates vasodilation. Inhibition of MLCP by Rho kinase (ROK) mediated phosphorylation or binding to phosphorylated CPI-17 increases phosphorylation of myosin light chain and causes Ca^{2+} sensitization.

cells and released upon endothelial cell stimulation by physicochemical factors such as elevated oxidized low-density lipoproteins and hypoxia (Rakugi et al. 1990; Unoki et al. 1999; Xie, Bevan 1999). Sphingosine-1-phosphate is stored in relatively high concentrations in platelets, is released on platelet activation and may play an aggravating role in vascular diseases (Ohmori et al. 2004; Yatomi 2006). Endothelin-1 and sphingosine-1-phosphate cause sustained vasospasms of the spiral modiolar artery that are mediate via ET_A and sphingosine-1-phosphate receptors, cause a transient increase in the cytosolic Ca^{2+} concentration and a Rho-kinase mediated increase in the Ca^{2+} sensitivity of the myofilaments (Scherer et al. 2001; Scherer et al. 2002; Scherer et al. 2006). Interestingly, endothelin antagonists can prevent vasospasm in the spiral modiolar artery but fail to release vasospasms, which reduced their potential as drugs to treat sudden hearing loss with suspected vascular origin. In contrast, inhibitors of Rho-kinase have been shown to release vasospasms in the spiral modiolar artery (Scherer et al. 2005). Vasospasms in cardiac and cerebral arteries have already been treated successfully with Rho-kinase inhibitors. The development of novel and improved Rho-kinase inhibitors is underway driven by the economic potential associated with the recreational enhancement of erectile function.

Insufficient vasodilation may be another pathobiological mechanism that leads to sudden hearing loss. Among the most potent vasodilators are calcitonin gene related peptide (CGRP), substance P, and •NO (Carlisle et al. 1990; McLaren et al. 1993; Qiu et al. 2001; Vass et al. 2004). CGRP and •NO increase cochlear blood flow and relax smooth muscle cells of the isolated vessel (Hillerdal and Andersson 1991; Quirk et al. 1994; Herzog et al. 2002). •NO is likely to mediate vasodilation via cGMP and activation of guanylyl cyclase (Fessenden and Schacht 1997). CGRP acts in the spiral modiolar artery via activation of CGRP receptors, an increase of the second messenger cAMP leading to a transient activation of K^+ channels, a transient decrease in the cytosolic Ca^{2+} concentration, and a long-lasting Rho-kinase mediated Ca^{2+} desensitization of the myofilaments (Herzog et al. 2002). This signaling mechanism observed in the spiral modiolar artery differs from signaling mechanism found in other vessels, where CGRP mediates vasodilation via cGMP and •NO. The observation that signaling mechanisms of potent vasoconstrictors and vasodilators converge in the spiral modiolar artery on the control of Rho-kinase supports the notion that this kinase is an important drug target.

Smooth muscle cells of the spiral modiolar artery integrate information from various sources. Vasoconstrictors and dilators may originate from the innervation surrounding the vessel, from endothelial cells lining the vascular lumen, or from the smooth muscle cells themselves. Signal transduction mechanisms, which mediate these neurogenic, local, and paracrine regulation of smooth muscle contractility are now beginning to be understood which opens perspectives for the development of novel drugs for the treatment of inner ear disorders with a vascular etiology.

5. Maintenance of Inner Ear Fluids

The inner ear houses several unusual extracellular fluids including endolymph and perilymph (Fig. 3.7). Homeostasis of the composition of these fluids is intimately intertwined with the generation of the endocochlear potential and with sensory transduction.

5.1 K^+ Cycling

Endolymph is an unusual extracellular fluid in that the major monovalent cation is K^+ (Smith et al. 1954) rather than Na^+, which is the major monovalent cation in perilymph and most other extracellular fluids (Fig. 3.7). The presence of K^+

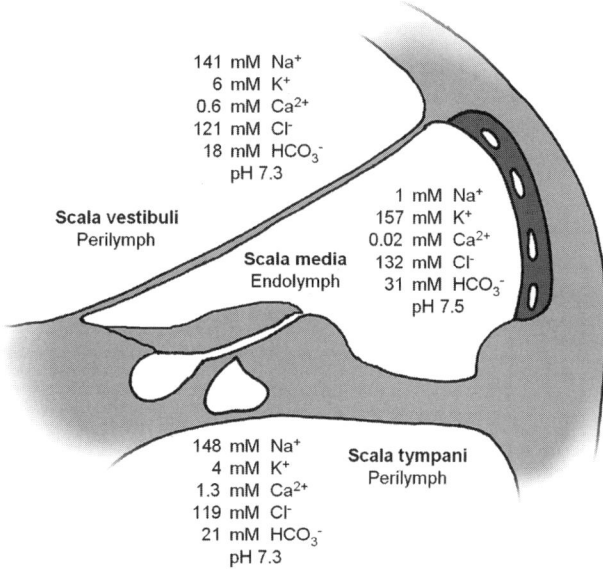

141 mM Na⁺
6 mM K⁺
0.6 mM Ca²⁺
121 mM Cl⁻
18 mM HCO₃⁻
pH 7.3

Scala vestibuli
Perilymph

Scala media
Endolymph

1 mM Na⁺
157 mM K⁺
0.02 mM Ca²⁺
132 mM Cl⁻
31 mM HCO₃⁻
pH 7.5

148 mM Na⁺
4 mM K⁺
1.3 mM Ca²⁺
119 mM Cl⁻
21 mM HCO₃⁻
pH 7.3

Scala tympani
Perilymph

FIGURE 3.7. Composition of cochlear fluids. Scala media of the cochlea is filled with endolymph and scala vestibule and scala tympani are filled with perilymph. Endolymph and perilymph differ greatly in their composition in particular in their cation composition.

in endolymph is of great importance, as K^+ provides the major charge carrier for sensory transduction. The choice of K^+ over other monovalent cations such as Na^+ has the advantage that an influx of K^+ into the sensory cells causes the least relative change in the cytosolic ion homeostasis compared to any other ion since K^+ is by far the most abundant ion in the cytosol (Wangemann 2002). Additional advantages of K^+ are that the mechanically gated transduction channel in the stereocilia of the sensory cells is highly K^+ permeable and that both influx into and efflux from the sensory cell can occur down the electrochemical gradient. This electrochemical gradient is produced by stria vascularis and by the K^+ selectivity of channels in the basolateral membrane of the sensory cells. Stria vascularis supported by the spiral ligament generates, at great energetic expense, a current that drives sensory transduction in the organ of Corti (Zidanic and Brownell 1990). The metabolic demand for the generation of this current requires a dense capillary network for the delivery of O_2 and glucose and the removal of CO_2. Spatial separation between stria vascularis as the source of the transduction current and organ of Corti as the sensory transducer has the advantage that the highly sensitive mechanosensor is somewhat isolated from the low-frequency vibrations associated with blood flow through capillaries.

Sensory transduction in the cochlea depends on the cycling of K^+ (Konishi et al. 1978) (Fig. 3.8). Endolymphatic K^+ is driven along the electrochemical gradient into the sensory hair cells via the apical transduction channel and out of the hair cells into perilymph via basolateral K^+ channels including KCNQ4 ($I_{K,n}$),

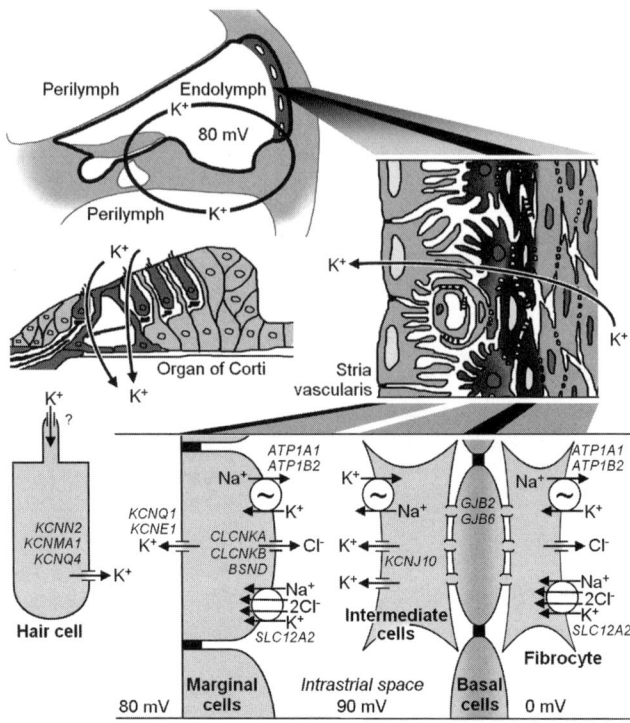

FIGURE 3.8. K^+ cycling and the endocochlear potential. Epithelial cells lining scala media and basal cells of stria vascularis are joined by tight junctions. The endocochlear potential is generated across the basal cell barrier and can be measured across the basal cell barrier as well as across the barrier between endolymph perilymph. K^+ is recycled between endolymph and perilymph. K^+ in endolymph enters the hair cells via the transduction channel in the apical membrane and leaves the hair cell via K^+ channels in the basolateral membrane. From perilymph, K^+ is taken up by fibrocytes in the lateral wall of the cochlea and transported from cell to cell via gap junctions into basal cells and from there into intermediate cells of stria vascularis. K^+ is released from the intermediate cells into the intrastrial space via the KCNJ10 K^+ channel that generates the endocochlear potential. Marginal cells take up K^+ from the intrastrial space and secrete K^+ into endolymph.

KCNN2 (I_{SK2}), and KCNMA1 ($I_{K,f}$) (Johnstone et al. 1989; Kros 1996). K^+ is taken up from perilymph by fibrocytes of the spiral ligament and transported from cell to cell via gap junctions GJB2 (Cx26) and GJB6 (Cx30) into strial intermediate cells (Kikuchi et al. 1995; Xia et al. 1999, 2000, 2001). K^+ is released from the intermediate cells into the intrastrial space via the KCNJ10 K^+ channel that generates the endocochlear potential (Marcus et al. 2002). Other K^+ channels may contribute to the delivery of K^+ into the intrastrial space (Takeuchi and Irimajiri 1996; Marcus et al. 2002; Nie et al. 2005). From the intrastrial space, K^+ is taken up by strial marginal cells and secreted into endolymph (Wangemann et al. 1995a; Wangemann 2002). K^+ is taken up across the basolateral membrane of strial marginal cells via the $Na^+/2Cl^-/K^+$ cotransporter SLC12A2 and the Na^+/K^+-ATPase ATP1A1/ATP1B2 and secreted across the apical membrane into endolymph via the K^+ channel KCNQ1/KCNE1 (Wangemann et al. 1995a; Shen et al. 1997). Na^+ and Cl^- taken up together with K^+ via the $Na^+/2Cl^-/K^+$ cotransporter SLC12A2 are recycled in the basolateral membrane via the Na^+/K^+-ATPase and Cl^- channels CLCNKA/BSND and CLCNKB/BSND that are associated with BSND (Barttin).

The pathway of K^+ through stria vascularis and through the hair cells is well established. In contrast, two concepts are promoted for the pathway from perilymph toward spiral ligament. One concept envisions that K^+ enters perilymph and flows through perilymph toward spiral ligament. This concept is supported by flux and current measurements (Zidanic and Brownell 1990; Salt and Ohyama 1993). The other concept envisions that K^+ released from the hair cells never enters perilymph but is taken up by adjacent supporting cells and funneled from cell to cell via a gap junction network toward the spiral ligament (Spicer and Schulte 1996). This concept is based on an interesting, but unproven, hypothesis derived from the finding of gap junctions and KCl transporters in supporting cells of the organ of Corti. The transport direction of KCl transporters, however, is outwardly directed given the usual stoichiometry and ionic gradients, which precludes their involvement in the uptake of K^+ (Warnock and Eveloff 1989; Adragna et al. 2004). Further, if the gap junctions GJB2 or GJB6 would be required for K^+ cycling, a disruption of K^+ delivery to stria vascularis and a collapse of scala media would have been expected in mice that lack these gap junction proteins (Cohen-Salmon et al. 2002; Teubner et al. 2003). The fact that no collapse was seen in mice lacking GJB2 or GJB6 argues against this concept that GJB2 or GJB6 are essential for K^+ cycling.

K^+ cycling in the cochlea is not limited to the loop through the sensory cells. Part of the current that is generated by stria vascularis is carried through outer sulcus cells and through Reissner's membrane (Konishi et al. 1978; Zidanic and Brownell 1990; Salt and Ohyama 1993). The significance of these parallel current loops may lie in the opportunity to fine tune the current through the sensory hair cells. Currents through Reissner's membrane are carried by Na^+ (Lee and Marcus 2003) and currents through outer sulcus epithelial cells are carried by K^+ and Na^+ (Marcus and Chiba 1999; Chiba and Marcus 1992).

K^+ cycling does not only occur in the cochlea but also in the vestibular labyrinth (Wangemann 1995). Endolymphatic K^+ flows into the sensory hair cells via the apical transduction channel and is released from the hair cells via basolateral K^+ channels (Valli et al. 1990). Fibrocytes connected by gap junctions including GJB2 may be involved in delivering K^+ to vestibular dark cells (Kikuchi et al. 1995). Extracellular K^+ is taken up into vestibular dark cells via SLC12A2 and ATP1A1/ATP1B2 (Marcus et al. 1987; Wangemann 1995). To conclude the cycle, K^+ is released into endolymph via the K^+ channel KCNQ1/KCNE1 (Marcus and Shen 1994).

Cochlear fluid homeostasis requires a well tuned balance between K^+ secretion and K^+ reabsorption. Failure to maintain this balance will result in an enlargement of the endolymphatic compartment as seen in Ménière's disease and Pendred's syndrome or result in a collapse of scala media as seen in Jervell–Lange-Nielsen syndrome. K^+ secretion by strial marginal cells and vestibular dark cells is sensitive to the basolateral K^+ concentration. Even small millimolar changes in the basolateral K^+ concentration alter the rate of K^+ secretion (Wangemann et al. 1995a; Wangemann et al. 1996b). It appears that the K^+ secretory cells are sensors for K^+ rivaling the exquisite sensitivity of the plasma K^+ sensor in the adrenal cortex that governs the release of aldosterone and thereby K^+ excretion in the kidney. The sensitivity of K^+ secretory cells in the cochlea and vestibular labyrinth ensures that K^+ does not accumulate in perilymph or in the intrastrial fluid space, which, in the vestibular system, would interfere with neuronal transmission and, in the cochlea, would interfere with the generation of the endocochlear potential.

K^+ cycling is subject control by acoustic stimulation. Intense acoustic stimulation causes a temporary threshold shift, reduces the endocochlear potential and endolymphatic K^+ concentration and alters the K^+ efflux pathway from endolymph to perilymph (Salt and Konishi 1979; Thorne et al. 2004). Sound stimulation releases ATP into cochlear fluids (Munoz et al. 2001). ATP is not only an energy equivalent but also a paracrine messenger. ATP released into perilymph reaches P2X receptors in the organ of Corti and reduces outer hair cell motility, which dampens the cochlear amplifier (Housley et al. 1999; Zhao et al. 2005). ATP released into endolymph stimulates P2X receptors in Reissner's membrane and the apical membrane of outer sulcus (King et al. 1998; Lee et al. 2001). P2X receptors are nonselective ion channels that in concert with basolateral K^+ channels support transcellular effluxes of K^+ from endolymph to perilymph. These cation effluxes generate currents that are parallel to the transduction current through the hair cells and effectively reduce the current density through the sensory hair cells. ATP released into endolymph activates not only P2X receptors but also P2Y4 receptors located in the apical membrane of strial marginal cells (Sage and Marcus 2002). P2Y4 receptors are G-protein-coupled receptors that reduce K^+ secretion across the apical membrane of strial marginal cells via protein kinase C-mediated phosphorylation and inhibition of the apical K^+ channel KCNQ1/KCNE1 (Marcus et al. 2005). A reduction in K^+ secretion can be expected to reduce the endocochlear potential and thereby the driving force for sensory transduction. A similar protective mechanism has been observed in the vestibular labyrinth. ATP inhibits K^+ secretion in vestibular dark cells via activation of apical P2Y4 receptor and activates parallel current loops via activation of P2X receptors in the apical membrane of vestibular transitional cells (Liu et al. 1995; Marcus et al. 1997; Lee et al. 2001).

K^+ cycling is also subject to systemic stimulation. Physical and emotional stresses leading to increased norepinephrine and epinephrine plasma levels enable a number of organ-specific responses that are mediated by β-adrenergic receptors and support a "fight–or-flight" reaction. For example, $β_1$-adrenergic receptors increase heart rate and force and $β_2$-adrenergic receptors open and moisturize airways and eyes. $β_1$-adrenergic receptors have been found in the organ of Corti, in the outer sulcus, and in stria vascularis (Fauser et al. 2004). β_1-Adrenergic receptors in stria vascularis and vestibular dark cells stimulate K^+ secretion via an increase in the cytosolic cAMP concentration (Wangemann et al. 1999, 2000). In addition, β_2-adrenergic receptors have been identified in the spiral ligament (Schimanski et al. 2001). Acceleration of K^+ cycling in the inner ear may be part of the preparation for a successful "fight-or-flight" reaction.

5.2 pH and HCO$_3^-$

The pH of cochlear endolymph of 7.5 is higher than the pH of perilymph (pH 7.3) or plasma (Wangemann et al. 2007). Similar observations have been made in the vestibular labyrinth (Nakaya et al. 2007). The higher pH of endolymph is consistent with a higher concentration of HCO_3^-, 31 versus 20 mM in perilymph (Sterkers et al. 1984; Ikeda et al. 1987c) (Fig. 3.7). The endolymphatic pH depends on H^+ and HCO_3^- secretion and on carbonic anhydrase activity (Sterkers et al. 1984; Ikeda et al. 1987c; Stankovic et al. 1997). Mechanisms of endolymphatic HCO_3^- homeostasis are not yet fully understood, but it is conceivable that the endolymphatic HCO_3^- concentration is maintained by secretory and absorptive mechanisms. Sites discussed for HCO_3^- secretion are spiral prominence and outer sulcus epithelial cells as well as spindle cells of stria vascularis that contain the Cl^-/HCO_3^- exchanger SLC26A4 (pendrin) in their apical membrane (Wangemann et al. 2004, 2007). Sites that may be engaged in the absorption of HCO_3^- may include the interdental cells of the spiral limbus. It is conceivable that HCO_3^-, which originates from metabolically generated CO_2, is cycled through endolymph before being eliminated from the cochlea via the capillary beds in stria vascularis, spiral ligament and spiral limbus. Support for this cycling of HCO_3^- comes from the observation that acoustic stimulation, which increases stria metabolism and CO_2 production, causes an alkalization of endolymph (Ikeda 1988).

The importance of pH homeostasis of the cochlear fluids is linked to cochlear function via the general pH sensitivity of ion channels, transporters, and metabolic enzymes. For example, Ca^{2+} absorption from endolymph via TRPV5 and TRPV6 Ca^{2+} channels is inhibited in Pendred syndrome, which is associated with an acidification of endolymph (Nakaya et al. 2007; Wangemann et al. 2007). Further, experimental maneuvers that cause acidification of cochlear fluids including inhibition of carbonic anhydrase, application of acidic fluids to the round window or flushing the round window membrane with CO_2 gas, have been shown to reduce the endocochlear potential (Sterkers et al. 1984; Ikeda et al. 1987c; Ikeda and Morizono 1989a,1989b). These short-term effects of an acidic pH may become detrimental under long-term conditions. Application of acidic fluids to the round window has been shown to be detrimental by enhancing free radical stress mediated hearing loss induced by platinum containing anticancer drugs whereas application of alkaline fluids has been shown to have a protective effect on hearing (Rybak et al. 1997; Tanaka et al. 2004). Consistently, some forms of systemic acidosis and impairment of cochlear pH regulation by mutations of essential subunits of H^+-ATPases, mutations of Cl^-/HCO_3^- exchangers SLC26A4 (pendrin) and the Na^+/HCO_3^- transporter SLC4A7 are associated with hearing loss (Karet et al. 1999a; Everett et al. 2001; Stover et al. 2002; Bok et al. 2003). It is conceivable that long-term effects of fluid and tissue acidosis are potentiated by promoting the generation of free radicals and activation of the innate immune system (Lardner 2001; Kellum et al. 2004).

5.3 Ca^{2+}

Endolymph is not only an unusual extracellular fluid for its high K^+ and low Na^+ concentration but also for its low Ca^{2+} concentration. Endolymph contains 20 µM Ca^{2+}, which is very low compared to other extracellular fluids such perilymph or plasma, which contain 1–2 mM Ca^{2+} (Bosher and Warren 1978; Ninoyu and Meyer zum Gottesberge 1986a; Ikeda et al. 1987d; Salt et al. 1989) (Fig. 3.7). Ca^{2+} enters the hair bundle together with K^+ and is necessary for the generation of the mechanoelectrical transduction current (Ohmori 1985; Lumpkin et al. 1997; Ricci and Fettiplace 1998). In addition, Ca^{2+} is responsible for fast (milliseconds) and slow (tenth of milliseconds) adaptations of the transduction mechanism (Holt and Corey 2000). Ca^{2+} that enters the hair bundle via the transduction channel is removed from the cytoplasm of the stereocilia by the Ca^{2+}-ATPase PMCA2 (Apicella et al. 1997; Yamoah et al. 1998).

For normal auditory function, the endolymphatic Ca^{2+} concentration can neither be too low nor too high. Elevated Ca^{2+} concentrations block transduction and the generation of microphonic potentials but reduced Ca^{2+} concentration suppress microphonic potentials as well (Tanaka et al. 1980; Marcus et al. 1982; Ohmori 1985; Ricci and Fettiplace 1998). Pathologically low endolymphatic Ca^{2+} concentration have been found in deaf-*waddler* mice and are suspected to be the cause of deafness (Wood et al. 2004a). Reduced endolymphatic Ca^{2+} concentrations are also suspected to be the cause for hearing loss associated with vitamin D deficiency and hypoparathyroidism, two conditions that are associated with low plasma Ca^{2+} concentrations (Brookes 1983; Ikeda et al. 1987a, 1987b, 1989). Conversely, elevated endolymphatic Ca^{2+} concentrations are thought to contribute to hearing loss in the guinea pig model of Ménière's disease (Ninoyu and Meyer zum Gottesberge 1986b), to transient threshold shifts following acoustic overstimulation (Ikeda and Morizono 1988), and to failure to aquire hearing in a mouse model of Pendred's syndrome (Wangemann et al. 2007).

Ca^{2+} absorption from endolymph appears to be driven at least in part by the endocochlear potential since the endolymphatic Ca^{2+} concentration is somewhat correlated with the magnitude of the endocochlear potential (Ikeda et al. 1987d). In analogy to Ca^{2+} absorption in proximal and distal tubules of the kidney, it is conceivable that Ca^{2+} absorption occurs partially through paracellular and partially through transcellular mechanisms. Transcellular pathways include uptake of Ca^{2+} across the apical membrane via TRPV5 and TRPV6 channels, buffering of Ca^{2+} in the cytosol by Ca^{2+} binding proteins and export via basolateral Ca^{2+}-ATPases (Yamauchi et al. 2005; Nakaya et al. 2007; Wangemann et al. 2007).

Ca^{2+} homeostasis in cochlear endolymph has long been assumed to involve active Ca^{2+}secretion based on the finding that the endolymphatic Ca^{2+} concentration is higher than what would be predicted for passive distribution across a semipermeable barrier. The finding that vanadate, an inhibitor of Ca^{2+}-ATPases, reduced the endolymphatic Ca^{2+} concentration more than would be predicted from the inhibition of the endocochlear potential suggested that vanadate-sensitive Ca^{2+}-ATPase are responsible for Ca^{2+} secretion into endolymph (Ikeda and Morizono 1988). This concept has been refined by the finding of very low endolymphatic Ca^{2+} concentrations in the deaf-*waddler* mouse, which bears a loss-of-function mutation in the PMCA Ca^{2+}-ATPase ATP2B2 (Street et al. 1998; Konrad-Martin et al. 2001; Wood, et al. 2004b).

6. Generation of the Endocochlear Potential

The endocochlear potential is generated by stria vascularis and provides the main driving force for sensory transduction in the organ of Corti (von Békésy 1950; Davis 1953; Wangemann and Schacht 1996; Wangemann 2002). Stria vascularis is functionally a two-layered epithelium consisting of the strial marginal cell layer, which provides a barrier between endolymph and intrastrial fluid and the basal cell layer that provides the a barrier between interstrial fluid and the extracellular fluid spaces in spiral ligament that are in open contact with perilymph (Fig. 3.8). Strial intermediate cells are functionally a part of the basal cell barrier, as they are connected to basal cells via a high density of gap junctions (Kikuchi et al. 1995). The endocochlear potential is essentially a K^+ equilibrium potential that is generated by the KCNJ10 K^+ channel located in the intermediate cells of stria vascularis in conjunction with a very low K^+ concentration in the intrastrial fluid spaces and a normal high K^+ concentration in the cytosol of intermediate cells (Takeuchi et al. 2000; Marcus et al. 2002). The endocochlear potential can be measured across the basal cell barrier, as intermediate cells are functionally a part of the basal cell barrier (Salt et al. 1987). Strial marginal cells contribute to the generation of the endocochlear potential in that they keep the K^+ concentration in the intrastrial fluid spaces low (Wangemann et al. 1995a). Similarly, spiral ligament, which consists of a large network of interconnected fibrocytes that are endowed with uptake mechanisms for K^+, contributes to the endocochlear potential in that it assists in maintaining the high cytosolic K^+ concentration in the intermediate cells.

7. Homestatic Disorders

Cochlear homeostasis is critical for all aspects of sensory transduction. Consequently, dysfunction of homeostatic mechanisms is a main cause of deafness (Fig. 3.9). Many hereditary forms deafness, including the most frequently occurring forms, as well as acquired forms of hearing loss, are due to failures to maintain cochlear homeostasis. These include connexin-related deafness,

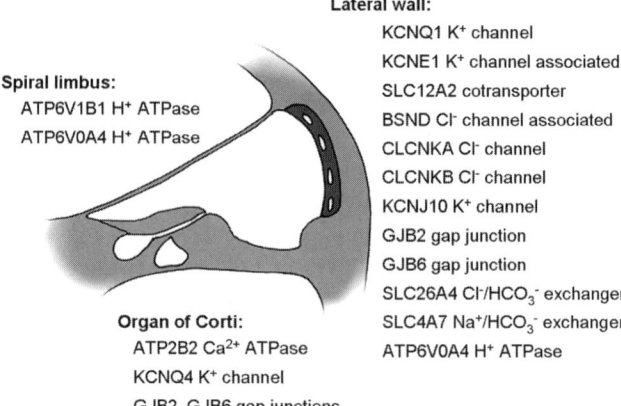

Lateral wall:
KCNQ1 K⁺ channel
KCNE1 K⁺ channel associated
SLC12A2 cotransporter
BSND Cl⁻ channel associated
CLCNKA Cl⁻ channel
CLCNKB Cl⁻ channel
KCNJ10 K⁺ channel
GJB2 gap junction
GJB6 gap junction
SLC26A4 Cl⁻/HCO₃⁻ exchanger
SLC4A7 Na⁺/HCO₃⁻ exchanger
ATP6V0A4 H⁺ ATPase

Spiral limbus:
ATP6V1B1 H⁺ ATPase
ATP6V0A4 H⁺ ATPase

Organ of Corti:
ATP2B2 Ca²⁺ ATPase
KCNQ4 K⁺ channel
GJB2, GJB6 gap junctions

FIGURE 3.9. Overview of homeostatic disorders. Dysfunction of genes that are expressed in the lateral wall, the organ of Corti or the spiral limbus cause deafness due to failure to maintain cochlear homeostasis.

Pendred syndrome, Jervell–Lange-Nielsen syndrome, Bartter syndrome, renal tubular acidosis with sensorineural deafness syndrome, and Ménière's disease.

7.1 Connexin-Related Deafness

Most cells in tissues are connected among each other by gap junctions to coordinate their behavior through electrical coupling and the exchange of metabolites and second messengers. Examples of cochlear cells that are interconnected by gap junction include supporting cells of the organ of Corti, inner and outer sulcus epithelial cells, strial intermediate and basal cells, spiral ligament fibrocytes, Reissner's membrane epithelial cells, and fibrocytes of the spiral limbus (Kikuchi et al. 2000). Exceptions to this rule are inner and outer hair cells and strial marginal cells. Gap junction are formed by docking two hemichannels expressed by adjacent cells. Each hemichannel consists of homo- or heteromeric connexin proteins. Five connexins are most prominently expressed in the cochlea, GJB2 (Cx26), GJB6 (Cx30), GJA1 (Cx43), GJA7 (Cx45), and GJE1 (Cx29) (Ahmad et al. 2003).

Mutations of GJB2 (Cx26) are the most prevalent cause of nonsyndromic deafness in childhood accounting for about one-half of all cases (Kelsell et al. 1997; Zelante et al. 1997; Rabionet et al. 2000). Mutations of GJB2 are responsible for recessive as well as dominant forms of non-syndromic deafness (Denoyelle et al. 1998). GJB2 is widely expressed in the cochlea in particular in strial basal cells and spiral ligament fibrocytes as well as in supporting cells of the organ of Corti and in inner and outer sulcus epithelial cells (Kikuchi et al. 1995). It had been suggested that cells in the gap junction networks surrounding the bases of inner and outer hair cells take up K⁺ and funnel it toward K⁺ secretory cells in the lateral wall and hypothetical K⁺ secretory cells in the spiral limbus

(Kikuchi et al. 1995; Spicer and Schulte 1998). Gap junctions would thereby be essential for K^+ cycling. Mice lacking expression of GJB2 from the organ of Corti (but not from lateral wall of the cochlea or from other organs) had a significant hearing loss and provided valuable insights into the role of gap junctions in the organ of Corti (Cohen-Salmon et al. 2002). The endocochlear potential was normal at the onset of hearing but failed to fully develop to adult levels, which is consistent with the observed hearing loss. The endolymphatic K^+ concentration was normal at all ages suggesting that GJB2 is not essential for K^+ cycling. The observed reduction in the endocochlear potential is likely due to a compromise in the epithelial barrier caused by progressive apoptosis of supporting cells in the organ of Corti that begins at the onset of hearing with the cells nearest to the inner hair cells. It is conceivable that GJB2 in these cells is essential for the sharing of metabolites (Matsunami et al. 2006). An import molecule, which is taken up by the cells nearest to the inner hair cells, is glutamate that is released from the inner hair cells in response to sound stimulation. Buffering of glutamate may depend on coupling between supporting cells, as glutamine synthase is mainly expressed in the neighbors of the cells that express the glutamate uptake transporter SLC1A3 (Eybalin et al. 1996; Ottersen et al. 1998). Consistent with glutamate toxicity as a key factor is the finding that gap junctions containing mutations of *GJB2*, which cause hearing loss, maintained ionic coupling but failed to exchange larger biochemicals (Zhang et al. 2005).

Mutations of *GJB6* (Cx30) are another prevalent cause of nonsyndromic deafness in childhood (Del Castillo et al. 2003). *GJB6* is expressed in a pattern similar to that of *GJB2* (Lautermann et al. 1998). Mice that lack functional expression of *GJB6* are profoundly deaf since they lack the endocochlear potential (Teubner et al. 2003). The endolymphatic K^+ concentration, however, was normal at the onset of hearing although reduced later. Cells in the organ of Corti underwent massive apoptosis at the time when the endocochlear potential failed to develop. These results illustrate that *GJB6* is required for the generation of the endocochlear potential but not essential for K^+ cycling. Interestingly, loss of GJB6 (CX30) renders the capillaries in stria vascularis leaky to the intrastrial space (Cohen-Salmon et al. 2007). This leak provides an electrical short for the endocochlear potential. Generation of the endocochlear potential depends directly on the electrical coupling between intermediate and basal cells of stria vascularis, however, the chemical coupling between intermediate and basal cells of stria vascularis and spiral ligament fibrocytes may be equally important in that the connected cells may provide a buffer system HCO_3^-, pH, free radicals, Ca^{2+}, and metabolites. Differences in the quality of coupling of gap junctions that consist of *GJB2* with or without *GJB6* are now beginning to be understood (Sun et al. 2005; Zhao 2005). Nevertheless, a failure to buffer metabolites may be the cause for the pathologic leakiness of capillaries in stria vascularis that leads to a collapse of the endocochlear potential and to deafness in mice and patients lacking functional GJB6.

Mutations of *GJA1* (Cx43) are also associated with non-syndromic deafness (Liu et al. 2001). *GJA1* expression in the cochlea of adult mice is limited to the spiral ligament, capillaries of stria vascularis, and bone of the otic capsule (Cohen-Salmon et al. 2004). However, *GJA1* is expressed in various cells of the developing mouse cochlea. The etiology of hearing loss is yet unclear.

Other connexins expressed in the cochlea are candidate genes for human deafness. *GJA7* (Cx45), as shown for *GJA1* (Cx43), is expressed in the developing mouse cochlea (Cohen-Salmon et al. 2004). Expression in the adult cochlea is limited to capillary endothelial cells. Whether mutations of *GJA7* cause deafness is currently unclear. In contrast, mice lacking *GJE1* (Cx29) are delayed in the maturation of hearing, have an early onset of a high-frequency hearing loss, and an elevated sensitivity to noise damage (Tang et al. 2006). The phenotype includes prolonged latencies of auditory brain stem responses which is consistent with the expression of *GJE1* in Schwann cells that provide myelination of the soma and dendritic fibers of the spiral ganglion cells.

7.2 Pendred Syndrome

Pendred syndrome is the most common syndromic form of deafness. Pendred syndrome is an autosomal recessive disorder associated with sensorineural hearing loss, developmental abnormalities of the cochlea, and euthyroid goiter (Pendred 1896). Deafness in Pendred's syndrome is generally profound although sometimes late in onset and provoked by stresses such as light head injury or infection (Cremers et al. 1998; Usami et al. 1999; Luxon et al. 2003). A positive perchlorate discharge test and an enlarged vestibular aqueduct appear to be the most reliable clinical signs of Pendred syndrome (Reardon et al. 2000). Vestibular dysfunction is uncommon. Goiter is variable and generally develops after puberty (Royaux et al. 2000).

The syndrome is caused by mutations of SLC26A4 (pendrin) (Everett et al. 1997). SLC26A4 is an anion exchanger that can transport Cl^-, I^-, and HCO_3^- (Scott et al. 1999; Scott and Karniski 2000). In the cochlea, SLC26A4 is expressed in apical membranes of spiral prominence epithelial cells and spindle-shaped cells of stria vascularis, in outer sulcus cells and root cells of the spiral ligament (Everett et al. 1999; Wangemann et al. 2004). SLC26A4 appears to mediate HCO_3^- secretion into endolymph and is responsible for the slightly alkaline pH (Fig. 3.7) (Wangemann et al. 2007). SLC26A4 may have a similar role in the vestibular system, where it is expressed in transitional cells (Nakaya et al. 2007). Loss of pendrin leads to the development of enlarged endolymphatic compartments in the cochlea and the vestibular labyrinth and to an acidification of endolymph (Everett et al. 2001). The enlarged endolymphatic compartment in the cochlea requires higher metabolic rates in stria vascularis to sustain higher rates of K^+ secretion necessary to maintain K^+ homeostasis (Wangemann et al. 2004). Acidification of endolymph coupled with higher metabolic rates causes free radical stress and redox imbalance in stria vascularis. Oxidative stress degenerates stria vascularis and causes the loss of the K^+ channel KCNJ10, which leads to deafness via the loss of the normal endocochlear potential (Wangemann et al. 2004, 2007; Singh and Wangemann 2008). Acidification of endolymph inhibits Ca^{2+} absorption and elevated Ca^{2+} concentrations in endolymph inhibit the transduction channel (Wangemann et al. 2007). In addition, progressive degeneration of stria vascularis is associated with an invasion of macrophages into stria vascularis (Jabba et al. 2006).

In the thyroid SLC26A4 is partially responsible for I^- transport across the apical membrane of thyroid follicular epithelial cells. Mutations of SLC26A4 appear to limit iodide fixation in the follicular lumen, which may lead to a compensatory enlargement of the thyroid gland (Bidart et al. 2000).

7.3 Jervell–Lange-Nielsen Syndrome

Jervell–Lange-Nielsen syndrome is a rare autosomal recessive disease characterized by profound sensorineural deafness and prolonged QT-intervals of the cardiac action potentials that in response to exercise or emotions can precipitate arrhythmias, syncope and even sudden death in otherwise healthy individuals (Jervell and Lange-Nielsen 1957). Jervell–Lange-Nielsen syndrome is due to severe mutations of the K^+ channel α-subunit KCNQ1 and/or the β-subunit KCNE1 that are required for proper K^+ channel function (Neyroud et al. 1997; Schulze-Bahr et al. 1997; Wang et al. 2002).

In the inner ear, the K^+ channel KCNQ1/KCNE1 is expressed in the apical membrane of strial marginal cells and vestibular dark cells (Sakagami et al. 1991; Marcus and Shen 1994; Wangemann et al. 1995a). This K^+ channel is essential for K^+ secretion across the apical membrane of strial marginal cells and vestibular dark cells. Without this channel, K^+ secretion fails and endolymphatic spaces in patients suffering from Jervell–Lange-Nielsen syndrome appear to be collapsed (Friedmann et al. 1966). Similarly, mice that lack KCNE1, KCNQ1, or harbor a spontaneous mutation in KCNE1 develop normally endolymphatic spaces until the onset of K^+ secretion. The onset of K^+ secretion is likely paralleled by the onset of absorptive processes. Failure of K^+ secretion in the presence of unimpeded absorptive processes leads to the apparent collapse of endolymphatic spaces in mice as seen in human patients (Vetter et al. 1996; Lee et al. 2000; Letts et al. 2000; Casimiro et al. 2001).

The KCNQ1/KCNE1 K^+ channel is not only responsible for K^+ secretion and the formation of endolymph in the cochlear and vestibular labyrinth but carries in cardiac myocytes the slowly activating I_{Ks} current that plays a major role in the repolarization phase of the cardiac action potential (Varnum et al. 1993; Barhanin et al. 1996; Sanguinetti et al. 1996). Although transcription of KCNQ1 in the heart is independent of KCNE1, it appears that KCNQ1 interacts at the translational or posttranslational level with KCNE1 (Drici et al. 1998). Support for a posttranslational interaction comes from the observation that KCNQ1 in vestibular dark cells of mice lacking KCNE1 failed to concentrate in the apical membrane and appeared to remain in the cytoplasm rather than being trafficked to the apical membrane (Nicolas et al. 2001). Thus, KCNE1 may be necessary for trafficking of KCNQ1 to the apical membrane.

Other mutations of KCNQ1 and/or KCNE1 lead to Romano–Ward syndrome, which is a more frequently observed autosomal dominantly disease consisting of long-QT syndrome without other abnormalities. More than 80 mutations in KCNQ1 and KCNE1 have so far been described (Splawski et al. 2000). It is still largely unclear why most heterozygous mutations of KCNQ1 and KCNE1 cause Romano–Ward syndrome with no apparent effect on the inner ear and why several homozygous mutations or compounding heterozygous mutations affect both the inner ear and the heart. At least some mutations that cause Jervell–Lange-Nielsen syndrome impair the organization of KCNQ1 into the required tetramers thus precluding the assembly of a functional K^+ channel (Schmitt et al. 2000; Tyson et al. 2000).

7.4 Bartter Syndrome

Bartter syndrome refers to a group of currently four types of autosomal recessive impairments of renal salt reabsorption (Bartter et al. 1962). Types 1–3 are caused by mutations of the $Na^+/2Cl^-/K^+$ cotransporter SLC12A1 (NKCC2), the K^+ channel KCNJ1 (ROMK), and the Cl^- channel (CLCNKB) that lead to renal salt wasting since these transporters and channels play essential roles in the kidney. Type 4 of Bartter syndrome is characterized by renal salt wasting and deafness. This type of Bartter syndrome is caused either by mutations of the Cl^- channel β-subunit BSND (barttin) or by coincidence of mutations of the Cl^- channel α-subunits CLCNKA and CLCNKB (Estevez et al. 2001; Birkenhager et al. 2001; Schlingmann et al. 2004).

In the inner ear, the Cl^- channels CLCNKA (CLCK-1) and CLCNKB (CLCK-2) are expressed in the organ of Corti, the spiral ligament, and stria vascularis (Kawasaki et al. 2000; Sage and Marcus 2001; Maehara et al. 2003; Qu et al. 2006). Conceivably, the expression of both channels protects the inner ear in type 3 Bartter syndrome. Cl^- channels support the uptake of K^+ across the basolateral membrane of strial marginal cells and thereby the secretion of K^+ and the generation of the endocochlear potential (Takeuchi et al. 1995; Ando and Takeuchi 2000; Qu et al. 2006). Failure to generate an endocochlear potential may be the mechanism of deafness in type 4 Bartter syndrome. Cl^- channels are expressed at a high density in the basolateral membrane of strial marginal cells and vestibular dark cells to support the recycling of Cl^- that is taken up together with Na^+ and K^+ via the $Na^+/2Cl^-/K^+$ cotransporter SLC12A2 (Wangemann and Marcus 1992; Marcus et al. 1993). Although CLCNKA and CLCNKB are the pore-forming α-subunit of these Cl^- channel, the β-subunit barttin is needed for trafficking from the endoplasmic reticulum to the plasma membrane (Waldegger et al. 2002).

7.5 Renal Tubular Acidosis with Sensorineural Deafness Syndrome

Renal tubular acidosis with sensorineural deafness syndrome is an autosomal recessive impairment of renal acid secretion and progressive hearing loss (Dunger et al. 1980). The syndrome is caused by mutations of ATP6V1B1 or ATPV0A4, which are subunits of vacuolar H^+-ATPase (Karet et al. 1999b; Stover et al. 2002; Vargas-Poussou et al. 2006). ATP6V1B1 is expressed in a limited number of tissues including the acid secreting intercalated cells of the renal distal tubule and interdental cells of the cochlea. Mutations of ATP6V1B1 are generally associated with an early onset of sensorineural hearing loss whereas mutations of ATP6V0A4, which is a more commonly expressed subunit of H^+-ATPases, is generally associated with a later onset in hearing loss. The etiology of hearing loss is unclear.

Mice lacking functional expression of ATP6V1B1 differ significantly from the human phenotype. Humans bearing mutations of ATP6V1B1 generate an abnormally alkaline urine; develop a metabolic acidosis; lose Ca^{2+} from bone; and suffer from growth retardation and rickets, from calciuria and the formation of renal calcium deposits, and from the loss of hearing. In contrast, mice lacking functional expression of ATP6V1B1 do not develop metabolic acidosis although they generated a significantly more alkaline urine. Mice do not suffer from calciuria, grow normally, and have normal hearing (Dou et al. 2003; Finberg et al. 2005). The difference in metabolic acidosis may be explainable by a greater alkali load in typical rodent diets compared to typical protein-rich human diets. The difference in the development of hearing loss may suggest, that mice express compensatory pH regulatory mechanism in the cochlea or that hearing loss is not directly related to the loss of H^+-ATPase function but rather a secondary event potentially related to changes in the cochlear Ca^{2+} homeostasis.

7.6 Ménière's Disease

Ménière's disease is an enigma characterized by episodic vertigo, fluctuating low-frequency hearing loss, tinnitus, the sensation of oral fullness, and endolymphatic hydrops (Ménière 1861; Yamakawa 1938). The etiology of Ménière's disease is unknown and whether endolymphatic hydrops is a cause, a consequence or an epiphenomenon of Ménière's disease is unclear (Kiang 1989; Rauch et al. 1989; Merchant et al. 2005). A wide variety of etiologies have been discussed including genetic predispositions, immune diseases, vascular defects, and viral infections as initiators of the disease that is associated with cochlear fluid imbalances. The difficulty deciphering this enigmatic disease may rest with the possibility that clinical symptoms may be manifestations of multiple diseases and that symptoms are episodic, at least in the early stages. Recent research has been concentrated on the elucidation of genetic defects, the role of intralabyrinthine fluid dynamics, the role of stress, the development of diagnostic tests for early diagnosis and the development of animal models.

Familial Ménière's disease suggestive of a genetic defect or predispositions may account for as many as 14% of cases of Ménière's disease (Birgerson et al. 1987). In particular, the HLA-A2 allele of the major histocompatibility complex, which regulate immune inflammatory responses, has been found in 90% of patient with familial Ménière's disease but only in 29% of the general population (Arweiler et al. 1995). This allele, however, is not specific to Ménière's disease but also associated with an earlier onset of Alzheimer disease and a poorer prognosis of ovarian cancer (Listi et al. 2006; Gamzatova et al. 2006).

Symptoms of Ménière's disease overlaps to some extend with autosomal dominant hearing loss DFNA9, which is due to mutations of COCH (Manolis et al. 1996). COCH (cochlin) is a secreted protein of unknown function, that is expressed in fibrocytes of the spiral limbus and spiral ligament (Robertson et al. 2001). Mutation of COCH have been found in a subset of patients that have symptoms consistent with Ménière's disease (Fransen et al. 1999; De Kok et al. 1999; Usami et al. 2003; Sanchez et al. 2004). Overexpression of cochlin has not only been associated with hearing loss but also with blindness. Inappropriate aggregation of cells due to extracellular cochlin depositions have been observed in the trabecular network of patients suffering from open-angle glaucoma, a leading cause for age-related blindness (Bhattacharya et al. 2005). Cochlin deposits appear to contribute to the obstruction of Schlemm's canal, which leads to an elevated intraocular pressure and glaucoma. In analogy, it can speculated that mutated cochlin limits diffusional pathways between cells of the inner ear and that this limitation contributes to the loss of inner ear function. Interestingly, cochlin appears to not be essential for normal cochlear and vestibular function since mice lacking cochlin have normal hearing and show no overt signs of vestibular dysfunction (Makishima et al. 2005).

An autoimmune-mediated etiology of some cases of Ménière's disease has been suggested based on the finding that plasma of some patients contains antibodies against inner ear antigens (Boulassel et al. 2001; Passali et al. 2004; Mouadeb and Ruckenstein 2005). Support for this etiology comes from the effectiveness of immunosuppressive treatments (Selivanova et al. 2005; Garduno-Anaya 2005) and from the observation that infusions of antibodies against inner ear antigens result in hearing loss and endolymphatic hydrops, symptoms that are associated with but not limited to Ménière's disease (Yoo et al. 1984; Soliman 1989).

Disturbances of cochlear fluid dynamics have historically been seen as a mediator of Ménière's disease in conjunction with the longitudinal flow hypothesis. The longitudinal flow hypothesis states that endolymph is generated in the cochlea and reabsorbed in the endolymphatic sac (Guild 1927). This hypothesis was based on a rather simple ink-injection experiment and received indirect support from the finding that surgical destruction of the endolymphatic sac leads to endolymph hydrops (Kimura 1967). With regard to controlling endolymph composition, the longitudinal flow hypothesis has since been replaced by the radial flow hypothesis. The radial flow hypothesis states that flow is insignificant and that the composition of endolymph is controlled within radial cross sections of cochlea by secretory and absorptive processes. Measurements of flow have revealed an absence of significant flow in endolymph and perilymph, which gave rise to the radial flow hypothesis (Salt et al. 1986; Ohyama et al. 1988; Salt and Thalmann 1988). With the rise of the radial flow hypothesis it became imperative to understand the secretory and absorptive mechanisms and their regulation in the epithelia lining cochlea and vestibular endolymph. Stria marginal cells and vestibular dark cells have been identified as the premier sites of K^+ secretion (Wangemann 1995). Outer sulcus epithelial cells and transitional cells have been shown to participate in cation reabsorption (Marcus and Chiba 1999) and the epithelium lining the semicircular canals has been shown to mediate Na^+ and Ca^{2+} absorption as well as Cl^- secretion (Milhaud et al. 2002; Yamauchi et al. 2005; Pondugula et al. 2006). Although endolymph does not flow in the sense of the longitudinal flow hypothesis, fluid movements do occur in response to pressure changes or in response to low-frequency sounds (Salt 2004). Communication between cochlear endolymph and endolymph of the endolymphatic sac, that differs greatly in composition, appears to be guarded by a membranous valve (Salt and Demott 2000; Salt and Rask-Andersen 2004). The role of this valve in endolymphatic hydrops and Ménière's disease is yet unclear.

Emotional stress has long been associated with the precipitation of Ménière's disease attacks (Horner and Cazals 2003). Stress hormones such as norepinephrine increase K^+ and Cl^- secretion via β-adrenergic receptors in strial marginal cells, vestibular dark cells and semicircular duct epithelial cells (Wangemann et al. 1999, 2000; Milhaud et al. 2002). Vasopressin induces endolymphatic hydrops whereas vasopressin antagonists cause a collapse of the endolymphatic compartment (Takeda et al. 2003). Although the mechanism of vasopressin-induced endolymphatic hydrops is unclear, it may be related to a stimulated expression of aquaporin 2 (Mhatre et al. 2002; Sawada et al. 2002). Further, plasma concentrations of prolactin are elevated in some patients suffering from Ménière's disease (Horner et al. 2001). In a subset of these Ménière's patients hyperprolactinemia was due to prolactinoma, a benign tumor of the pituitary gland that secretes prolactin. The role of prolactin on cochlear homeostasis is yet unclear. Stimulation of ion transport by stress hormones such norepinephrine may not only contribute to endolymphatic hydrops but also the elevated metabolic demands to ensure homeostasis of larger volumes may increase free radical stress (Labbe et al. 2005).

The difficulty deciphering the enigmatic Ménière's disease is not the least due to the lack of an ideal animal model. The traditional surgically derived guinea pig model, unlike Ménière's disease, does not show vestibular symptoms, does not show the typical low-frequency hearing loss and does not improve with glycerol treatment (Kimura 1967; Horner and Cazals 1987). Other guinea pig models develop endolymphatic hydrops (Dunnebier et al. 1997; Lohuis et al. 1999; Matsuoka et al. 2002). Guinea pig models in general may not be ideal because the majority of molecular research tools are available for mouse models. Several mouse models with endolymphatic hydrops have recently described including mice lacking SLC26A4 (pendrin), FOX1, and BRN4 (Everett et al. 2001; Xia et al. 2002; Hulander et al. 2003; Wangemann et al. 2005). These models may not be ideal either but may prove useful to investigate certain aspects of Ménière's disease.

Acknowledgment. Support by NIH-R01-DC01098 and NIH-R01-DC04280 is gratefully acknowledged.

References

Adragna NC, Fulvio MD, Lauf PK (2004) Regulation of K-Cl cotransport: from function to genes. J Membr Biol 201:109–137.

Agrup C, Bagger-Sjoback D, Fryckstedt J (1999) Presence of plasma membrane-bound Ca^{2+}-ATPase in the secretory epithelia of the inner ear. Acta Otolaryngol 119:437–445.

Ahmad S, Chen S, Sun J, Lin X (2003) Connexins 26 and 30 are co-assembled to form gap junctions in the cochlea of mice. Biochem Biophys Res Commun 307:362–368.

Ando M, Takeuchi S (2000) mRNA encoding ClC-K1, a kidney Cl^- channel is expressed in marginal cells of the stria vascularis of rat cochlea: its possible contribution to Cl^- currents. Neurosci Lett 284:171–174.

Apicella S, Chen S, Bing R, Penniston JT, Llinas R, Hillman DE (1997) Plasmalemmal ATPase calcium pump localizes to inner and outer hair bundles. Neuroscience 79: 1145–1151.

Arweiler DJ, Jahnke K, Grosse-Wilde H (1995) [Meniere disease as an autosome dominant hereditary disease]. Laryngorhinootologie 74:512–515.

Banfi B, Malgrange B, Knisz J, Steger K, Dubois-Dauphin M, Krause KH (2004) NOX3, a superoxide-generating NADPH oxidase of the inner ear. J Biol Chem 279: 46065–46072.

Barhanin J, Lesage F, Guillemare E, Fink M, Lazdunski M, Romey G (1996) KVLQT1 and IsK (minK) proteins associate to form the IKs cardiac potassium current. Nature 384:78–80.

Bartter FC, Pronove P, Gill JR, Jr., MacCardle RC (1962) Hyperplasia of the juxta-glomerular complex with hyperaldosteronism and hypokalemic alkalosis. A new syndrome. Am J Med 33:811–828.

Beckman JS, Beckman TW, Chen J, Marshall PA, Freeman BA (1990) Apparent hydroxyl radical production by peroxynitrite: implications for endothelial injury from nitric oxide and superoxide. Proc Natl Acad Sci USA 87:1620–1624.

Bernard C (1878) Leçons sur les phénomènes de la vie communs aux animaux es aux végétaux. Cours de Physiologie générale du Museum d'Histoire Naturelle.

Bhattacharya SK, Rockwood EJ, Smith SD, Bonilha VL, Crabb JS, Kuchtey RW, Robertson NG, Peachey NS, Morton CC, Crabb JW (2005) Proteomics reveal Cochlin deposits associated with glaucomatous trabecular meshwork. J Biol Chem 280: 6080–6084.

Bidart JM, Mian C, Lazar V, Russo D, Filetti S, Caillou B, Schlumberger M (2000) Expression of pendrin and the Pendred syndrome (PDS) gene in human thyroid tissues. J Clin Endocrinol Metab 85:2028–2033.

Birgerson L, Gustavson KH, Stahle J (1987) Familial Ménière's disease: a genetic investigation. Am J Otol 8:323–326.

Birkenhager R, Otto E, Schurmann MJ, Vollmer M, Ruf EM, Maier-Lutz I, Beekmann F, Fekete A, Omran H, Feldmann D, Milford DV, Jeck N, Konrad M, Landau D, Knoers NV, Antignac C, Sudbrak R, Kispert A, Hildebrandt F (2001) Mutation of BSND causes Bartter syndrome with sensorineural deafness and kidney failure. Nat Genet 29:310–314.

Bok D, Galbraith G, Lopez I, Woodruff M, Nusinowitz S, BeltrandelRio H, Huang W, Zhao S, Geske R, Montgomery C, Van S, I, Friddle C, Platt K, Sparks MJ, Pushkin A,

Abuladze N, Ishiyama A, Dukkipati R, Liu W, Kurtz I (2003) Blindness and auditory impairment caused by loss of the sodium bicarbonate cotransporter NBC3. Nat Genet 34:313–319.

Bond BR, Ng LL, Schulte BA (1998) Identification of mRNA transcripts and immuno-histochemical localization of Na/H exchanger isoforms in gerbil inner ear. Hear Res 123:1–9.

Bosher SK, Warren RL (1978) Very low calcium content of cochlear endolymph, an extracellular fluid. Nature 273:377–378.

Boulassel MR, Deggouj N, Tomasi JP, Gersdorff M (2001) Inner ear autoantibodies and their targets in patients with autoimmune inner ear diseases. Acta Otolaryngol 121:28–34.

Brookes GB, (1983) Vitamin D deficiency—-a new cause of cochlear deafness. J Laryngol Otol 97:405–420.

Brown JN, Nuttall AL (1994) Autoregulation of cochlear blood flow in guinea pigs. Am J Physiol 266:H458–H467.

Cannon W (1929) Organization for physiological homeostasis. Physiol Rev 9:399–431.

Carlisle L, Aberdeen J, Forge A, Burnstock G (1990) Neural basis for regulation of cochlear blood flow: peptidergic and adrenergic innervation of the spiral modiolar artery of the guinea pig. Hear Res 43:107–113.

Casimiro MC, Knollmann BC, Ebert SN, Vary JC Jr, Greene AE, Franz MR, Grinberg A, Huang SP, Pfeifer K (2001) Targeted disruption of the *Kcnq1* gene produces a mouse model of Jervell and Lange-Nielsen syndrome. Proc Natl Acad Sci USA 98: 2526–2531.

Cecola RP, Bobbin RP (1992) Lowering extracellular chloride concentration alters outer hair cell shape. Hear Res 61:65–72.

Chiba T, Marcus DC (2000) Nonselective cation and BK channels in apical membrane of outer sulcus epithelial cells. J Membr Biol 174:167–179.

Cohen-Salmon M, Ott T, Michel V, Hardelin JP, Perfettini I, Eybalin M, Wu T, Marcus DC, Wangemann P, Willecke K, Petit C (2002) Targeted ablation of connexin26 in the inner ear epithelial gap junction network causes hearing impairment and cell death. Curr Biol 12:1106–1111.

Cohen-Salmon M, Maxeiner S, Kruger O, Theis M, Willecke K, Petit C (2004) Expression of the *connexin43-* and *connexin45-*encoding genes in the developing and mature mouse inner ear. Cell Tissue Res 316:15–22.

Cohen-Salmon M, Regnault B, Cayet N, Caille D, Demuth K, Hardelin JP, Janel N, Meda P, Petit C (2007) Connexin30 deficiency causes instrastrial fluid-blood barrier disruption within the cochlear stria vascularis. Proc Natl Acad Sci USA 104:6229–6234.

Conlon BJ, Smith DW (1998) Supplemental iron exacerbates aminoglycoside ototoxicity in vivo. Hear Res 115:1–5.

Cremers CW, Admiraal RJ, Huygen PL, Bolder C, Everett LA, Joosten FB, Green ED, Van Camp G, Otten BJ (1998) Progressive hearing loss, hypoplasia of the cochlea and widened vestibular aqueducts are very common features in Pendred's syndrome. Int J Pediatr Otorhinolaryngol 45:113–123.

D'Ambrosio R, Gordon DS, Winn HR (2002) Differential role of KIR channel and Na^+/K^+-pump in the regulation of extracellular K^+ in rat hippocampus. J Neurophysiol 87:87–102.

Danbolt NC (2001) Glutamate uptake. Prog Neurobiol 65:1–105.

Davis H (1953) Energy into nerve impulses: the inner ear. Adv Sci 9:420–425.

Dechesne CJ, Winsky L, Kim HN, Goping G, Vu TD, Wenthold RJ, Jacobowitz DM (1991) Identification and ultrastructural localization of a calretinin-like calcium-binding protein (protein 10) in the guinea pig and rat inner ear. Brain Res 560:139–148.

De Kok YJ, Bom SJ, Brunt TM, Kemperman MH, van Beusekom E, van der Velde–Visser SD, Robertson NG, Morton CC, Huygen PL, Verhagen WI, Brunner HG, Cremers CW, Cremers FP (1999) A Pro51Ser mutation in the *COCH* gene is associated with late onset autosomal dominant progressive sensorineural hearing loss with vestibular defects. Hum Mol Genet 8:361–366.

Del Castillo I, Moreno–Pelayo MA, Del Castillo FJ, Brownstein Z, Marlin S, Adina Q, Cockburn DJ, Pandya A, Siemering KR, Chamberlin GP, Ballana E, Wuyts W, Maciel-Guerra AT, Alvarez A, Villamar M, Shohat M, Abeliovich D, Dahl HH, Estivill X, Gasparini P, Hutchin T, Nance WE, Sartorato EL, Smith RJ, Van Camp G, Avraham KB, Petit C, Moreno F (2003) Prevalence and evolutionary origins of the del (GJB6–D13S1830) mutation in the *DFNB1* locus in hearing-impaired subjects: a multicenter study. Am J Hum Genet 73:1452–1458.

Denoyelle F, Lina-Granade G, Plauchu H, Bruzzone R, Chaib H, Levi-Acobas F, Weil D, Petit C (1998) *Connexin 26* gene linked to a dominant deafness. Nature 393:319–320.

Dou H, Finberg K, Cardell EL, Lifton R, Choo D (2003) Mice lacking the B1 subunit of H^+-ATPase have normal hearing. Hear Res 180:76–84.

Drici MD, Arrighi I, Chouabe C, Mann JR, Lazdunski M, Romey G, Barhanin J (1998) Involvement of IsK-associated K^+ channel in heart rate control of repolarization in a murine engineered model of Jervell and Lange-Nielsen syndrome. Circ Res 83:95–102.

Dunger DB, Brenton DP, Cain AR (1980) Renal tubular acidosis and nerve deafness. Arch Dis Child 55:221–225.

Dunnebier EA, Segenhout JM, Wit HP, Albers FW (1997) Two-phase endolymphatic hydrops: a new dynamic guinea pig model. Acta Otolaryngol (Stockh) 117:13–19.

Eisenstein RS (2000) Iron regulatory proteins and the molecular control of mammalian iron metabolism. Annu Rev Nutr 20:627–662.

El Barbary A, Altschuler RA, Schacht J (1993) Glutathione *S*-transferases in the organ of Corti of the rat: enzymatic activity, subunit composition and immunohistochemical localization. Hear Res 71:80–90.

Estevez R, Boettger T, Stein V, Birkenhager R, Otto E, Hildebrandt F, Jentsch TJ (2001) Barttin is a Cl^- channel beta-subunit crucial for renal Cl^- reabsorption and inner ear K^+ secretion. Nature 414:558–561.

Everett LA, Glaser B, Beck JC, Idol JR, Buchs A, Heyman M, Adawi F, Hazani E, Nassir E, Baxevanis AD, Sheffield VC, Green ED (1997) Pendred syndrome is caused by mutations in a putative sulphate transporter gene (*PDS*). Nat Genet 17:411–422.

Everett LA, Morsli H, Wu DK, Green ED (1999) Expression pattern of the mouse ortholog of the Pendred's syndrome gene (*Pds*) suggests a key role for pendrin in the inner ear. Proc Natl Acad Sci USA 96:9727–9732.

Everett LA, Belyantseva IA, Noben-Trauth K, Cantos R, Chen A, Thakkar SI, Hoogstraten-Miller SL, Kachar B, Wu DK, Green ED (2001) Targeted disruption of mouse Pds provides insight about the inner-ear defects encountered in Pendred syndrome. Hum Mol Genet 10:153–161.

Eybalin M, Norenberg MD, Renard N (1996) Glutamine synthetase and glutamate metabolism in the guinea pig cochlea. Hear Res 101:93–101.

Fauser C, Schimanski S, Wangemann P (2004) Localization of beta1-adrenergic receptors in the cochlea and the vestibular labyrinth. J Membr Biol 201:25–32.

Ferrary E, Sterkers O, Saumon G, Tran Ba Huy P, Amiel C (1987) Facilitated transfer of glucose from blood into perilymph in the rat cochlea. Am J Physiol 253:F59–F65.

Fessenden JD, Schacht J (1997) Localization of soluble guanylate cyclase in the guinea pig cochlea suggests involvement in regulation of blood flow and supporting cell physiology. J Histochem Cytochem 45:1401–1408.

Finberg KE, Wagner CA, Bailey MA, Paunescu TG, Breton S, Brown D, Giebisch G, Geibel JP, Lifton RP (2005) The B1-subunit of the H^+ ATPase is required for maximal urinary acidification. Proc Natl Acad Sci USA 102:13616–13621.

Foster JD, Drescher MJ, Hatfield JS, Drescher DG (1994) Immunohistochemical localization of S-100 protein in auditory and vestibular end organs of the mouse and hamster. Hear Res 74:67–76.

Fransen E, Verstreken M, Verhagen WI, Wuyts FL, Huygen PL, D'Haese P, Robertson NG, Morton CC, McGuirt WT, Smith RJ, Declau F, Van de Heyning PH, Van Camp G (1999) High prevalence of symptoms of Meniere's disease in three families with a mutation in the COCH gene. Hum Mol Genet 8:1425–1429.

Friedmann I, Fraser GR, Froggatt P (1966) Pathology of the ear in the cardioauditory syndrome of Jervell and Lange-Nielsen (recessive deafness with electrocardiographic abnormalities). J Laryngol Otol 80:451–470.

Furness DN, Lehre KP (1997) Immunocytochemical localization of a high-affinity glutamate-aspartate transporter, GLAST, in the rat and guinea–pig cochlea. Eur J Neurosci 9:1961–1969.

Furuta H, Luo L, Hepler K, Ryan AF (1998) Evidence for differential regulation of calcium by outer versus inner hair cells: plasma membrane Ca-ATPase gene expression. Hear Res 123:10–26.

Gamzatova Z, Villabona L, Dahlgren L, Dalianis T, Nillson B, Bergfeldt K, Masucci GV (2006) Human leucocyte antigen (HLA) A2 as a negative clinical prognostic factor in patients with advanced ovarian cancer. Gynecol Oncol 103:145–150.

Garduno-Anaya MA, Couthino DT, Hinojosa-Gonzalez R, Pane-Pianese C, Rios-Castaneda LC (2005) Dexamethasone inner ear perfusion by intratympanic injection in unilateral Méniére's disease: a two-year prospective, placebo-controlled, double-blind, randomized trial. Otolaryngol Head Neck Surg 133:285–294.

Grant L, Slapnick S, Kennedy H, Hackney C (2006) Ryanodine receptor localisation in the mammalian cochlea: an ultrastructural study. Hear Res 219:101–109.

Gruber DD, Dang H, Shimozono M, Scofield MA, Wangemann P (1998) Alpha1A adrenergic receptors mediate vasoconstriction of the isolated spiral modiolar artery in vitro. Hear Res 119:113–124.

Guild SR (1927) The circulation of the endolymph. Am J Anat 39:57–81.

Hackney CM, Mahendrasingam S, Penn A, Fettiplace R (2005) The concentrations of calcium buffering proteins in mammalian cochlear hair cells. J Neurosci 25:7867–7875.

Hakuba N, Koga K, Gyo K, Usami SI, Tanaka K (2000) Exacerbation of noise-induced hearing loss in mice lacking the glutamate transporter GLAST. J Neurosci 20: 8750–8753.

Herzog M, Scherer EQ, Albrecht B, Rorabaugh B, Scofield MA, Wangemann P (2002) CGRP receptors in the gerbil spiral modiolar artery mediate a sustained vasodilation via a transient cAMP-mediated Ca^{2+}-decrease. J Membr Biol 189:225–236.

Hillerdal M, Andersson SE (1991) The effects of calcitonin gene-related peptide (CGRP) on cochlear and mucosal blood flow in the albino rabbit. Hear Res 52:321–328.

Holt JR, Corey DP (2000) Two mechanisms for transducer adaptation in vertebrate hair cells. Proc Natl Acad Sci USA 97:11730–11735.

Horner KC, Cazals Y (1987) Glycerol-induced changes in the cochlear responses of the guinea pig hydropic ear. Arch Otorhinolaryngol 244:49–54.

Horner KC, Cazals Y (2003) Stress in hearing and balance in Ménière's disease. Noise Health 5:29–34.

Horner KC, Giraudet F, Lucciano M, Cazals Y (2001) Sympathectomy improves the ear's resistance to acoustic trauma—could stress render the ear more sensitive? Eur J Neurosci 13:405–408.

Housley GD, Kanjhan R, Raybould NP, Greenwood D, Salih SG, Jarlebark L, Burton LD, Setz VC, Cannell MB, Soeller C, Christie DL, Usami S, Matsubara A, Yoshie H, Ryan AF, Thorne PR (1999) Expression of the P2X(2) receptor subunit of the ATP-gated ion channel in the cochlea: implications for sound transduction and auditory neurotransmission. J Neurosci 19:8377–8388.

Hulander M, Kiernan AE, Blomqvist SR, Carlsson P, Samuelsson EJ, Johansson BR, Steel KP, Enerback S (2003) Lack of pendrin expression leads to deafness and expansion of the endolymphatic compartment in inner ears of *Foxi1* null mutant mice. Development 130:2013–2025.

Ichimiya I, Adams JC, Kimura RS (1994) Immunolocalization of Na^+, K^+-ATPase, Ca^{++}-ATPase, calcium-binding proteins, and carbonic anhydrase in the guinea pig inner ear. Acta Otolaryngol (Stockh) 114:167–176.

Ikeda K, Morizono T (1988) Calcium transport mechanism in the endolymph of the chinchilla. Hear Res 34:307–311.

Ikeda K, Morizono T (1989a) Effects of carbon dioxide in the middle ear cavity upon the cochlear potentials and cochlear pH. Acta Otolaryngol (Stockh) 108:88–93.

Ikeda K, Morizono T (1989b) The preparation of acetic acid for use in otic drops and its effect on endocochlear potential and pH in inner ear fluid. Am J Otolaryngol 10:382–385.

Ikeda K, Kobayashi T, Kusakari J, Takasaka T, Yumita S, Furukawa Y (1987a) Sensorineural hearing loss associated with hypoparathyroidism. Laryngoscope 97:1075–1079.

Ikeda K, Kusakari J, Kobayashi T, Saito Y (1987b) The effect of vitamin D deficiency on the cochlear potentials and the perilymphatic ionized calcium concentration of rats. Acta Otolaryngol Suppl Stockh 435:64–72.

Ikeda K, Kusakari J, Takasaka T, Saito Y (1987c) Early effects of acetazolamide on anionic activities of the guinea pig endolymph: evidence for active function of carbonic anhydrase in the cochlea. Hear Res 31:211–216.

Ikeda K, Kusakari J, Takasaka T, Saito Y (1987d) The Ca^{2+} activity of cochlear endolymph of the guinea pig and the effect of inhibitors. Hear Res 26:117–125.

Ikeda K, Kusakari J, Takasaka T (1988) Ionic changes in cochlear endolymph of the guinea pig induced by acoustic injury. Hear Res 32:103–110.

Ikeda K, Kobayashi T, Itoh Z, Kusakari J, Takasaka T (1989) Evaluation of vitamin D metabolism in patients with bilateral sensorineural hearing loss. Am J Otol 10:11–13.

Ikeda K, Saito Y, Nishiyama A, Takasaka T (1992a) Intracellular pH regulation in isolated cochlear outer hair cells of the guinea-pig. J Physiol Lond 447:627–648.

Ikeda K, Saito Y, Nishiyama A, Takasaka T (1992b) Na^+-Ca^{2+} exchange in the isolated cochlear outer hair cells of the guinea-pig studied by fluorescence image microscopy. Pflügers Arch 420:493–499.

Imamura S, Adams JC (1996) Immunolocalization of peptide 19 and other calcium–binding proteins in the guinea pig cochlea. Anat Embryol (Berl) 194:407–418.

Ito M, Spicer SS, Schulte BA (1993) Immunohistochemical localization of brain type glucose transporter in mammalian inner ears: comparisons of developmental and adult stages. Hear Res 71:230–238.

Jabba SV, Oelke A, Singh R, Maganti RJ, Feming S, Wall SM, Everett LA, Green ED, Wangemann P (2006) Macrophage invasion contributes to degeneration of stria vascularis in Pendred syndrome mouse model. BMC Med 4:37–ff.

Jacono AA, Hu B, Kopke RD, Henderson D, Van De Water TR, Steinman HM (1998) Changes in cochlear antioxidant enzyme activity after sound conditioning and noise exposure in the chinchilla. Hear Res 117:31–38.

Jervell A, Lange-Nielsen F (1957) Congenital deaf–mutism, functional heart disease with prolongation of the Q–T interval and sudden death. Am Heart J 54:59–68.

Johnstone BM, Patuzzi R, Syka J, Sykova E (1989) Stimulus-related potassium changes in the organ of Corti of guinea-pig. J Physiol (Lond) 408:77–92.

Kambayashi J, Kobayashi T, Marcus NY, Demott JE, Thalmann I, Thalmann R (1982) Minimal concentrations of metabolic substrates capable of supporting cochlear potentials. Hear Res 7:105–114.

Karet FE, Finberg KE, Nayir A, Bakkaloglu A, Ozen S, Hulton SA, Sanjad SA, Al Sabban EA, Medina JF, Lifton RP (1999a) Localization of a gene for autosomal recessive distal renal tubular acidosis with normal hearing (rdRTA2) to 7q33–34. Am J Hum Genet 65:1656–1665.

Karet FE, Finberg KE, Nelson RD, Nayir A, Mocan H, Sanjad SA, Rodriguez-Soriano J, Santos F, Cremers CW, Di Pietro A, Hoffbrand BI, Winiarski J, Bakkaloglu A, Ozen S, Dusunsel R, Goodyer P, Hulton SA, Wu DK, Skvorak AB, Morton CC, Cunningham MJ, Jha V, Lifton RP (1999b) Mutations in the gene encoding B1 subunit of H^+-ATPase cause renal tubular acidosis with sensorineural deafness. Nat Genet 21:84–90.

Kawasaki E, Hattori N, Miyamoto E, Yamashita T, Inagaki C (2000) mRNA expression of kidney-specific ClC-K1 chloride channel in single-cell reverse transcription-polymerase chain reaction analysis of outer hair cells of rat cochlea. Neurosci Lett 290:76–78.

Kellum JA, Song M, Li J (2004) Science review: extracellular acidosis and the immune response: clinical and physiologic implications. Crit Care 8:331–336.

Kelsell DP, Dunlop J, Stevens HP, Lench NJ, Liang JN, Parry G, Mueller RF, Leigh IM (1997) Connexin 26 mutations in hereditary non-syndromic sensorineural deafness. Nature 387:80–83.

Kiang NY (1989) An auditory physiologist's view of Ménière's syndrome. In: Nadol JB, Jr. (ed) Second International Symposium on Ménière's Disease. Amsterdam: Kugler and Ghedini, pp. 13–24.

Kikuchi T, Kimura RS, Paul DL, Adams JC (1995) Gap junctions in the rat cochlea: immunohistochemical and ultrastructural analysis. Anat Embryol (Berl) 191:101–118.

Kikuchi T, Kimura RS, Paul DL, Takasaka T, Adams JC (2000) Gap junction systems in the mammalian cochlea. Brain Res Brain Res Rev 32:163–166.

Kimura K, Ito M, Amano M, Chihara K, Fukata Y, Nakafuku M, Yamamori B, Feng J, Nakano T, Okawa K, Iwamatsu A, Kaibuchi K (1996) Regulation of myosin phosphatase by Rho and Rho-associated kinase (Rho-kinase). Science 273:245–248.

Kimura RS (1967) Experimental blockage of the endolymphatic duct and sac and its effect on the inner ear of the guinea pig. A study on endolymphatic hydrops. Ann Otol Rhinol Laryngol 76:664–687.

King M, Housley GD, Raybould NP, Greenwood D, Salih SG (1998) Expression of ATP-gated ion channels by Reissner's membrane epithelial cells. NeuroReport 9:2467–2474.

Kofuji P, Biedermann B, Siddharthan V, Raap M, Iandiev I, Milenkovic I, Thomzig A, Veh RW, Bringmann A, Reichenbach A (2002) Kir potassium channel subunit expression in retinal glial cells: implications for spatial potassium buffering. Glia 39:292–303.

Konishi K, Yamane H, Iguchi H, Takayama M, Nakagawa T, Sunami K, Nakai Y (1998) Local substances regulating cochlear blood flow. Acta Otolaryngol Suppl (Stockh) 538:40–46.

Konishi T, Butler RA, Fernández C (1961) Effect of anoxia on cochlear potentials. J Acoust Soc Am 33:349–356.

Konishi T, Hamrick PE, Walsh PJ (1978) Ion transport in guinea pig cochlea. I. Potassium and sodium transport. Acta Otolaryngol (Stockh) 86:22–34.

Konrad-Martin D, Norton SJ, Mascher KE, Tempel BL (2001) Effects of *PMCA2* mutation on DPOAE amplitudes and latencies in deafwaddler mice. Hear Res 151:205–220.

Kros CJ (1996) Physiology of mammalian hair cells. In: Dallos P, Popper AN, Fay R (eds)The Cochlea. New York: Springer, pp. 319–385.

Kuffler SW, Nicholls JG, Orkand RK (1966) Physiological properties of glial cells in the central nervous system of amphibia. J Neurophysiol 29:768–787.

Kuijpers W, Bonting SL (1969) Studies on (Na^+-K^+)-activated ATPase. XXIV. Localization and properties of ATPase in the inner ear of the guinea pig. Biochim Biophys Acta 173:477–485.

Labbe D, Teranishi MA, Hess A, Bloch W, Michel O (2005) Activation of caspase-3 is associated with oxidative stress in the hydropic guinea pig cochlea. Hear Res 202: 21–27.

Lamm K, Zajic G, Schacht J (1994) Living isolated cells from inner ear vessels: a new approach for studying the regulation of cochlear microcirculation and vascular permeability. Hear Res 81:83–90.

Lang F, Busch GL, Ritter M, Volkl H, Waldegger S, Gulbins E, Haussinger D (1998) Functional significance of cell volume regulatory mechanisms. Physiol Rev 78: 247–306.

Lardner A (2001) The effects of extracellular pH on immune function. J Leukoc Biol 69:522–530.

Lautermann J, ten Cate WJ, Altenhoff P, Grummer R, Traub O, Frank H, Jahnke K, Winterhager E (1998) Expression of the gap-junction connexins 26 and 30 in the rat cochlea. Cell Tissue Res 294:415–420.

Lee JH, Marcus DC (2003) Endolymphatic sodium homeostasis by Reissner's membrane. Neuroscience 119:3–8.

Lee JH, Chiba T, Marcus DC (2001) P2X2 receptor mediates stimulation of parasensory cation absorption by cochlear outer sulcus cells and vestibular transitional cells. J Neurosci 21:9168–9174.

Lee MP, Ravenel JD, Hu RJ, Lustig LR, Tomaselli G, Berger RD, Brandenburg SA, Litzi TJ, Bunton TE, Limb C, Francis H, Gorelikow M, Gu H, Washington K, Argani P, Goldenring JR, Coffey RJ, Feinberg AP (2000) Targeted disruption of the *Kvlqt1* gene causes deafness and gastric hyperplasia in mice. J Clin Invest 106:1447–1455.

Letts VA, Valenzuela A, Dunbar C, Zheng QY, Johnson KR, Frankel WN (2000) A new spontaneous mouse mutation in the *Kcne1* gene. Mamm Genome 11:831–835.

Li HS, Niedzielski AS, Beisel KW, Hiel H, Wenthold RJ, Morley BJ (1994) Identification of a glutamate/aspartate transporter in the rat cochlea. Hear Res 78:235–242.

Lim DJ, Karabinas C, Trune DR (1983) Histochemical localization of carbonic anhydrase in the inner ear. Am J Otolaryngol 4:33–42.

Listi F, Candore G, Balistreri CR, Grimaldi MP, Orlando V, Vasto S, Colonna-Romano G, Lio D, Licastro F, Franceschi C, Caruso C (2006) Association between the HLA-A2 allele and Alzheimer disease. Rejuvenation Res 9:99–101.

Liu J, Kozakura K, Marcus DC (1995) Evidence for purinergic receptors in vestibular dark cell and strial marginal cell epithelia of the gerbil. Audit Neurosci 1:331–340.

Liu XZ, Xia XJ, Adams J, Chen ZY, Welch KO, Tekin M, Ouyang XM, Kristiansen A, Pandya A, Balkany T, Arnos KS, Nance WE (2001) Mutations in *GJA1* (connexin 43) are associated with non-syndromic autosomal recessive deafness. Hum Mol Genet 10:2945–2951.

Lohuis PJ, Klis SF, Klop WM, van Emst MG, Smoorenburg GF (1999) Signs of endolymphatic hydrops after perilymphatic perfusion of the guinea pig cochlea with cholera toxin; a pharmacological model of acute endolymphatic hydrops. Hear Res 137: 103–113.

Lumpkin EA, Marquis RE, Hudspeth AJ (1997) The selectivity of the hair cell's mechanoelectrical-transduction channel promotes Ca^{2+} flux at low Ca^{2+} concentrations. Proc Natl Acad Sci USA 94:10997–11002.

Luxon LM, Cohen M, Coffey RA, Phelps PD, Britton KE, Jan H, Trembath RC, Reardon W (2003) Neuro-otological findings in Pendred syndrome. Int J Audiol 42: 82–88.

Lymar SV, Jiang Q, Hurst JK (1996) Mechanism of carbon dioxide-catalyzed oxidation of tyrosine by peroxynitrite. Biochemistry 35:7855–7861.

Maehara H, Okamura HO, Kobayashi K, Uchida S, Sasaki S, Kitamura K (2003) Expression of CLC-KB gene promoter in the mouse cochlea. NeuroReport 14: 1571–1573.

Makishima T, Rodriguez CI, Robertson NG, Morton CC, Stewart CL, Griffith AJ (2005) Targeted disruption of mouse *Coch* provides functional evidence that DFNA9 hearing loss is not a *COCH* haploinsufficiency disorder. Hum Genet 118:29–34.

Mammano F, Frolenkov GI, Lagostena L, Belyantseva IA, Kurc M, Dodane V, Colavita A, Kachar B (1999) ATP–Induced $Ca^{(2+)}$ release in cochlear outer hair cells: localization of an inositol triphosphate-gated $Ca^{(2+)}$ store to the base of the sensory hair bundle. J Neurosci 19:6918–6929.

Manolis EN, Yandavi N, Nadol JB, Jr., Eavey RD, McKenna M, Rosenbaum S, Khetarpal U, Halpin C, Merchant SN, Duyk GM, MacRae C, Seidman CE, Seidman JG (1996) A gene for non-syndromic autosomal dominant progressive postlingual sensorineural hearing loss maps to chromosome 14q12–13. Hum Mol Genet 5: 1047–1050.

Marcaggi P, Coles JA (2001) Ammonium in nervous tissue: transport across cell membranes, fluxes from neurons to glial cells, and role in signalling. Prog Neurobiol 64:157–183.

Marcus DC, Chiba T (1999) K^+ and Na^+ absorption by outer sulcus epithelial cells. Hear Res 134:48–56.

Marcus DC, Shen Z (1994) Slowly activating, voltage-dependent K^+ conductance is apical pathway for K^+ secretion in vestibular dark cells. Am J Physiol 267:C857–C864.

Marcus DC, Thalmann R, Marcus NY (1978a) Respiratory quotient of stria vascularis of guinea pig in vitro. Arch Otorhinolaryngol 221:97–103.

Marcus DC, Thalmann R, Marcus NY (1978b) Respiratory rate and ATP content of stria vascularis of guinea pig in vitro. Laryngoscope 88:1825–1835.

Marcus DC, Ge XX, Thalmann R (1982) Comparison of the non-adrenergic action of phentolamine with that of vanadate on cochlear function. Hear Res 7:233 246.

Marcus DC, Marcus NY, Greger R (1987) Sidedness of action of loop diuretics and ouabain on nonsensory cells of utricle: a micro-Ussing chamber for inner ear tissues. Hear Res 30:55–64.

Marcus DC, Takeuchi S, Wangemann P (1993) Two types of chloride channel in the basolateral membrane of vestibular dark cell epithelium. Hear Res 69:124–132.

Marcus DC, Sunose H, Liu J, Shen Z, Scofield MA (1997) P2U purinergic receptor inhibits apical IsK/KvLQT1 channel via protein kinase C in vestibular dark cells. Am J Physiol 273:C2022–C2029.

Marcus DC, Wu T, Wangemann P, Kofuji P (2002) KCNJ10 (Kir4.1) potassium channel knockout abolishes endocochlear potential. Am J Physiol Cell Physiol 282:C403–C407.

Marcus DC, Liu J, Lee JH, Scherer EQ, Scofield MA, Wangemann P (2005) Apical membrane P2Y4 purinergic receptor controls K^+ secretion by strial marginal cell epithelium. Cell Commun Signal 3(13):1–8.

Matsunami T, Suzuki T, Hisa Y, Takata K, Takamatsu T, Oyamada M (2006) Gap junctions mediate glucose transport between GLUT1-positive and –negative cells in the spiral limbus of the rat cochlea. Cell Commun Adhes 13:93–102.

Matsuoka H, Kwon SS, Yazawa Y, Barbieri M, Yoo TJ (2002) Induction of endolymphatic hydrops by directly infused monoclonal antibody against type II collagen CB11 peptide. Ann Otol Rhinol Laryngol 111:587–592.

McLaren GM, Quirk WS, Laurikainen E, Coleman JK, Seidman MD, Dengerink HA, Nuttall AL, Miller JM, Wright JW (1993) Substance P increases cochlear blood flow without changing cochlear electrophysiology in rats. Hear Res 71:183–189.

Ménière P (1861) Mémoire sur des lésions de l'orielle interne donnant lieu à des symptômes de cogestion cérébrale apoplectiforme. Gaz Med Paris 16:597–601.

Merchant SN, Adams JC, Nadol JB Jr (2005) Pathophysiology of Ménière's syndrome: are symptoms caused by endolymphatic hydrops? Otol Neurotol 26:74–81.

Mhatre AN, Jero J, Chiappini I, Bolasco G, Barbara M, Lalwani AK (2002) Aquaporin-2 expression in the mammalian cochlea and investigation of its role in Ménière's disease. Hear Res 170:59–69.

Milhaud PG, Pondugula SR, Lee JH, Herzog M, Lehouelleur J, Wangemann P, Sans A, Marcus DC (2002) Chloride secretion by semicircular canal duct epithelium is stimulated via beta 2-adrenergic receptors. Am J Physiol Cell Physiol 283: C1752–C1760.

Moriyama Y, Maeda M, Futai M (1992) The role of V-ATPase in neuronal and endocrine systems. J Exp Biol 172:171–178.

Mouadeb DA, Ruckenstein MJ (2005) Antiphospholipid inner ear syndrome. Laryngoscope 115:879–883.

Mount DB, Romero MF (2004) The SLC26 gene family of multifunctional anion exchangers. Pflugers Arch 447:710–721.

Munoz DJ, Kendrick IS, Rassam M, Thorne PR (2001) Vesicular storage of adenosine triphosphate in the guinea-pig cochlear lateral wall and concentrations of ATP in the endolymph during sound exposure and hypoxia. Acta Otolaryngol 121:10–15.

Nakaya K, Harbidge DG, Wangemann P, Schultz BD, Green E, Wall SM, Marcus DC (2007) Lack of pendrin HCO_3^- transport elevates vestibular endolymphatic $[Ca^{2+}]$ by inhibition of acid-sensitive TRPV5 and TRPV6. Am J Physiol Renal Physiol 292:F1314–F1321.

Nakazawa K (2001) Ultrastructural localization of calmodulin in gerbil cochlea by immunogold electron microscopy. Hear Res 151:133–140.

Nakazawa K, Spicer SS, Schulte BA (1995) Postnatal expression of the facilitated glucose transporter, GLUT 5, in gerbil outer hair cells. Hear Res 82:93–99.

Newman EA, Frambach DA, Odette LL (1984) Control of extracellular potassium levels by retinal glial cell K^+ siphoning. Science 225:1174–1175.

Neyroud N, Tesson F, Denjoy I, Leibovici M, Donger C, Barhanin J, Faure S, Gary F, Coumel P, Petit C, Schwartz K, Guicheney P (1997) A novel mutation in the potassium channel gene *KVLQT1* causes the Jervell and Lange-Nielsen cardioauditory syndrome. Nat Genet 15:186–189.

Nicolas M, Dememes D, Martin A, Kupershmidt S, Barhanin J (2001) KCNQ1/KCNE1 potassium channels in mammalian vestibular dark cells. Hear Res 153:132–145.

Nie L, Gratton MA, Mu KJ, Dinglasan JN, Feng W, Yamoah EN (2005) Expression and functional phenotype of mouse ERG K^+ channels in the inner ear: potential role in K^+ regulation in the inner ear. J Neurosci 25:8671–8679.

Ninoyu O, Meyer zum Gottesberge AM (1986a) Ca^{++} activity in the endolymphatic space. Arch Otorhinolaryngol 243:141–142.

Ninoyu O, Meyer zum Gottesberge AM (1986b) Changes in Ca^{++} activity and DC potential in experimentally induced endolymphatic hydrops. Arch Otorhinolaryngol 243:106–107.

Ohmori H (1985) Mechano-electrical transduction currents in isolated vestibular hair cells of the chick. J Physiol (Lond) 359:189–217.

Ohmori T, Yatomi Y, Osada M, Ozaki Y (2004) Platelet-derived sphingosine 1-phosphate induces contraction of coronary artery smooth muscle cells via S1P2. J Thromb Haemost 2:203–205.

Ohyama K, Salt AN, Thalmann R (1988) Volume flow rate of perilymph in the guinea-pig cochlea. Hear Res 35:119–129.

Okamura HO, Sugai N, Suzuki K, Ohtani I (1996) Enzyme-histochemical localization of carbonic anhydrase in the inner ear of the guinea pig and several improvements of the technique. Histochem Cell Biol 106:425–430.

Okamura H, Spicer SS, Schulte BA (2001) Developmental expression of monocarboxylate transporter in the gerbil inner ear. Neuroscience 107:499–505.

Oliveira JA, Canedo DM, Rossato M, Andrade MH (2004) Self-protection against amino-glycoside ototoxicity in guinea pigs. Otolaryngol Head Neck Surg 131:271–279.

Orkand RK, Nicholls JG, Kuffler SW (1966) Effect of nerve impulses on the membrane potential of glial cells in the central nervous system of amphibia. J Neurophysiol 29:788–806.

Oshima T, Ikeda K, Furukawa M, Takasaka T (1997) Alternatively spliced isoforms of the Na^+/Ca^{2+} exchanger in the guinea pig cochlea. Biochem Biophys Res Commun 233:737–741.

Ottersen OP, Takumi Y, Matsubara A, Landsend AS, Laake JH, Usami S (1998) Molecular organization of a type of peripheral glutamate synapse: the afferent synapses of hair cells in the inner ear. Prog Neurobiol 54:127–148.

Pack AK, Slepecky NB (1995) Cytoskeletal and calcium-binding proteins in the mammalian organ of Corti: cell type-specific proteins displaying longitudinal and radial gradients. Hear Res 91:119–135.

Passali D, Damiani V, Mora R, Passali FM, Passali GC, Bellussi L (2004) P0 antigen detection in sudden hearing loss and Ménière's disease: a new diagnostic marker? Acta Otolaryngol 124:1145–1148.

Payne JA (1997) Functional characterization of the neuronal-specific K-Cl cotransporter: implications for $[K^+]o$ regulation. Am J Physiol 273:C1516–C1525.

Payne JA, Xu JC, Haas M, Lytle CY, Ward D, Forbush B, III (1995) Primary structure, functional expression, and chromosomal localization of the bumetanide-sensitive Na-K-Cl cotransporter in human colon. J Biol Chem 270:17977–17985.

Pendred V (1896) Deaf-mutism and goitre. Lancet 11:532.

Pierson MG, Gray BH (1982) Superoxide dismutase activity in the cochlea. Hear Res 6:141–151.

Pondugula SR, Raveendran NN, Ergonul Z, Deng Y, Chen J, Sanneman JD, Palmer LG, Marcus DC (2006) Glucocorticoid regulation of genes in the amiloride-sensitive sodium transport pathway by semicircular canal duct epithelium of neonatal rat. Physiol Genomics 24:114–123.

Qiu J, Steyger PS, Trune DR, Nuttall AL (2001) Co-existence of tyrosine hydroxylase and calcitonin gene-related peptide in cochlear spiral modiolar artery of guinea pigs. Hear Res 155:152–160.

Qu C, Liang F, Hu W, Shen Z, Spicer SS, Schulte BA (2006) Expression of CLC-K chloride channels in the rat cochlea. Hear Res 213:79–87.

Quirk WS, Dengerink HA, Coleman JK, Wright JW (1989) Cochlear blood flow autoregulation in Wistar-Kyoto rats. Hear Res 41:53–60.

Quirk WS, Avinash G, Nuttall AL, Miller JM (1992) The influence of loud sound on red blood cell velocity and blood vessel diameter in the cochlea. Hear Res 63:102–107.

Quirk WS, Seidman MD, Laurikainen EA, Nuttall AL, Miller JM (1994) Influence of calcitonin-gene related peptide on cochlear blood flow and electrophysiology. Am J Otol 15:56–60.

Rabionet R, Gasparini P, Estivill X (2000) Molecular genetics of hearing impairment due to mutations in gap junction genes encoding beta connexins. Hum Mutat 16:190–202.

Rakugi H, Tabuchi Y, Nakamaru M, Nagano M, Higashimori K, Mikami H, Ogihara T, Suzuki N (1990) Evidence for endothelin-1 release from resistance vessels of rats in response to hypoxia. Biochem Biophys Res Commun 169:973–977.

Rauch SD, Merchant SN, Thedinger BA (1989) Ménière's syndrome and endolymphatic hydrops. Double-blind temporal bone study. Ann Otol Rhinol Laryngol 98:873–883.

Reardon W, OMahoney CF, Trembath R, Jan H, Phelps PD (2000) Enlarged vestibular aqueduct: a radiological marker of pendred syndrome, and mutation of the PDS gene. QJM 93:99–104.

Ricci AJ, Fettiplace R (1998) Calcium permeation of the turtle hair cell mechanotransducer channel and its relation to the composition of endolymph. J Physiol 506 (Pt 1):159–173.

Rimaniol AC, Mialocq P, Clayette P, Dormont D, Gras G (2001) Role of glutamate transporters in the regulation of glutathione levels in human macrophages. Am J Physiol Cell Physiol 281:C1964–C1970.

Robertson NG, Resendes BL, Lin JS, Lee C, Aster JC, Adams JC, Morton CC (2001) Inner ear localization of mRNA and protein products of COCH, mutated in the sensorineural deafness and vestibular disorder, DFNA9. Hum Mol Genet 10:2493–2500.

Romero MF, Fulton CM, Boron WF (2004) The SLC4 family of HCO_3^{--} transporters. Pflugers Arch 447:495–509.

Royaux IE, Suzuki K, Mori A, Katoh R, Everett LA, Kohn LD, Green ED (2000) Pendrin, the protein encoded by the Pendred syndrome gene (PDS), is an apical porter of iodide in the thyroid and is regulated by thyroglobulin in FRTL-5 cells. Endocrinology 141:839–845.

Rubanyi GM, Polokoff MA (1994) Endothelins: molecular biology, biochemistry, pharmacology, physiology, and pathophysiology. Pharmacol Rev 46:325–415.

Ryan AF, Goodwin P, Woolf NK, Sharp F (1982) Auditory stimulation alters the pattern of 2-deoxyglucose uptake in the inner ear. Brain Res 234:213–225.

Rybak LP, Husain K, Evenson L, Morris C, Whitworth C, Somani SM (1997) Protection by 4-methylthiobenzoic acid against cisplatin-induced ototoxicity: antioxidant system. Pharmacol Toxicol 81:173–179.

Sadanaga M, Liu J, Wangemann P (1997) Endothelin-A receptors mediate endothelin-induced vasoconstriction in the spiral ligament of the inner ear. Hear Res 112:106–114.

Sage CL, Marcus DC (2001) Immunolocalization of ClC-K chloride channel in strial marginal cells and vestibular dark cells. Hear Res 160:1–9.

Sage CL, Marcus DC (2002) Immunolocalization of P2Y4 and P2Y2 purinergic receptors in strial marginal cells and vestibular dark cells. J Membr Biol 185:103–115.

Sakagami M, Fukazawa K, Matsunaga T, Fujita H, Mori N, Takumi T, Ohkubo H, Nakanishi S (1991) Cellular localization of rat Isk protein in the stria vascularis by immunohistochemical observation. Hear Res 56:168–172.

Sakaguchi N, Henzl MT, Thalmann I, Thalmann R, Schulte BA (1998) Oncomodulin is expressed exclusively by outer hair cells in the organ of Corti. J Histochem Cytochem 46:29–40.

Salt AN (2004) Acute endolymphatic hydrops generated by exposure of the ear to nontraumatic low-frequency tones. JARO 5:203–214.

Salt AN, Demott JE (2000) Ionic and potential changes of the endolymphatic sac induced by endolymph volume changes. Hear Res 149:46–54.

Salt AN, Konishi T (1979) Effects of noise on cochlear potentials and endolymph potassium concentration recorded with potassium-selective electrodes. Hear Res 1: 343–363.

Salt AN, Ohyama K (1993) Accumulation of potassium in scala vestibuli perilymph of the mammalian cochlea. Ann Otol Rhinol Laryngol 102:64–70.

Salt AN, Rask-Andersen H (2004) Responses of the endolymphatic sac to perilymphatic injections and withdrawals: evidence for the presence of a one-way valve. Hear Res 191:90–100.

Salt AN, Thalmann R (1988) Rate of longitudinal flow of cochlear endolymph. In: Nadol Jr JB (ed) Second International Symposium on Ménière's Disease. Amsterdam. Kugler and Ghedini, pp. 69–73.

Salt AN, Thalmann R, Marcus DC, Bohne BA (1986) Direct measurement of longitudinal endolymph flow rate in the guinea pig cochlea. Hear Res 23:141–151.

Salt AN, Melichar I, Thalmann R (1987) Mechanisms of endocochlear potential generation by stria vascularis. Laryngoscope 97:984–991.

Salt AN, Inamura N, Thalmann R, Vora A (1989) Calcium gradients in inner ear endolymph. Am J Otolaryngol 10:371–375.

Sanchez E, Lopez-Escamez JA, Lopez-Nevot MA, Lopez-Nevot A, Cortes R, Martin J (2004) Absence of COCH mutations in patients with Ménière disease. Eur J Hum Genet 12:75–78.

Sanguinetti MC, Curran ME, Zou A, Shen J, Spector PS, Atkinson DL, Keating MT (1996) Coassembly of KVLQT1 and minK (IsK) proteins to form cardiac IKs potassium channel. Nature 384:80–83.

Sato M, Tani E, Fujikawa H, Kaibuchi K (2000) Involvement of Rho-kinase-mediated phosphorylation of myosin light chain in enhancement of cerebral vasospasm. Circ Res 87:195–200.

Sawada S, Takeda T, Kitano H, Takeuchi S, Kakigi A, Azuma H (2002) Aquaporin-2 regulation by vasopressin in the rat inner ear. NeuroReport 13:1127–1129.

Scheibe F, Haupt H, Ludwig C (1992) Intensity-dependent changes in oxygenation of cochlear perilymph during acoustic exposure. Hear Res 63:19–25.

Scherer EQ, Wonneberger K, Wangemann P (2001) Differential desensitization of Ca^{2+} mobilization and vasoconstriction by ETA receptors in the gerbil spiral modiolar artery. J Membr Biol 182:183–191.

Scherer EQ, Herzog M, Wangemann P (2002) Endothelin-1-induced vasospasms of spiral modiolar artery are mediated by rho-kinase-induced Ca^{2+} sensitization of contractile apparatus and reversed by calcitonin gene-related peptide. Stroke 33:2965–2971.

Scherer EQ, Arnold W, Wangemann P (2005) Pharmacological reversal of endothelin-1 mediated constriction of the spiral modiolar artery: a potential new treatment for sudden sensorineural hearing loss. BMC Ear Nose Throat Disord 5:10.

Scherer EQ, Lidington D, Oestreicher E, Arnold W, Pohl U, Bolz SS (2006) Sphingosine-1-phosphate modulates spiral modiolar artery tone: a potential role in vascular-based inner ear pathologies? Cardiovasc Res

Schimanski S, Scofield MA, Wangemann P (2001) Functional b2-adrenergic receptors are present in non-strial tissues of the lateral wall in the gerbil cochlea. Audiol Neurootol 6:124–136.

Schlingmann KP, Konrad M, Jeck N, Waldegger P, Reinalter SC, Holder M, Seyberth HW, Waldegger S (2004) Salt wasting and deafness resulting from mutations in two chloride channels. N Engl J Med 350:1314–1319.

Schmitt N, Schwarz M, Peretz A, Abitbol I, Attali B, Pongs O (2000) A recessive C–terminal Jervell and Lange-Nielsen mutation of the KCNQ1 channel impairs subunit assembly. EMBO J 19:332–340.

Schulze-Bahr E, Wang Q, Wedekind H, Haverkamp W, Chen Q, Sun Y, Rubie C, Hordt M, Towbin JA, Borggrefe M, Assmann G, Qu X, Somberg JC, Breithardt G, Oberti C, Funke H (1997) *KCNE1* mutations cause jervell and Lange-Nielsen syndrome. Nat Genet 17:267–268.

Scott DA, Karniski LP (2000) Human pendrin expressed in *Xenopus laevis* oocytes mediates chloride/formate exchange. Am J Physiol Cell Physiol 278:C207–C211.

Scott DA, Wang R, Kreman TM, Sheffield VC, Karniski LP (1999) The Pendred syndrome gene encodes a chloride-iodide transport protein. Nat Genet 21:440–443.

Selivanova OA, Gouveris H, Victor A, Amedee RG, Mann W (2005) Intratympanic dexamethasone and hyaluronic acid in patients with low-frequency and Ménière's-associated sudden sensorineural hearing loss. Otol Neurotol 26:890–895.

Shen Z, Marcus DC, Sunose H, Chiba T, Wangemann P (1997) IsK channel in strial marginal cell: voltage-dependence, ion selectivity, inhibition by 293B and sensitivity to clofilium. Audit Neurosci 3:215–230.

Shibata T, Hibino H, Doi K, Suzuki T, Hisa Y, Kurachi Y (2006) Gastric type H^+,K^+-ATPase in the cochlear lateral wall is critically involved in formation of the endocochlear potential. Am J Physiol Cell Physiol 291:C1038–C1048.

Shimokawa H, Takeshita A (2005) Rho-kinase is an important therapeutic target in cardiovascular medicine. Arterioscler Thromb Vasc Biol 25:1767–1775.

Shimozono M, Scofield MA, Wangemann P (1997) Functional evidence for a monocarboxylate transporter (MCT) in strial marginal cells and molecular evidence for MCT1 and MCT2 in stria vascularis. Hear Res 114:213–222.

Singh, R, Wangemann, P (2008, 2007 epub) Free radical stress mediated loss of Kcnj10 protein expression in stria vascularis contributes to deafness in Pendred syndrome mouse model. Am J Physiol Renal Physiol 294:F139–F148.

Smith CA, Lowry OH, Wu ML (1954) The electrolytes of the labyrinthine fluids. Laryngoscope 64:141–153.

Soliman AM (1989) Experimental autoimmune inner ear disease. Laryngoscope 99: 188–193.

Somlyo AP, Somlyo AV (2003) Ca^{2+} sensitivity of smooth muscle and nonmuscle myosin II: modulated by G proteins, kinases, and myosin phosphatase. Physiol Rev 83:1325–1358.

Song BB, Sha SH, Schacht J (1998) Iron chelators protect from aminoglycoside-induced cochleo- and vestibulo-toxicity. Free Radic Biol Med 25:189–195.

Spector GJ, Carr C (1979) The ultrastructural cytochemistry of peroxisomes in the guinea pig cochlea: a metabolic hypothesis for the stria vascularis. Laryngoscope 89:1–38.

Spicer SS, Schulte BA (1991) Differentiation of inner ear fibrocytes according to their ion transport related activity. Hear Res 56:53–64.

Spicer SS, Schulte BA (1996) The fine structure of spiral ligament cells relates to ion return to the stria and varies with place-frequency. Hear Res 100:80–100.

Spicer SS, Schulte BA (1998) Evidence for a medial K^+ recycling pathway from inner hair cells. Hear Res 118:1–12.

Splawski I, Shen J, Timothy KW, Lehmann MH, Priori S, Robinson JL, Moss AJ, Schwartz PJ, Towbin JA, Vincent GM, Keating MT (2000) Spectrum of mutations in long-QT syndrome genes. KVLQT1, HERG, SCN5A, KCNE1, and KCNE2. Circulation 102:1178–1185.

Stankovic KM, Brown D, Alper SL, Adams JC (1997) Localization of pH regulating proteins H^+ ATPase and Cl^-/HCO^{3-} exchanger in the guinea pig inner ear. Hear Res 114:21–34.

Sterkers O, Saumon G, Tran Ba Huy P, Ferrary E, Amiel C (1984) Electrochemical heterogeneity of the cochlear endolymph: effect of acetazolamide. Am J Physiol 246:F47–F53.

Sterling D, Reithmeier RA, Casey JR (2001) A transport metabolon. Functional interaction of carbonic anhydrase II and chloride/bicarbonate exchangers. J Biol Chem 276: 47886–47894.

Stover EH, Borthwick KJ, Bavalia C, Eady N, Fritz DM, Rungroj N, Giersch AB, Morton CC, Axon PR, Akil I, Al Sabban EA, Baguley DM, Bianca S, Bakkaloglu A, Bircan Z, Chauveau D, Clermont MJ, Guala A, Hulton SA, Kroes H, Li VG, Mir S, Mocan H, Nayir A, Ozen S, Rodriguez SJ, Sanjad SA, Tasic V, Taylor CM, Topaloglu R, Smith AN, Karet FE (2002) Novel *ATP6V1B1* and *ATP6V0A4* mutations in autosomal recessive distal renal tubular acidosis with new evidence for hearing loss. J Med Genet 39:796–803.

Street VA, McKee-Johnson JW, Fonseca RC, Tempel BL, Noben-Trauth K (1998) Mutations in a plasma membrane Ca^{2+}-ATPase gene cause deafness in deafwaddler mice. Nat Genet 19:390–394.

Sun J, Ahmad S, Chen S, Tang W, Zhang Y, Chen P, Lin X (2005) Cochlear gap junctions coassembled from Cx26 and 30 show faster intercellular Ca^{2+} signaling than homomeric counterparts. Am J Physiol Cell Physiol 288:C613–C623.

Sweadner KJ (1985) Enzymatic properties of separated isozymes of the Na,K-ATPase. Substrate affinities, kinetic cooperativity, and ion transport stoichiometry. J Biol Chem 260:11508–11513.

Tabuchi K, Tsuji S, Asaka Y, Hara A, Kusakari J (2001) Ischemia-reperfusion injury of the cochlea: effects of an iron chelator and nitric oxide synthase inhibitors. Hear Res 160:31–36.

Takeda T, Sawada S, Takeda S, Kitano H, Suzuki M, Kakigi A, Takeuchi S (2003) The effects of V2 antagonist (OPC–31260) on endolymphatic hydrops. Hear Res 182:9–18.

Takeuchi S, Ando M (1997) Marginal cells of the stria vascularis of gerbils take up glucose via the facilitated transporter GLUT: application of autofluorescence. Hear Res 114:69–74.

Takeuchi S, Irimajiri A (1994) Cl$^-$ and nonselective cation channels in the basolateral membrane of strial marginal cells. Proc Sendai Symp 4:35–37.

Takeuchi S, Irimajiri A (1996) Maxi–K+ channel in plasma membrane of basal cells dissociated from the stria vascularis of gerbils. Hear Res 95:18–25.

Takeuchi S, Ando M, Kozakura K, Saito H, Irimajiri A (1995) Ion channels in basolateral membrane of marginal cells dissociated from gerbil stria vascularis. Hear Res 83: 89–100.

Takeuchi S, Ando M, Kakigi A (2000) Mechanism generating endocochlear potential: role played by intermediate cells in stria vascularis. Biophys J 79:2572–2582.

Takumi Y, Matsubara A, Danbolt NC, Laake JH, Storm-Mathisen J, Usami S, Shinkawa H, Ottersen OP (1997) Discrete cellular and subcellular localization of glutamine synthetase and the glutamate transporter GLAST in the rat vestibular end organ. Neuroscience 79:1137–1144.

Takumi Y, Matsubara A, Laake JH, Ramirez-Leon V, Roberg B, Torgner I, Kvamme E, Usami S, Ottersen OP (1999) Phosphate activated glutaminase is concentrated in mitochondria of sensory hair cells in rat inner ear: a high resolution immunogold study. J Neurocytol 28:223–237.

Takumi Y, Matsubara A, Tsuchida S, Ottersen OP, Shinkawa H, Usami S (2001) Various glutathione S-transferase isoforms in the rat cochlea. NeuroReport 12:1513–1516.

Tanaka Y, Asanuma A, Yanagisawa K (1980) Potentials of outer hair cells and their membrane properties in cationic environments. Hear Res 2:431–438.

Tanaka F, Whitworth CA, Rybak LP (2004) Round window pH manipulation alters the ototoxicity of systemic cisplatin. Hear Res 187:44–50.

Tang W, Zhang Y, Chang Q, Ahmad S, Dahlke I, Yi H, Chen P, Paul DL, Lin X (2006) Connexin29 is highly expressed in cochlear Schwann cells, and it is required for the normal development and function of the auditory nerve of mice. J Neurosci 26:1991–1999.

Teubner B, Michel V, Pesch J, Lautermann J, Cohen-Salmon M, Sohl G, Jahnke K, Winterhager E, Herberhold C, Hardelin JP, Petit C, Willecke K (2003) Connexin30 (Gjb6)–deficiency causes severe hearing impairment and lack of endocochlear potential. Hum Mol Genet 12:13–21.

Thalmann I, Matschinsky FM, Thalmann R (1970) Quantitative study of selected enzymes involved in energy metabolism of the cochlear duct. Ann Otol Rhinol Laryngol 79: 12–29.

Thalmann R, Miyoshi T, Thalmann I (1972) The influence of ischemia upon the energy reserves of inner ear tissues. Laryngoscope 82:2249–2272.

Thorne PR, Munoz DJ, Housley GD (2004) Purinergic modulation of cochlear partition resistance and its effect on the endocochlear potential in the guinea pig. J Assoc Res Otolaryngol 5:58–65.

Tyson J, Tranebjaerg L, McEntagart M, Larsen LA, Christiansen M, Whiteford ML, Bathen J, Aslaksen B, Sorland SJ, Lund O, Pembrey ME, Malcolm S, Bitner-Glindzicz M (2000) Mutational spectrum in the cardioauditory syndrome of Jervell and Lange-Nielsen. Hum Genet 107:499–503.

Unoki H, Fan J, Watanabe T (1999) Low-density lipoproteins modulate endothelial cells to secrete endothelin-1 in a polarized pattern: a study using a culture model system simulating arterial intima. Cell Tissue Res 295:89–99.

Usami S, Hjelle OP, Ottersen OP (1996) Differential cellular distribution of glutathione—an endogenous antioxidant—in the guinea pig inner ear. Brain Res 743:337–340.

Usami S, Abe S, Weston MD, Shinkawa H, Van Camp G, Kimberling WJ (1999) Non-syndromic hearing loss associated with enlarged vestibular aqueduct is caused by *PDS* mutations. Hum Genet 104:188–192.

Usami S, Takahashi K, Yuge I, Ohtsuka A, Namba A, Abe S, Fransen E, Patthy L, Otting G, Van Camp G (2003) Mutations in the *COCH* gene are a frequent cause of autosomal dominant progressive cochleo-vestibular dysfunction, but not of Ménière's disease. Eur J Hum Genet 11:744–748.

Valli P, Zucca G, Botta L (1990) Perilymphatic potassium changes and potassium homeostasis in isolated semicircular canals of the frog. J Physiol (Lond) 430: 585–594.

Vanhoutte PM (1978) Heterogeneity in vascular smooth muscle. In: Kaley G, Altura BM (eds) Microcirculation, Vol. II. Baltimore: University Park Press, pp. 181–309.

Vargas-Poussou R, Houillier P, Le Pottier N, Strompf L, Loirat C, Baudouin V, Macher MA, Dechau M, Ulinski T, Nobili F, Eckart P, Novo R, Cailliez M, Salomon R, Nivet H, Cochat P, Tack I, Fargeot A, Bouissou F, Kesler GR, Lorotte S, Godefroid N, Layet V, Morin G, Jeunemaitre X, Blanchard A (2006) Genetic investigation of autosomal recessive distal renal tubular acidosis: evidence for early sensorineural hearing loss associated with mutations in the *ATP6V0A4* gene. J Am Soc Nephrol 17:1437–1443.

Varnum MD, Busch AE, Bond CT, Maylie J, Adelman JP (1993) The min K channel underlies the cardiac potassium current IKs and mediates species-specific responses to protein kinase. Proc Natl Acad Sci USA 90:11528–11532.

Vass Z, Dai CF, Steyger PS, Jancso G, Trune DR, Nuttall AL (2004) Co-localization of the vanilloid capsaicin receptor and substance P in sensory nerve fibers innervating cochlear and vertebro-basilar arteries. Neuroscience 124:919–927.

Vetter DE, Mann JR, Wangemann P, Liu Z, McLaughlin KJ, Lesage F, Marcus DC, Lazdunski M, Heinemann SF, Barhanin J (1996) Inner ear defects induced by null mutation of isk gene. Neuron 17:1251–1264.

Vince JW, Reithmeier RA (1998) Carbonic anhydrase II binds to the carboxyl terminus of human band 3, the erythrocyte Cl^-/HCO_3^{--} exchanger. J Biol Chem 273: 28430–28437.

von Békésy G (1950) DC potentials and energy balance of the cochlear partition. J Acoust Soc Am 22:576–582.

Wagner CA, Finberg KE, Breton S, Marshansky V, Brown D, Geibel JP (2004) Renal vacuolar H^+-ATPase. Physiol Rev 84:1263–1314.

Waldegger S, Jeck N, Barth P, Peters M, Vitzthum H, Wolf K, Kurtz A, Konrad M, Seyberth HW (2002) Barttin increases surface expression and changes current properties of ClC-K channels. Pflugers Arch 444:411–418.

Wang S, Schacht J (1990) Insulin stimulates protein synthesis and phospholipid signaling systems but does not regulate glucose uptake in the inner ear. Hear Res 47:53–61.

Wang Z, Li H, Moss AJ, Robinson J, Zareba W, Knilans T, Bowles NE, Towbin JA (2002) Compound heterozygous mutations in *KvLQT1* cause Jervell and Lange-Nielsen syndrome. Mol Genet Metab 75.308–316.

Wangemann P (1995) Comparison of ion transport mechanisms between vestibular dark cells and strial marginal cells. Hear Res 90:149–157.

Wangemann P (2002) K^+ cycling and the endocochlear potential. Hear Res 165:1–9.

Wangemann P, Gruber DD (1998) The isolated in-vitro perfused spiral modiolar artery: pressure dependence of vasoconstriction. Hear Res 115:113–118.

Wangemann P, Marcus DC (1992) The membrane potential of vestibular dark cells is controlled by a large Cl^{--} conductance. Hear Res 62:149–156.

Wangemann P, Schacht J (1996) Homeostasic mechanisms in the cochlea. In: Dallos P, Popper AN, Fay R (eds) The Cochlea. New York: Springer, pp. 130–185.

Wangemann P, Shiga N (1994) Cell volume control in vestibular dark cells during and after a hyposmotic challenge. Am J Physiol Cell Physiol 266:C1046–C1060.

Wangemann P, Wonneberger K (2005) Neurogenic regulation of cochlear blood flow occurs along the basilar artery, the anterior inferior cerebellar artery and at branch points of the spiral modiolar artery. Hear Res 209:91–96.

Wangemann P, Liu J, Marcus DC (1995a) Ion transport mechanisms responsible for K^+ secretion and the transepithelial voltage across marginal cells of stria vascularis in vitro. Hear Res 84:19–29.

Wangemann P, Liu J, Shen Z, Shipley A, Marcus DC (1995b) Hypo-osmotic challenge stimulates transepithelial K^+ secretion and activates apical IsK channel in vestibular dark cells. J Membr Biol 147:263–273.

Wangemann P, Liu J, Shiga N (1996a) Vestibular dark cells contain the Na^+/H^+ exchanger NHE-1 in the basolateral membrane. Hear Res 94:94–108.

Wangemann P, Shen Z, Liu J (1996b) K^+-induced stimulation of K^+ secretion involves activation of the IsK channel in vestibular dark cells. Hear Res 100:201–210.

Wangemann P, Cohn ES, Gruber DD, Gratton MA (1998) Ca^{2+}-dependence and nifedipine-sensitivity of vascular tone and contractility in the isolated superfused spiral modiolar artery in vitro. Hear Res 118:90–100.

Wangemann P, Liu J, Shimozono M, Scofield MA (1999) b1-adrenergic receptors but not b2-adrenergic or vasopressin receptors regulate K^+ secretion in vestibular dark cells of the inner ear. J Membr Biol 170:67–77.

Wangemann P, Liu J, Shimozono M, Schimanski S, Scofield MA (2000) K^+ secretion in strial marginal cells is stimulated via b1-adrenergic receptors but not via b2-adrenergic or vasopressin receptors. J Membr Biol 175:191–202.

Wangemann P, Itza EM, Albrecht B, Wu T, Jabba SV, Maganti RJLJH, Everett LA, Wall SM, Royaux IE, Green ED, Marcus DC (2004) Loss of KCNJ10 protein expression abolishes endocochlear potential and causes deafness in Pendred syndrome mouse model. BMC Medicine 2:30.

Wangemann P, Jabba SV, Singh R, Wu T, Oelke A, Gollapudi ASB, Marcus DC (2005) Deafness in Pendred syndrome is related to free radical stress in stria vascularis. In: David J. Lim (ed.) Proc 5th Int Sym Menière's Dis Inner Ear Homeostasis Disorders, House Ear Institute, Los Angeles, CA.

Wangemann P, Nakaya K, Wu T, Maganti R, Itza EM, Sanneman J, Harbidge D, Billings S, Marcus DC (2007) Loss of cochlear HCO^{3-} secretion causes deafness via endolymphatic acidification and inhibition of Ca^{2+} reabsorption in a Pendred syndrome mouse model. Am J Physiol Renal Physiol 292:F1345–1353.

Warnock DG, Eveloff J (1989) K-Cl cotransport systems. Kidney Int 36:412–417.

Wood JD, Muchinsky SJ, Filoteo AG, Penniston JT, Tempel BL (2004a) Low endolymph calcium concentrations in deafwaddler2J mice suggest that PMCA2 contributes to endolymph calcium maintenance. J Assoc Res Otolaryngol 5:99–110.

Wood JD, Muchinsky SJ, Filoteo AG, Penniston JT, Tempel BL (2004b) Low endolymph calcium concentrations in deafwaddler2J mice suggest that PMCA2 contributes to endolymph calcium maintenance. J Assoc Res Otolaryngol 5:99–110.

Xia A, Katori Y, Oshima T, Watanabe K, Kikuchi T, Ikeda K (2001) Expression of connexin 30 in the developing mouse cochlea. Brain Res 898:364–367.

Xia A, Kikuchi T, Hozawa K, Katori Y, Takasaka T (1999) Expression of connexin 26 and Na,K-ATPase in the developing mouse cochlear lateral wall: functional implications. Brain Res 846:106–111.

Xia AP, Ikeda K, Katori Y, Oshima T, Kikuchi T, Takasaka T (2000) Expression of connexin 31 in the developing mouse cochlea. NeuroReport 11:2449–2453.

Xia AP, Kikuchi T, Minowa O, Katori Y, Oshima T, Noda T, Ikeda K (2002) Late–onset hearing loss in a mouse model of DFN3 non-syndromic deafness: morphologic and immunohistochemical analyses. Hear Res 166:150–158.

Xie H, Bevan JA (1999) Oxidized low-density lipoprotein enhances myogenic tone in the rabbit posterior cerebral artery through the release of endothelin-1. Stroke 30: 2423–2429.

Yada T, Shimokawa H, Kajiya F (2006) Cardioprotective effect of hydroxyfasudil as a specific Rho-kinase inhibitor, on ischemia-reperfusion injury in canine coronary microvessels in vivo. Clin Hemorheol Microcirc 34:177–183.

Yamakawa K (1938) Über pathologische Veränderungen bei einem Menière-Kranken. J Otolaryngol Soc Jpn 44:181–182.

Yamashita H, Sekitani T, Bagger-Sjöbäck D (1992) Expression of carbonic anhydrase isoenzyme-like immunoreactivity in the limbus spiralis of the human fetal cochlea. Hear Res 64:118–122.

Yamauchi D, Raveendran NN, Pondugula SR, Kampalli SB, Sanneman JD, Harbidge DG, Marcus DC (2005) Vitamin D upregulates expression of ECaC1 mRNA in semicircular canal. Biochem Biophys Res Commun 331:1353–1357.

Yamoah EN, Lumpkin EA, Dumont RA, Smith PJ, Hudspeth AJ, Gillespie PG (1998) Plasma membrane Ca^{2+}–ATPase extrudes Ca^{2+} from hair cell stereocilia. J Neurosci 18:610–624.

Yatomi Y (2006) Sphingosine 1–phosphate in vascular biology: possible therapeutic strategies to control vascular diseases. Curr Pharm Des 12:575–587.

Yoo TJ, Tomoda K, Hernandez AD (1984) Type II collagen-induced autoimmune inner ear lesions in guinea pigs. Ann Otol Rhinol Laryngol Suppl 113:3–5.

Zelante L, Gasparini P, Estivill X, Melchionda S, D'Agruma L, Govea N, Mila M, Monica MD, Lutfi J, Shohat M, Mansfield E, Delgrosso K, Rappaport E, Surrey S, Fortina P (1997) Connexin26 mutations associated with the most common form of non-syndromic neurosensory autosomal recessive deafness (DFNB1) in Mediterraneans. Hum Mol Genet 6:1605–1609.

Zhang Y, Tang W, Ahmad S, Sipp JA, Chen P, Lin X (2005) Gap junction-mediated intercellular biochemical coupling in cochlear supporting cells is required for normal cochlear functions. Proc Natl Acad Sci USA 102:15201–15206.

Zhao HB (2005) Connexin26 is responsible for anionic molecule permeability in the cochlea for intercellular signalling and metabolic communications. Eur J Neurosci 21:1859–1868.

Zhao HB, Yu N, Fleming CR (2005) Gap junctional hemichannel-mediated ATP release and hearing controls in the inner ear. Proc Natl Acad Sci USA 102:18724–18729.

Zidanic M, Brownell WE (1990) Fine structure of the intracochlear potential field. I. The silent current. Biophys J 57:1253–1268.

Ziegler EA, Brieger J, Heinrich UR, Mann WJ (2004) Immunohistochemical localization of cyclooxygenase isoforms in the organ of Corti and the spiral ganglion cells of guinea pig cochlea. ORL J Otorhinolaryngol Relat Spec 66:297–301.

Zimmermann U, Kopschall I, Rohbock K, Bosman GJ, Zenner HP, Knipper M (2000) Molecular characterization of anion exchangers in the cochlea. Mol Cell Biochem 205:25–37.

4
Tinnitus: Theories, Mechanisms, and Treatments

Carol A. Bauer and Thomas J. Brozoski

1. Introduction

Tinnitus is an auditory percept that originates in the head and not from an external sound source. This phantom sound is a symptom of an underlying abnormality in the auditory pathway. It is instructive to think of tinnitus as analogous to the symptom of pain. Tinnitus, like pain, occurs after peripheral trauma, is presumed to derive from peripheral deafferentation and associated processes of central compensation, may be distinguished by acute and chronic sensory states, is refractory to management and has no direct objective correlate.

2. History and Epidemiology

Tinnitus is nearly as common in human experience as hearing loss. The term "tinnitus" originated with Pliny the Elder (C.E 23–79), and appears in his description of "ears ringing and singing, or having in them any unnaturall (sic) sound and noise," in his work *Natural History* (Morgenstern 2005). Michelangelo described his tinnitus in a poem referring to his physical decline: "A spider's web is hidden in one ear, in the other a cricket sings throughout the night" (Girardi 1965). Perhaps because chronic tinnitus is not only persistent but virtually inescapable, it often significantly degrades the sufferer's quality of life. Scientific interest in tinnitus has expanded in recent decades, in parallel with advances in the field of auditory neuroscience.

The human experience of tinnitus is complex, comprising both the sensory features of the condition and the associated affective reactions. The sensory component comprises perceptual features such as tonality, laterality or location, loudness, and constancy. The reactive component derives from the emotional, cognitive and functional responses to the perception of tinnitus. This higher-order reactive component is unique to each individual and is modulated by factors not directly related to the sensory features of the tinnitus or the associated auditory pathology. In some cases, tinnitus can be disabling resulting in depression, anxiety, disordered sleep, and impaired concentration. These factors can be a significant source of morbidity in chronic tinnitus. Clinical studies that examine the effects of intervention to mitigate tinnitus must carefully address the dichotomous nature of the problem and apply appropriate instruments to measure relevant factors comprehensively.

Tinnitus demographics have been studied world-wide, with some disparity in the findings. Many of the disparities likely stem from the difficulty in surveying large groups of people, the reliance on anamnestic data, and the imprecise nature of language in accurately capturing the sensory features of tinnitus. Relatively few studies have quantified the sensory features of tinnitus using either psycho-acoustic measurements or validated questionnaires or both. Demographic data relevant to tinnitus include age, gender, hearing loss, history of cardiovascular disease, head injury, and exposure to noise or ototoxins.

Tinnitus is estimated to affect 8% to 30% of adults worldwide. In a national study of hearing in the United Kingdom, 10% of adults reported experiencing tinnitus for longer than 5 min, unrelated to tinnitus of immediate onset after noise exposure. In this study, 5% of adults described their tinnitus as moderate or severely annoying, 1% as severely affecting their quality of life, and 0.5% as prohibiting a normal life (Evered and Lawrenson 1981). A recent Australian study of tinnitus in a large population-based sample of 2015 adults, ages 55 to 99 years, combined detailed tinnitus questionnaires with audiologic assessment (Sindhusake et al. 2003): 30% of the sample reported experiencing tinnitus. Tinnitus prevalence was not related to age or gender but related to audiometric thresholds, and the association between tinnitus and hearing loss was greater in subjects younger than age 65. Tinnitus prevalence in people with normal hearing was lower (26.6%) than in people with hearing loss (35.1%). Mildly annoying tinnitus was reported by 50% of those with tinnitus, while extremely annoying tinnitus was reported by 16% of sufferers. Similar results were reported in a survey of 674 70-year-olds in Sweden (Rosenhall and Karlsson 1991). A recent prospective population based study of hearing loss, in Beaver Dam, Wisconsin, examined 3753 adults 48 to 92 years of age. On enrollment, subjects were questioned about 'significant tinnitus,' with a reported prevalence of 8.2%. The 5-year incidence of tinnitus, among those not reporting tinnitus at enrollment, was 5.7% (Nondahl et al. 2002).

2.1 Types of Tinnitus

The most common type of tinnitus is idiopathic subjective tinnitus. Although this form of tinnitus can be descriptively characterized by its sensory–perceptual properties, it does not have acoustic properties, in that it cannot be measured or detected with sound pressure instruments. This stands in contrast to objective tinnitus, which does have acoustic properties that can be measured. Examples of objective tinnitus include somatosensory sounds such as vascular bruits, the muscle contractions of palatal myoclonus and tensor tympani spasm, and the pathologic airflow via a patulous eustachian tube. An uncommon source of objective tinnitus is spontaneous otoacoustic emissions (SOAEs), related to an estimated to cause about 4% of bothersome tinnitus (Penner 1990).

Idiopathic subjective tinnitus is usually associated with auditory pathology and measurable deficits in auditory functioning. The two most common etiologies for chronic subjective tinnitus are noise-induced hearing loss and age-related hearing loss. Less common causes of tinnitus are specific pathologies including autoimmune disease, endolymphatic hydrops, ototoxin exposure, barotrauma, ischemia, infections, and neoplasms. All these etiologies imply pathology in the auditory system as the source of tinnitus.

However, it is important to note that an estimated 10% of chronic subjective tinnitus occurs in the absence of identifiable auditory pathology on routine clinical testing (Barnea et al. 1990; Borchgrevink et al. 2001). Conversely, an estimated 20% of people with profound hearing loss do not experience tinnitus (Levine 1999). Further, the incidence of tinnitus in population studies of noise-induced hearing loss (NIHL) ranges between 40% and 80% (Man and Naggan 1981; Axelsson and Sandh 1985). These observations suggest that the pathology that results in tinnitus is perhaps unique from the pathology of hearing loss.

Tinnitus is often described as tonal, buzzing, hissing, or noise-like. Although many of the sensory features and the affective reactions to tinnitus are not unique to or determined by tinnitus etiology, some qualitative aspects are nevertheless associated with particular etiologies. Tinnitus associated with endolymphatic hydrops is described as roaring or machine-like; tinnitus related to presbycusis or a high-frequency hearing loss is said to resemble crickets or cicadas on a summer night; tinnitus associated with acoustic trauma characterized by a focal notched hearing loss between 4 and 6 kHz, is typically tonal (perhaps the classic "ringing in the ears"). Despite years of inquiry, it is not known if, or how, the qualitative features of tinnitus are related to either the patterns of peripheral auditory pathology or possible compensatory changes in central auditory function induced by the peripheral pathology.

Clinical observations have identified several unique forms of tinnitus (see Cacace 2003 for review). These include typewriter tinnitus, somatic tinnitus, and cutaneous and gaze-evoked tinnitus. These less common forms of tinnitus have distinct features that may derive from pathological processes that are not shared by the more typical forms of subjective tinnitus. Nevertheless, these uncommon tinnitus types have been instructive in advancing understanding of how a phantom sound is generated.

A unique "somatic" form of tinnitus has been observed in individuals with the ability to modulate the loudness, laterality, or tonality of their tinnitus, using either head-and-neck maneuvers or stimulation of head and neck areas. This type of tinnitus was first noted in a small group of patients who underwent surgery for treatment of large vestibular schwannomas. Postoperatively, these patients noted the ability to modulate their chronic tinnitus by exaggerated eye movements, so-called gaze-evoked tinnitus. Subsequently, a more general form of somatosensory modulation has been described (Sanchez et al. 2002; Abel and

Levine 2004). In 65% to 80% of people with mild tinnitus, the loudness and pitch of the perception can be modified by forceful isometric contractions of head and neck muscles. Fifty-eight percent of study subjects without preexisting tinnitus could induce tinnitus by strong contractions of muscles in the jaw, head, or neck (Levine et al. 2003).

Animal research has established that a variety of multimodal and somatic inputs are integrated into the auditory pathway, (Itoh et al. 1987; Ryugo et al. 2003; Shore et al. 2003; Shore 2005; Shore and Zhou 2006). The parallels between the observed neuroanatomical projections and the clinical observations in humans are intriguing and may lead to important developments in understanding how tinnitus develops and is modulated. For example, it is possible that reduction of normal afferent input to brain stem auditory nuclei, such as following acoustic trauma or in presbycusis, results in inappropriate up-regulation of somatosensory inputs to the auditory system. In the normal auditory brain stem, somatic inputs may modulate auditory processing so that the hearing system is informed of relevant somatic events such as head position or movement-generated sound. After partial deafferentation, normally modulatory somatic inputs could partially replace lost auditory input and thus become part of the auditory stream, and be heard as "sound."

3. Animal Models

Advances in understanding the pathology of tinnitus were highly constrained for many years because tinnitus research was limited to studying the disorder in humans. In the clinical setting, it is very difficult to distinguish factors relevant to tinnitus from factors relevant to the typical coexisting hearing loss. Although recent advances in diagnostic audiometric testing and functional imaging have expanded capabilities for studying tinnitus in people, these methods still do not match the analytic power provided by reliable and valid animal models.

There are several critical features that determine the utility, reliability, and validity of an animal model of tinnitus. First, the metric representing tinnitus should be a measurement of tinnitus perception, that is, the animal's response to what it hears, as opposed to systemic measures presumed to reflect tinnitus. Before the development of behavioral animal models of tinnitus, surrogate measures presumed to represent tinnitus were the primary measures available. Examples of surrogate measures include the neural ensemble of spontaneous activity, bursting activity, and synchronized activity (Eggermont 1990). The problem with using these associative measures is that there is no evidence that these measures are reflecting the phenomenon of tinnitus rather than hearing loss or other phenomena resulting from the experimental manipulation. A useful animal model of tinnitus relies on a measure that cannot be easily interpreted as reflecting auditory dysfunction other than that of hearing a sound of endogenous origin. To maximize utility, a model should lend itself to repeated measurement.

A durable model can assess the chronicity of tinnitus and the effect of interventions. Because susceptibility to tinnitus most likely varies in animals as it does in people, the model should quantify tinnitus in individuals, or at least discriminate the presence or absence of tinnitus in individuals. Finally, the model should detect tinnitus regardless of etiology and duration. That is, the ideal model should be capable of detecting new onset, acute tinnitus, as well as chronic tinnitus, and tinnitus resulting from a variety of interventions, including noise trauma and ototoxicity. Currently there is no single model possessing all of these attributes. Existing models have selective strengths and weaknesses. The choice of a particular animal model is dictated by the experimental objectives of the investigator. A summary of the models is presented in Table 4.1 and a detailed review of different models is presented in the Addendum at the end of this chapter.

4. Mechanisms of Tinnitus Generation and Persistence

The primary objective of current tinnitus research is to understand the physiological basis of the disorder. Theories of tinnitus have been proposed to explain both its sensory features and the reactive components, such as the affective response of distress. Some theories have focused on specific brain regions that may act as tinnitus generators, while others have taken a systems approach and viewed tinnitus within existing frameworks, such as pain (Moller 1997; Tonndorf 1987) or aging (Milbrandt et al. 2000). A caveat for all theories and models of tinnitus is that the underlying pathology may be unique for different types and etiologies of tinnitus. New onset tinnitus may involve anatomic pathways and mechanisms that are different from those involved in chronic tinnitus. The pathophysiology of tinnitus from ototoxins (e.g., salicylate, carboplatin, cisplatin), acoustic trauma, cochlear ablation, and aging, may be similar or significantly different. This section reviews existing theories of tinnitus within the organizational framework of anatomical focus. It should be kept in mind that many of the theories presented here have not been evaluated via experimental methods that differentiate between the related, but distinct phenomena of hearing loss and tinnitus.

4.1 Cochlear Damage as the Source of Tinnitus

The complex structure, organization, and physiology of the cochlea is arguably the best understood and most extensively studied region of the auditory system. Although significant knowledge gaps remain, theories linking cochlear damage to tinnitus represent the earliest attempts to explain and understand tinnitus.

Tonndorf (1981) was the first to hypothesize that dysfunctional stereocilia might be responsible for a variety of auditory pathologies, including tinnitus. Partial loss of stereocilia function would lead to a partial or complete decoupling

TABLE 4.1. Comparative features of various models for inducing, detecting and measuring tinnitus in animals.

Model	Behavioral method	Tinnitus induction	Tinnitus metric	Experimental investment	Advantages	Disadvantages
Jastreboff et al. (1988)	Conditioned suppression of licking	Salicylate Noise Quinine Unilateral Bilateral	Extinction of conditioned suppression	Training : 5 days Testing : 5 days	Rapid	Not adapted to long term assessment
Bauer and Brozoski (1999)	Conditioned suppression of lever pressing	Noise Topical round window ototoxins Unilateral	Relative performance on psychophysical discrimination functions	Training : 8 weeks Testing : 4–6 weeks	Long term assessment	Lengthy training and testing time periods
Heffner and Harrington (2002)	Conditioned suppression of licking	Noise Unilateral	Extinction of conditioned suppression	Training : 8 weeks Testing : 5 days	Accurate Individual assessment	Acute measurements only
Heffner and Koay (2005)	Forced two choice procedure	10 kHz 4 hour exposure	Lick localization on silent trials	Training : 47 days Testing : 5 days	Individual assessment No extinction of effect	Detects only lateralized tinnitus
Lobarinas et al. (2004)	Schedule induced polydipsia (SIP) avoidance conditioning	Salicylate Unilateral Bilateral	Lick rate	SIP training : 4 days AC training : 8 days	Individual assessment	Potential instability of lick rate measurement
Turner et al. (2006)	Gap detection and startle suppression	Noise Unilateral	Sound gap inhibition of startle reflex	No training required Testing : 30 minutes	Individual assessment	Loss of sensitivity with extended testing
Guitton et al. (2003)	Active avoidance conditioning	Systemic salicylate or mefenamate	Detection of sound stimulus	Training : 7 days Testing : 9 days	Rapid Individual assessment	Unknown if sensitive for assessing long-term tinnitus

of the hair cells from the tectorial membrane. Tight coupling between hair cells and tectorial membrane results in an intrinsic physiologic noise level that is 6 dB below threshold. Tonndorf estimated that a loose coupling would increase the intrinsic noise at the hair cell synapse by 55 dB. Narrow bands of decoupled hair cells would result in tonal tinnitus; broader areas of involvement might correspond to other qualitative forms of tinnitus such as hissing or roaring (Tonndorf 1981).

Recognizing that there would be many instances of tinnitus in which stereo-cilia decoupling would be irrelevant, as in the case of cochlear degeneration with loss of hair cells, other hypotheses about peripheral causes of tinnitus have been developed. Selective loss of populations of hair cells and loss of efferent control are two possible mechanisms that might contribute to tinnitus. Kaltenbach has studied patterns of hair cell loss and stereocilia damage in hamsters after acoustic trauma (Kaltenbach et al. 1992). A common observation in these subjects, all of which showed changes in the tonotopic map of the dorsal cochlear nucleus (DCN; see section 4.3), is the absence of significant inner or outer hair cell loss. Subsequent work combining behavioral assessment using the extinction model of Heffner with detailed cochlear histology suggested that outer hair cell lesions may be more relevant to tinnitus (Kaltenbach and Heffner 1999), although there has been some disagreement over the interpretation of these results (Heffner and Koay 2005). Bauer et al. (2007) studied cochlear histology in rats that displayed behavioral evidence of tinnitus after acoustic trauma. Exposed subjects displayed minimal evidence of inner or outer hair cell loss or stereocilia damage. Interestingly, there was a significant loss of the large diameter primary afferent dendrites within the osseous spiral lamina throughout the cochleas of subjects with evidence of tinnitus. The selective loss of this population of fibers was not limited to the tonotopic region that corresponded to the frequency of the tonal tinnitus detected with the behavioral tests but was highly correlated with behavioral evidence of tinnitus (Bauer et al. 2007).

Glutamate is the primary neurotransmitter at the synapse between inner hair cells and auditory nerve dendrites (Eybalin 1993). Glutamate excitotoxi-city, altered spontaneous release of glutamate from damaged hair cells, or hair cells no longer appropriately regulated by efferent control, are possible mechanisms for tinnitus generation in the cochlea. Blockade of N-methyl-D-aspartate (NMDA)-induced, spontaneous activity of primary auditory fiber dendrites at the inner hair cell synapse has been demonstrated using the NMDA receptor antagonist 3,5-dimethyl-1-adamantamine hydrochloride (Oestreicher et al. 1998). In a conditioned avoidance task, behavioral evidence of salicylate-induced tinnitus in rats was blocked by various NMDA antagonists (Guitton et al. 2003, 2005). These results suggest that, at least for salicylate-induced tinnitus, cochlear NMDA receptors may serve as modulators of neural excitation that is perceived as tinnitus.

4.2 Auditory Nerve as Site of Tinnitus Generation

Tonndorf (1987) and others (Kiang et al. 1976) have suggested that altered spontaneous activity within the auditory nerve may play a role in generating a tinnitus signal. Specifically, deviation from the normal random activity that is present within the nerve toward more synchronous spontaneous activity may be interpreted by higher auditory centers as cochlear stimulation. Indirect support for this hypothesis was found in the altered firing rate and firing pattern of single auditory nerve fibers in cat during salicylate infusion (Evans and Borerwe 1982). Additional support was provided by the finding that spectrally averaged spontaneous auditory nerve activity after acute salicylate infusion showed an increase near 200 Hz in acute cat preparations (Martin et al. 1993). Spectral averaging of the auditory nerve spontaneous activity [also referred to as ensemble spontaneous activity (ESA), ensemble background activity, and the spectrum of neural noise] is thought to reflect the summed spontaneous activity of the entire population of auditory nerve fibers. Altered ESA has been demonstrated in an awake guinea pig preparation under conditions of chronic salicylate exposure (Cazals et al. 1998). Similar to the findings in cat, a spectral peak in the ESA at 200 Hz was observed in guinea pigs after cochlear perfusion with the glutamate receptor antagonist 6-7-dinitroquinoxaline-2,3-dione (DNQX) and the purinergic receptor agonist adenosine $5'$-O-(3-thiotriphosphate) (ATP γS). Anecdotal observations from eighth nerve recordings in humans with tinnitus suggest that altered ESA of the auditory nerve may be relevant to the neural code for tinnitus, but little additional work in this area has been reported (Feldmeier and Lenarz 1996).

4.3 Cochlear Nucleus and Tinnitus

An intuitive and simple physiological explanation of tinnitus invokes elevated spontaneous neural activity in the auditory pathway. The elevated neural activity may occur at one or more levels of the auditory system, and may emerge as a consequence of peripheral damage or altered function elsewhere in the system. Studies from several laboratories have reported elevated spontaneous activity in the cochlear nucleus, primarily the DCN, after manipulations similar to those associated with permanent or reversible tinnitus in humans. A few of these studies have directly associated elevated DCN activity with evidence of tinnitus in the same animal subjects.

Kaltenbach and Afman (2000) have shown that noise exposure results in elevated multiunit spontaneous activity (MSA) in the DCN. Hamsters were exposed to unilateral acoustic trauma that resulted in well-defined hair cell lesions. MSA recorded from the surface of the DCN was obtained 30 to 58 days after exposure. Spectral response plots and spontaneous activity were recorded and mapped topographically, rather than tonotopically (threshold elevations above 7 kHz prevented tonotopic mapping). DCN MSA was significantly higher in exposed animals compared to unexposed control subjects. The maximum average rate occurred in a region of the DCN topographic map close to the

estimated locus of the exposure-tone frequency. The mean increase in spontaneous rate was paralleled by the increase in the mean neural threshold shift.
Similar findings were reported in rats after a unilateral exposure to acoustic
trauma, although the increase in spontaneous rate at the 10 kHz locus of the
DCN topographic map was not as profound as in the hamster (Zhang and
Kaltenbach 1998). The source of the elevated spontaneous rate was not immediately evident, as there was no correspondence between the widths or magnitude
of the cochlear lesions and the widths of the DCN topographic map with elevated
spontaneous activity.

The same group determined onset latency of the elevated spontaneous activity
and the relationship to threshold elevations (Kaltenbach et al. 2000). A nonlinear
fluctuation in MSA was found, initially decreasing and then increasing over time.
These results could be interpreted as reflecting an initial hearing loss, followed
by the delayed emergence of tinnitus. Hamsters were exposed to a 10-kHz tone
at 127 dB SPL for 4 h. The MSA of exposed subjects 2 days after acoustic
exposure was less than 14 counts per second (CPS); across the mediolateral DCN
axis, compared to 34 CPS in control subjects. However, 30 days after exposure
the MSA had increased to 78 CPS for the exposed group. It was notable that
the mean multiunit threshold shift 2 days after exposure exceeded the threshold
shifts at 30 days, suggesting that the observed MSA increase at 30 days was not
simply reflecting hearing loss.

Several studies have examined the possible contribution of different patterns
and magnitudes of hair cell damage to the reorganization of the DCN topographic
map. Cisplatin, a chemotherapeutic agent, is ototoxic and commonly produces
hearing loss and tinnitus in humans. Cisplatin predominantly degrades the
cochlear outer hair cell system; the attendant tinnitus may arise because of
imbalanced neural activity in the type 1 afferents innervating the intact inner hair
cells and the type 2 afferents innervating damaged outer hair cells. Hamsters were
evaluated for changes in MSA surface recordings from the DCN after a range of
systemic cisplatin doses that produced hair cell damage (Kaltenbach et al. 2002).
Although the level of cochlear damage did not correspond to cisplatin dose, in
general the inner hair cell loss was limited to less than 1% in the basal turn of
the cochlea, while a greater level of outer hair cell loss was evident in the apical
and basal turns of the cochlea. Subjects with only mild cochlear lesions did not
exhibit a significant increase in DCN MSA. However, subjects with selective
outer hair cell lesions in the basal turn did show evidence of increased activity.
Intermediate and severe outer hair cell loss in the basal half of the cochlea,
without associated inner hair cell lesions, correlated with increased activity in
the high-frequency region of the DCN ($r = 0.89$). However, extensive outer hair
cell damage with associated inner hair cell damage was not as strongly correlated
($r = 0.51$) with MSA. Kaltenbach concluded that outer hair cell damage was
an important factor in the development of increased spontaneous activity in the
DCN after cisplatin ototoxicity.

Caution should be used in interpreting the previously cited DCN–spontaneous
activity research: Although elevated DCN neural activity appears to be reliably

produced by various types of acoustic trauma, conditions commonly associated with tinnitus in humans, tinnitus itself was not measured in the animal subjects of these studies. Caution should also be exercised in interpreting the observed changes in multiunit activity. DCN surface recordings of multiunit activity do not discriminate neural type or the direction of activity in the auditory pathway. Some of these issues were experimentally addressed in a study reporting elevated single-unit spontaneous activity in DCN fusiform cells of chinchillas with psychophysical evidence of tinnitus (Brozoski et al. 2002). Fusiform neurons are responsible for the major rostral output of the DCN. Tinnitus was measured in the chinchillas 18 months after a unilateral exposure to a 4 kHz tone at 80 dB SPL. All of the animals were behaviorally trained and psychophysically tested (see Table 4.1). Exposed subjects displayed evidence of tinnitus that was tonal and centered at 1 kHz. Both spontaneous activity and stimulus-driven activity at a frequency identical to the psychophysical tinnitus frequency were significantly elevated. Auditory brain stem response thresholds were minimally elevated at the tinnitus frequency, indicating that the DCN changes were not reflecting hearing loss.

The role of the DCN in either the development or the persistence of tinnitus is still in question. Kaltenbach proposed a neural circuit in which type II fiber deafferentation from selective outer hair cell damage resulted in decreased input to granule cells. Loss of input to these excitatory neurons would decrease synaptic drive on inhibitory interneurons such as cartwheel cells and stellate cells, with resulting loss of inhibition at the level of the fusiform cells (Kaltenbach 2006; Kaltenbach et al. 2005). Bauer proposed an alternate route of tinnitus development. The selective loss of high spontaneous rate (SR), large-diameter primary afferent fibers (ANF) will selectively reduce the input to small cells within the deep DCN. Small cells are physiologically characterized by type II responses, which include low spontaneous activity and high thresholds, and morphologically include a mixture of cell types, including interneurons and vertical cells (Young and Voigt 1982). Type II units provide inhibitory input to type IV units, the DCN principal cells. If, following acoustic trauma, a major source of inhibition of the fusiform cells is lost via the pathway of high-SR-ANF to the vertical cells of the deep DCN, the elevated SR of fusiform cells is predicted. Elevated DCN fusiform output could serve either as a trigger for tinnitus-related neural activity rostral to the cochlear nucleus, or as a generator of the chronic neural signal for tinnitus.

The role of DCN fusiform cells as the sole source of chronic tinnitus was not supported by recent work showing persistent psychophysical evidence of tinnitus in rats after DCN ablation (Brozoski and Bauer 2005). Rats were behaviorally trained and tested for acoustic-trauma-induced tinnitus. After the tinnitus was psychoacoustically characterized, both experimental and control subjects had unilateral and bilateral DCN ablations, and then were retested for tinnitus. Bilateral dorsal DCN ablation did not significantly affect the psychophysical evidence of tinnitus and ipsilateral DCN ablation increased the evidence of tinnitus compared to pre-ablation performance. These results suggest that the DCN does not act as a simple feed-forward source of chronic tinnitus.

Nevertheless, it is possible that initial DCN hyperactivity after cochlear trauma triggers persistent pathological neuroplastic changes distributed across more than one level of the auditory system. The DCN may also contribute to the chronic pathology of tinnitus, albeit not exclusively, since tinnitus was enhanced by ipsilateral DCN ablations. The ipsilateral ablation data suggest that asymmetric input from the DCN to higher centers may also be an important factor in tinnitus pathophysiology.

4.4 Inferior Colliculus

The inferior colliculus (IC) has been suggested as a possible tinnitus generator. Known alterations in glutamic acid decarboxylase (GAD), glutamate neurotrans-mitters, and neural coding occur under conditions of decreased or abnormal auditory input such as aging, salicylate toxicity, and acoustic trauma (Willott and Lu 1981; Jastreboff and Sasaki 1986; Salvi et al. 1990; Caspary et al. 1995). Very few studies have directly implicated the IC with neurophysiological correlates of the behavioral evidence of tinnitus.

Jastreboff and Brennan's work established that salicylate treatment could produce tinnitus in rats (Jastreboff et al. 1988a,b). Subsequent work has examined salicylate-induced effects along the auditory pathway. Increased c-fos expression was noted in rat IC but not in cochlear nucleus, after moderate salicylate exposure (Wu et al. 2003). The c-fos staining was most prominent in the 9- to 10-kHz region of the IC, corresponding to the reported frequency of tinnitus in psychophysical testing (Bauer et al. 1999) and the frequency of increased sponta-neous activity in electrophysiologic studies (Chen and Jastreboff 1995; Wang et al. 2002).

4.5 Auditory Cortex

Because tinnitus is a conscious perception, that is, one must be aware of it for tinnitus to exist, the auditory cortex (AC) is a logical site for investigation of potential neurophysiological correlates. Several hypotheses about the role of the AC in tinnitus have been offered: They include increased spontaneous activity and changes in the temporal pattern of neural activity, such as bursting, and alterations of tonotopic representation that lead to enhanced synchrony. A number of animal studies have demonstrated altered neural activity and processing within the primary and secondary AC after auditory damage similar to that causing tinnitus in humans. However, these studies of cortical neurophysiology have not directly measured tinnitus in their animal subjects.

Salicylate has been shown to increase burst-firing in secondary AC (AII) neurons, but not in primary AC (AI) (Ochi and Eggermont 1996). This is consistent with the observation that salicylate produced bursting in the external nucleus of the IC, a region with significant input to AII. However, the salicylate results are not consistent with the central effects of other types of cochlear insult, suggesting that the pathophysiology of salicylate-induced and trauma-induced tinnitus might be quite different. The significance of neural bursting for chronic tinnitus, as opposed to acute tinnitus, and salicylate-induced tinnitus, is not certain. For example, immediately after acoustic trauma, Norena and Eggermont (2003) reported AI activity in cats that did not parallel salicylate effects: Synchronized firing was evident in neurons with characteristic frequencies CFs one and two octaves above the trauma frequency, and bursting was not specific to any frequency region. More importantly, hours after trauma, bursting was no longer evident but neural synchrony increased in frequency regions both below and one to two octaves above the trauma frequency. Elevated spontaneous activity was also reported, but it was restricted to frequency bands above and below the trauma frequency, a finding that contrasts with the immediate effects of acoustic trauma on spontaneous activity in subcortical structures such as the CN and IC (Salvi et al. 1978; Wang et al. 1996; Kaltenbach et al. 2000).

The inconsistencies between salicylate-induced and trauma-induced tinnitus may be real, particularly in light of the likely difference between the peripheral effects of each treatment: Salicylate ototoxicity produces a short-term increase in cochlear output via the auditory nerve; in contrast, auditory trauma generally produces a long-term decrease in cochlear output and a partial de-afferentation. The perception of tinnitus may be similar in each instance, but the pathophysiology may be quite different. Inconsistencies between the cortical effects and subcortical effects of auditory trauma, on the other hand, may be more apparent than real. Following trauma, the frequency locus of effect, and the time course of the effects, may differ at each level of the auditory system. The pathophysiology of chronic tinnitus produced by auditory trauma may reflect a cascade of events, beginning with cochlear damage, unfolding with neuroplastic alterations at each level in the central auditory system, and concluding with changes in cortical processing.

Advances in imaging technology, including positron emission tomography (PET) and functional magnetic resonance imaging (fMRI), have enabled the study of ongoing neural activity in awake humans under normal and pathologic conditions, including tinnitus. Whether the imaged regions of abnormal activity are in fact serving a causal role in the perception of tinnitus is not yet clear. Several imaging studies of human tinnitus have identified auditory cortex as a potentially critical site. It remains to be established whether the identified regions are active because of an intrinsic pathological process or if they have been indirectly activated by a primary generator located elsewhere.

The recently developed diagnostic and treatment technique of transcranial magnetic stimulation (TMS) has opened a novel avenue for investigating the causal and associational aspects of tinnitus-related cortical activity. In TMS, a brief intense magnetic field is applied to the scalp. The magnetic pulse induces a temporary focal disruption of neural activity in a discrete area of adjacent cortex. This "virtual lesion" briefly, and reversibly, disrupts cortical activity and allows the investigator to determine if the cortical region of interest contributes to a specific behavior or perception. This technique may be useful for investigating cortical mechanisms of tinnitus and may also be of clinical value as a treatment for some forms of tinnitus.

5. Treatment Principles

Development of targeted effective therapies for chronic tinnitus has been hampered by poor understanding of tinnitus pathophysiology and limited hypotheses about mechanisms. Consequently, many clinical trials have been derived from anecdotal reports of tinnitus treatment successes, or adopted from other clinical indications such as pain, epilepsy, and depression. Not surprisingly, such blind empiricism has had limited success. The excellent review of randomized placebo-controlled trials by Dobie is recommended for anyone interested in therapeutic interventions (pharmacologic, surgical, electrical, and behavioral) for tinnitus (Dobie 1999).

There are many challenges inherent in studying a complex phenomenon such as tinnitus in humans. The impact of genetic diversity and the timing and extent of cochlear damage that occurs over a lifetime are two factors that are outside experimental control. Although theoretically amenable to experimental control, the powerful placebo effect that accompanies any intervention for a subjective condition must be addressed with particular care and skill. Finally, the dichotomy between the perception of tinnitus and the associated reactive response to tinnitus (anxiety, depression, impaired concentration, sleep disruption) lends an additional complexity to unraveling mechanisms that are responsible for tinnitus. Clinical studies must address these complex features to accurately determine effective treatments for specific patient populations and types of tinnitus.

Principle-driven tinnitus treatments, based on likely mechanisms, include pharmacologic interventions directed at enhancing inhibitory neurotransmitter function (loss of inhibition), acoustic modulation of the auditory pathway using sound stimulation (peripheral deafferentation), electrical stimulation of the auditory pathway, and magnetic stimulation of auditory cortex (deafferentation and abnormal pattern of neural activity). Clinical research evaluating these treatments is either preliminary or has not yet been implemented in placebo-controlled trials.

5.1 Pharmacologic Interventions for Tinnitus

Pharmacologic interventions for tinnitus, derived from experimentally determined mechanisms, are few. Several independent lines of laboratory research have suggested that tinnitus arises from a loss of inhibition within the auditory pathway. A likely target for therapeutic intervention is the neurotransmitter γ-aminobutyric acid (GABA). Basic research indicates GABA is a widely distributed, almost exclusively inhibitory, neurotransmitter; the functional role of GABA in central auditory processing has been described in some detail (Caspary et al. 1987, 1994); laboratory studies investigating acute or slow, progressive deafferentation caused by cochlear damage, have identified alterations in GABA at several levels in the auditory pathway (Eggermont 2005).

In a well-controlled trial, alprazolam effectively reduced the subjective loudness of chronic tinnitus (Johnson et al. 1993). Alprazolam is a benzodiazepine and the site of action is the GABA receptor. Gabapentin, a GABA analogue, reduced the perceptual loudness and annoyance of chronic tinnitus in a dose-dependent manner, as established in a placebo-controlled study (Bauer and Brozoski 2006). The drug was most effective for a clinical subpopulation whose tinnitus was associated with acoustic trauma, and was less effective for a subpopulation without evidence of acoustic trauma. The effect was dose dependent and reversible with discontinuation of the active medication. It is also notable that a test of gabapentin using an animal model, showed that it significantly decreased, but did not eliminate, psychophysical evidence of tinnitus (Bauer and Brozoski 2001).

5.2 Reorganization of Auditory Cortex Through Sound Stimulation

Norena and Eggermont (2003) demonstrated the cortical reorgani-zation that occurs in cat after acoustic trauma. The reorganization was presumed to occur as a result of peripheral deafferentation leading to altered tonotopic representation of critical frequency bands within AI and AII. Cats with trauma-induced cortical reorganization were subsequently exposed to chronic acoustic stimulation with either low-frequency or high-frequency broad-band sound. Only the cats that were stimulated with high-frequency sound displayed a normalization of the cortical tonotopic reorganization (Norena and Eggermont 2003). These results suggest that pathology resulting from deafferentation can be modulated by acoustic stimulation.

Sound stimulation using various delivery techniques is currently in clinical use for the treatment of tinnitus. While derived from an experimentally determined mechanism, data on the efficacy of sound-stimulation therapy for reducing the loudness and annoyance of tinnitus is mixed. Flor used a protocol of daily auditory discrimination training that was tailored to the specific tonality of the individual's tinnitus (Flor et al. 2004). The most benefit was obtained by subjects who used daily regular training. Hiller (Hiller and Haerkotter 2005) examined the effect of sound stimulation provided by a white noise generator combined with cognitive therapy directed at reducing the reactive components of disturbing tinnitus. The addition of sound stimulation did not enhance the improvement that occurred with cognitive therapy alone.

5.3 Reorganization of Auditory Cortex Through Electric and Magnetic Stimulation

Several studies have investigated the effect of cortical stimulation on the perception and annoyance of chronic tinnitus. Plewnia et al. (2003) transiently suppressed the perception of chronic tinnitus in a group of subjects by applying TMS to secondary auditory cortex. Kleinjung et al. (2005a, b) used PET and MRI to identify regions of increased activity in auditory cortex associated with chronic severe tinnitus. Neuronavigational techniques were employed to target these areas for repetitive TMS. In a placebo-controlled crossover design, improvement in tinnitus annoyance occurred in the actively treated group but not in the sham treated group. Improvement in tinnitus severity occurred in 11 of 14 subjects within 1 to 6 days after treatment. However, at 6 months follow-up tinnitus was improved in 8 of 14 and had worsened from baseline in the remaining 6 subjects.

Using TMS, DeRidder et al. (2006) identified subjects for implantation of stimulating electrodes in AI and AII. Interestingly, implanted subjects with unilateral tonal tinnitus experienced the largest reduction in tinnitus loudness (97%) compared to 24% reduction in subjects with tinnitus characterized as white noise. Important caveats were that the tinnitus had to be responsive to TMS and be of recent onset. Inclusion of a placebo control would have strengthened the conclusions, as subjects responsive to TMS may be particularly strong placebo responders and therefore biased toward a positive response after electrode implantation. It is also known that for many people new onset tinnitus gradually improves over time through presumably a normal habituation process. The positive results observed by De Ridder, et al., although highly intriguing, are potentially confounded by the phenomenon of spontaneous recovery.

6. Summary

Chronic tinnitus is a complex phenomenon that affects millions of people and poses a significant challenge to both the scientific community and clinical practice. Until the pathophysiology of this heterogeneous disorder is understood in greater detail, effective treatments will remain elusive. The use of existing animal models, and the development of new ones, will play an important role in future research directed at understanding the mechanisms responsible for tinnitus. Current challenges that can be addressed by appropriate animal models include identification of the site(s) in the auditory, and perhaps nonauditory, pathway that are involved in generating the phantom perception. Several regions have been linked to the tinnitus signal, such as the DCN, the IC, and auditory cortex. Much work needs to be done to determine if these sites are critical triggers or generators of tinnitus and the neural mechanisms that result in tinnitus.

The development of new treatments derived from experimental principles is exciting and holds great potential for the millions of people who suffer from tinnitus. Targeted pharmacotherapy, based on plausible mechanisms, and related to specific etiologies, needs to be further developed. Rigorously conducted, well-controlled clinical studies will serve a critical role as emerging technologies, such as magnetic transcranial stimulation and direct brain stimulation, are applied to the treatment of tinnitus. Chronic tinnitus will yield to effective treatment as both the basic and applied research progresses.

Acknowledgments. The research of the authors in the areas of animal psychophysics, cochlear anatomy and brainstem electrophysiology was sponsored by the National Institutes of Health RO1 NIDCD RO1 DC04830. The clinical research investigating gabapentin was sponsored by the Tinnitus Research Consortium.

Addendum: Animal Models of Tinnitus

A.1 Primer of Conditioning Principles

The principles on which all of the animal models of tinnitus rely are derived from conditioning paradigms in experimental psychology. This large, well-codified body of knowledge describes the fundamental principles of conditioning and learning in many species. Perhaps the most fundamental principle of instrumental, or operant, conditioning, is the law of effect, which states that behavior is modified by its consequences. More specifically, a positively reinforcing stimulus (Sr^+) such as water for a water-deprived subject or food for a nutrient-deprived subject, when contingent on an emitted behavior (such as licking a spout, pressing a lever, or moving to a specific location), will increase, or maintain, the frequency of the behavior. Similarly, a punishing, or aversive, stimulus (Sr^-), such as an applied electric current, when contingent on an emitted behavior, will decrease the frequency of that behavior. Behavior can also be modified by informative or discriminative stimuli (Sd). Discriminative stimuli are contextual, typically continuously present, and inform the subject of stimulus contingencies, e.g., the likelihood of an Sr^+ or Sr^-, if an appropriate response is emitted. An Sd signaling a contingent Sr^+ is called a positive discriminative stimulus (often, S+), and an Sd signaling a contingent Sr^- is a negative discriminative stimulus (often, S–). A rat can be trained (conditioned) to press a lever for food (Sr^+) when audible sound (S+) is present. In addition, the rat can be conditioned to stop lever pressing in the absence of sound (S–) if a mild electric foot shock (Sr^-) is made contingent on lever pressing. Because S+ and S– differentially affect behavior, testing animals with Sd's of different values enables an experimenter to objectively quantify what the animal hears. Finally, extinction refers to a procedure of removing contingent stimuli, either Sr^+ or Sr^-, or both. Extinction permits an experimenter to measure behavior without immediate contingencies, in which case the influence of previous contingencies may be assessed.

A.2 The Jastreboff–Brennan Model

Jastreboff and Brennan established that salicylate-induced tinnitus could be objectively measured in rats, by applying the previously described conditioning principles (Jastreboff et al. 1988a, b). Using a conditioned-suppression psychophysical method (Smith 1970), Jastreboff and Brennan trained water-deprived rats to lick from a spout to obtain their daily water ration. Licking is a natural behavior for a rodent, and does not require lengthy training. A free-field auditory stimulus (the S+) was present in the test chamber when water was available (Sr^+). The same stimulus was also continually presence in the rats' home cage, as well as in the experimental chamber. The S– was offset of the background sound, which signaled the contingency of a mild footshock (Sr^-). In the conditioned suppression paradigm the rat can avoid the Sr^- by not making contact with the water spout during S–. Suppression of licking indicates the rat's detection of S–. Typically, suppression is quantified by a relative measure expressed as the ratio of lick rate during S– (B) compared to the behavior during S+ (A). The suppression ratio, $R = B/A + B$, is a standard metric for quantifying behavior in this paradigm and can be used to quantify the effect of S–. When the conditioned suppression paradigm is applied to stimulus-controlled behavior, R provides a running index of the subject's detection and interpretation of *S–*.

It is important to note that the Jastreboff–Brennan experiments used the conditioned suppression method to quantify salicylate-induced tinnitus by determining R during *extinction* of the conditioned suppression. Extinction was used so that the experimenters could assess the rats' interpretation of S– as affected by immediately prior conditions. For this purpose subjects were divided into three groups: controls, which did not receive salicylate; an SA group, which received Na-salicylate only *after* suppression training (Sr^- removed); an SB group, that received Na-salicylate *during and after* suppression training (Sr^-contingent, then removed). All subjects were water deprived and trained to lick from a spout with a background of audible noise (Sr^+). Suppression training was identical for all groups, the Sr^- being foot shock, delivered at the end of the S– presentation. The objective of the Na-salicylate treatments was to induce acute salicylate-induced tinnitus either during training (SB group), where it would be associated with the S– Sr^- suppression contingency, or after training (SA group), where it would *not* be associated with the S– Sr^- suppression contingency.

Different suppression ratios were obtained for each of the three groups during extinction: the SA group had the highest R (little suppression), the SB group had the lowest R (deep suppression), and the control group had an intermediate value of R. The following interpretation was given: All groups have the same S+, but not the same functional S–, which was determined by their suppression training. For all groups the nominal S– in extinction was sound off, as it was in training. However, functionally the S– in extinction was different for each group. SA subjects had the highest extinction R because with Na-salicylate on board they were incapable of hearing silence, their functional S–. In contrast, SB subjects had salicylate-induced tinnitus all along, and when tested in extinction, SB subjects suppressed deeply (low R) because tinnitus was their training S–. Finally, control extinction was intermediate between the SA and the SB groups because for the control group the S– was both functionally and nominally the same (no sound) during training and extinction testing.

The Jastreboff et al. (1988a, b) experiment was the first to demonstrate the perceptual consequences of salicylate ototoxicity in animals. They established that animals could experience tinnitus and that the effects of tinnitus could be objectively quantified. Nevertheless, a substantial limitation of their model was the reliance on the extinction test. This necessarily restricts the time frame in which an animal with tinnitus can be studied. Extinction is a transition state which is complete after four to five experimental sessions, at which point all differences between treatment groups disappear. This limits opportunities for manipulating tinnitus or studying changes that occur over time, such as the effects of aging on tinnitus. To a lesser extent, the Jastreboff–Brennan model is also hampered by reliance on licking behavior as the dependent measure. Licking in the rodent is episodic as it is in most mammals, thereby providing a baseline with considerable variability. Although correction methods exist for high variability baselines, corrections add a layer of complexity and potentially introduce additional artifacts.

A.3 The Bauer–Brozoski Model

Bauer and Brozoski (Bauer et al. 1999) developed an animal model of tinnitus using the Jastreboff–Brennan model as a point of departure. In the Bauer–Brozoski model, which has been successfully applied using both rats and chinchillas, tinnitus is induced by a single unilateral sound exposure sufficient to produce a temporary threshold shift in the exposed ear without contralateral effects. The Bauer–Brozoski model is a derivative conditioned suppression method, but there are substantial differences between it and Jastreboff–Brennan method. In the Bauer–Brozoski method, subjects are required to perform a running operant task that establishes a steady-state baseline and continuously trains and tests stimulus-controlled behavior. In this way the model indicates the presence of tinnitus over any duration of time; using rats, for example, tinnitus was measured continuously over a 17-month period with no apparent loss of sensitivity (Bauer and Brozoski 2001).

Subjects are trained to press a lever for food pellets. Restricted food availability combined with a variable-interval reinforcement schedule in the conditioning chambers ensures stable and high rates of lever pressing throughout a test session. Free-field, low-intensity broad-band noise is always present in each conditioning chamber. This background sound is also the S+. Although food reinforcement is always available (there is no extinction), there are randomly inserted interruptions in the background sound. Typically there are 10 such interruptions per session, each 1 min in length. For two of the 10 interruptions the sound is turned off (S–). These may be considered suppression-training periods, as they conclude with a 1-s duration electric foot shock (Sr^-). An Sr^- occurs only if the subject lever presses in the sound-off period above a performance criterion level (typically $R \geq 0.2$). The suppression-training periods insure that subjects pay close attention to the acoustic environment; this they communicate to the run-time computer by strongly suppressing their lever pressing, thereby avoiding a footshock. The eight remaining background sound interruptions may be considered stimulus test periods (Sd): Typically tones of variable intensity, extending over the subjects' audible range, are played over the overhead speaker. Food reinforcement is available, but not Sr⁻ (i.e., footshock). This procedure requires the animals to make a three-way discrimination: Sd (potentially novel stimulus: contingency unknown) versus S+ (contingency: Sr^+) vs. S- (contingency: Sr^-). Behavior in the presence of Sd is primarily driven by this three-way comparison, and it is this behavior that permits the derivation of the psychophysical functions used to indicate tinnitus, independent of unilateral hearing loss.

Comparison of psychophysical functions between experimental and control subjects, for a series of auditory test stimuli, demonstrates that tonal tinnitus in rats resulted from a single unilateral exposure to octave band noise. Maximum separation of the pure-tone discrimination function between trauma and control subjects was at 20 kHz. The interpretation is that trauma-exposed subjects have tonal tinnitus resembling a 20-kHz tone. The trauma-exposed subjects hear their tinnitus during the S- suppression training periods and therefore suppress more than control subjects when a 20-kHz test tone is presented. While there is some generalization to other test tones, as seen in Fig. 4.1, the greatest effect surrounds 20 kHz. A more analytic way to explain the results depicted in Fig. 4.1 is to say that for trauma subjects $S^{20kHz} \approx S-$, while for control subjects $S^{20kHz} \neq S-$.

The Bauer–Brozoski model can be used to detect tinnitus induced by any procedure that spares hearing thresholds in at least one ear. The model cannot be used with procedures that compromise hearing thresholds bilaterally, because bilateral loss of hearing sensitivity significantly affects free-field discrimination performance. The results depicted in Fig. 4.2 show the effect of acute severe bilateral hearing loss that developed in a rat that had been well trained and tested with only mild unilateral loss. After the onset of bilateral severe hearing loss, all auditory stimulus conditions were equivalent for this subject and the animal pressed the lever indiscriminately, which is indicated by the horizontal function. Control experiments examined the effect of temporary unilateral threshold elevation produced an ear plug. The plugs produced a mean unilateral threshold elevation of 40 dB SPL and did not alter the configuration of the psychophysical functions compared with normal-hearing subjects (Bauer and Brozoski 2001). This critical control condition demonstrates that tinnitus, and not hearing loss, is reflected in the psychophysical functions.

Because this model does not rely on extinction to detect tinnitus, conditioned suppression and reliable discrimination functions can be obtained over the entire lifetime of the subject, providing a powerful opportunity to study the course of chronic tinnitus. In humans, tinnitus that occurs after a cochlear insult can resolve, persist, or worsen throughout life. Animal models that permit repeated testing over months are required for assessing this aspect of tinnitus.

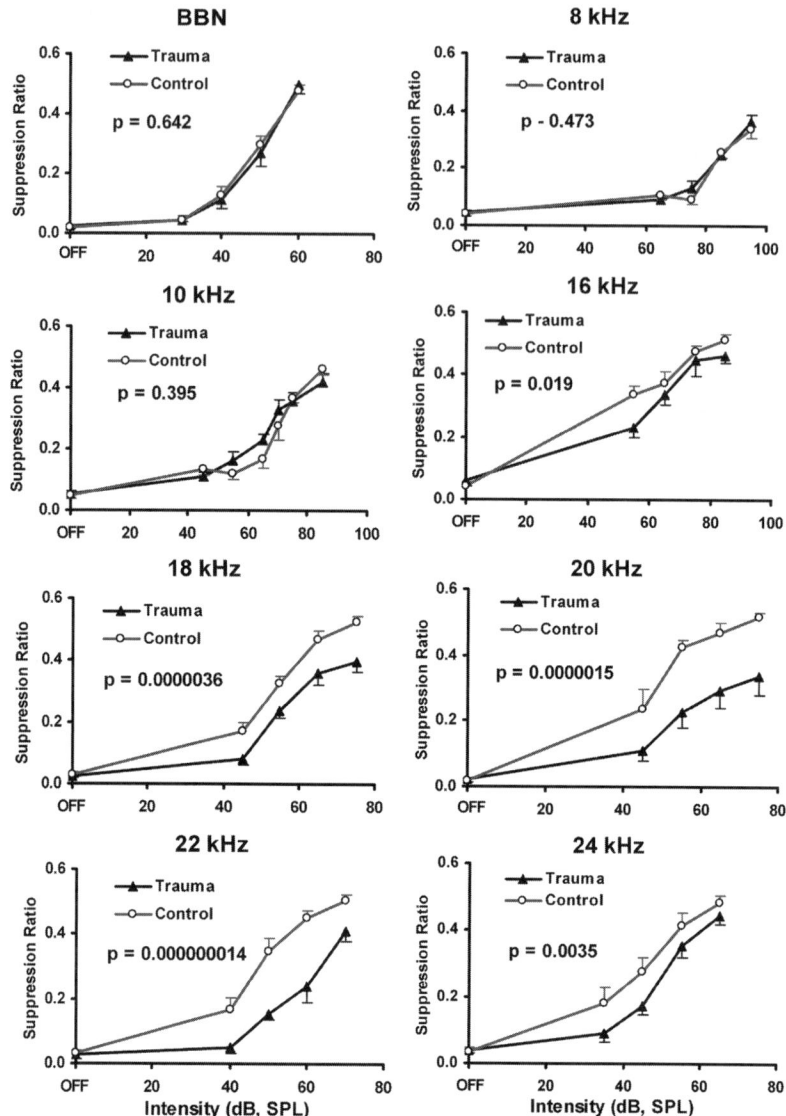

FIGURE 4.1. Psychophysical group data from rats ($n = 7$) traumatized before training. Tonal tinnitus that resembles a 20-kHz tone is induced by exposure to a 16-kHz tone at 105 dB SPL. The maximum separation in the psychophysical functions between the control and trauma-exposed subjects occurs with presentation of the 20-kHz test tone. Generalization to other tones results in a lesser degree of suppression with presentation of surrounding tones (18 and 22 kHz). (From Bauer 2003. Reprinted by permission of Elsevier.)

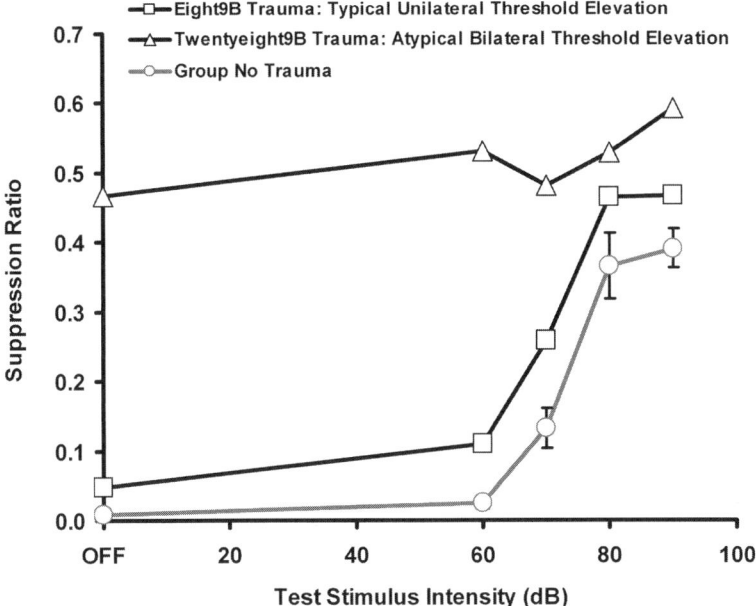

FIGURE 4.2. Bilateral severe hearing loss prevents the detection and discrimination of auditory stimulus conditions. Psychophysical functions are flat and reflect the absence of suppression under all test conditions. The psychophysical function of subject Eight9B shows the typical decrease in suppression with respect to unexposed (no trauma) controls, reflecting tinnitus developed after suppression training. The psychophysical function of subject Twentyeight9B shows the performance of a well-trained rat that developed an atypical bilateral hearing loss after months of testing demonstrating well-established tinnitus. The function of the subject with bilateral threshold elevation is approximately horizontal, because none of the stimuli can be discriminated from the speaker off condition. (Unpublished observations by Brozoski and Bauer.)

A.4 The Heffner and Harrington Model

A limiting feature of the two previously described animal models is their reliance on group data, thus making it difficult to draw conclusions about individual subjects. On the basis of personal histories and audiometric data, humans display an apparent range of susceptibility to tinnitus. This may also be true for animals, and as such a method for reliably detecting tinnitus in individual animals would be useful. Heffner and Harrington (2002) devised a model with this capability, deriving once again from the seminal work of Jastreboff et al. In the Heffner–Harrington model, Syrian golden hamsters were trained to lick from a water spout and obtain their daily water ration in a single 20-min experimental session. Continuous sound was present in the experimental chamber and subjects were trained to stop drinking in the absence of sound to avoid a foot shock. Subjects learned to discriminate between silence, when they break contact with the water spout, and a broad range of sounds, when they lick for water.

Tinnitus that results from a range of trauma parameters has been studied using the Heffner–Harrington model. For example, subjects were unilaterally exposed to a 10 kHz tone at 124 dB SPL for 0.5, 1, or 4 h, or 127 dB SPL for 2 h, immediately after completion of training. During testing, no tones were presented and no shock was given, making this a short-term extinction test, similar to that in the Jastreboff–Brennan model. Subjects exposed to unilateral acoustic trauma for 4 h contacted the water spout more during silent trials than control subjects. Performance degraded over repeated test sessions for both groups, becoming random by session 5. It is important to note that, as in the case of the Jastreboff–Brennan model, the exposed subjects most likely continue to experience tinnitus, but extinction methods rapidly lose sensitivity with repeated testing.

A significant feature of the Heffner–Harrington model, although obtained at the expense of extensive training periods and short-term sensitivity, is the capability of detecting tinnitus in individual subjects. As previously mentioned, acoustic trauma would not be expected to produce tinnitus in all subjects. An animal model that permits individual tinnitus assessment has the potential to distinguish pathologic factors that accompany hearing loss from those that accompany tinnitus. The overlap in behavioral performance between control subjects and subjects exposed to varying amounts of acoustic energy is detected via this model.

A.4.1 A Forced Two-Choice Procedure

Heffner and Koay (2005) developed and tested an animal model with improved sensitivity in characterizing qualitative features of a subject's tinnitus. In this model subjects were trained in a two-choice procedure to discriminate the presence, absence, and localization of sound stimuli. The test chamber was equipped with a center water spout and two side spouts, each with a corresponding loud speaker that presented an audible signal. Licking from the correct side spout corresponding to the active sound source delivered a reinforcing drop of water. Incorrect responses, such as licking a spout other than that at the sound source, were punished by a foot shock. After the hamsters learned to select correctly the side spout corresponding to the sound source, silent trials were randomly inserted. The silent trials were used to characterize tinnitus. On such trials subjects were required to make a side-spout response to enable the next trial, but the side response was never rewarded or punished.

Tinnitus was induced by unilateral exposure to a 10-kHz tone for 4 h at various intensity levels. It was hypothesized that unilateral cochlear damage would produce tinnitus perceptually lateralized to the traumatized ear. The hypothesis would be confirmed by preferential licking, on silent trails, from the spout ipsilateral to the trauma-exposed ear. Tinnitus was therefore measured as the difference in the average choice score on silent trials before trauma and after trauma. The largest shift in response choice during silent trials was in subjects trauma exposed at 125 dB SPL. Smaller shifts in response choice occurred after 110 dB SPL and 80 dB SPL exposures, suggesting a gradient in the effectiveness of acoustic trauma in inducing tinnitus. A control condition demonstrated that the observed results did not reflect hearing loss. These experiments demonstrated the successful use of a forced two-choice procedure to detect localized tinnitus in individual subjects after unilateral acoustic trauma.

Heffner (Heffner and Koay 2005) also used the model to assess the early time course of tinnitus development immediately after acoustic trauma. Subjects had sufficient motor coordination and motivation to perform the two-choice task within 3 to 17 min after trauma exposure. Two notable results were reported: First, hearing thresholds from trauma sufficient to produce tinnitus were assessed in individual subjects. From these data, the trauma dose sufficient to induce temporary or permanent tinnitus in half the subjects (TD_{50}) was determined. Second, tinnitus onset immediately after trauma exposure was demonstrated. These techniques will be useful in future studies that examine the relationship among temporary threshold shift, acute tinnitus, and central neural activity. Currently it is not known if pathologic changes in central neural activity are similar under conditions of transient tinnitus associated with temporary threshold shift and chronic tinnitus with or without threshold shift.

A.5 Schedule-Induced Polydipsia

Lobarinas et al. (2004) developed a novel model for detecting salicylate- and noise-induced tinnitus in rats termed schedule-induced polydipsia avoidance conditioning (SIP-AC). Rodents will naturally seek and drink water, even when not water deprived, after eating. Time-scheduled delivery of food pellets to food-deprived rats produces a high rate of licking for water (polydipsia) between pellet deliveries. The polydipsia can be brought under acoustic control using shock-avoidance (avoidance conditioning). This technique has been adapted to tinnitus assessment in individual subjects over long periods of time without extinction. In addition, this method can potentially detect the presence of acute tinnitus immediately after cochlear treatments.

Food-deprived rats are tested in acoustically isolated chambers containing a floor grid for shock delivery, a center-ceiling-mounted loudspeaker, a pellet dispenser, and a monitored lick spout. During training, subjects associate presentation of a sound stimulus with delivery of a food pellet. Pellet delivery is immediately followed by either 30 s of quiet or 30 s of sound. During this training phase there was no foot shock.

Lick suppression training was accomplished by making a foot shock contingent on licking in the presence of sound. Subjects were trained with a variety of sounds and intensity levels to maximize the generalization of suppression. Under these conditions, high stable high lick rates (2000–4000 per session) in quiet and near-zero lick rates in sound are typical. Systemic salicylate resulted in a dose-dependent decrease in licks during quiet intervals, without altering lick rates during sound intervals. This attenuation of lick suppression was interpreted as reflecting the presence of tinnitus. Control injections of saline and low-dose salicylate did not affect lick suppression. The salicylate effect obtained from higher doses (100–350 mg/kg) was significant and reversible, consistent with known effects of salicylate on cochlear function and tinnitus reported in other studies (Jastreboff and Brennan 1994).

A.6 Sound-Gap Inhibition of Acoustic Startle

The acoustic startle reflex is a well known unconditioned reflex in humans and animals: Presentation of a loud brief acoustic signal will elicit a short-latency motor response. The magnitude of startle to a sound stimulus can be decreased by preceding the startle stimulus with another stimulus. Effective stimuli for startle inhibition include events such as discrete tone bursts or gaps of silence interrupting an otherwise constant acoustic background. A gap of silence embedded in an acoustic background can effectively decrease the amplitude of a subjects' subsequent response to a startle stimulus. In this paradigm it has been established that the magnitude of startle inhibition is directly proportional to parameters such as gap width or depth, which affect gap salience (Allen et al. 2002; Forrest and Green 1987; Green and Forrest 1989; Ison et al. 2005). The advantages of this technique for studying tinnitus are: Food or water deprivation are unnecessary; there are no learning, memory, or motivational requirements; the assessment is rapid, allowing acute manipulations to be tested immediately; and the reflex habituates slowly and therefore can be used for chronic assessment.

In this model tinnitus was induced with a single 1-h unilateral exposure to octave-band noise centered at 16 kHz (Turner et al. 2006). The hypothesis was that tinnitus would fill the gap interval when the tinnitus and background sound were sufficiently similar (achieved through systematic variation of background sound composition), thereby decreasing gap inhibition.

Testing was conducted using a commercially available system that permitted stimulus parameters to be varied over a broad range and measurement of startle force applied to the chamber floor. The acoustic background within the chamber was systematically varied, testing gaps embedded in either broadband noise (BBN) or narrow-band noise centered at either 10 kHz or 16 kHz. Confirming the experimental hypothesis, startle inhibition in rats exposed to acoustic trauma was least effective for gaps embedded in the 10-kHz background compared to backgrounds of either BBN or 16 kHz. Further, this frequency-specific loss of startle inhibition was not evident in rats unexposed to acoustic trauma but with a unilateral conductive hearing loss produced by an ear plug. This control observation demonstrated that stimulus-specific loss of inhibition does not result from hearing loss, but rather another cause, with tinnitus as the most likely candidate. The gap-inhibition performance of subjects was compared to their operant stimulus discrimination scores obtained using the Bauer–Brozoski method. A significant positive correlation was obtained ($r = +0.753$, $F_{1,25} = 32.78$, $p < 0.001$), strongly supporting the conclusion that that the observed behavior reflects tinnitus (Turner et al. 2006).

A.7 Active Conditioned Avoidance

Avoidance of an aversive stimulus can be conditioned using a variety of behaviors. Guitton et al. (2003) trained rats in a paradigm in which a footshock could be avoided by climbing a pole. The conditioned stimulus was a tone paired with a footshock. Performance was scored as the number of times the rats correctly climbed the pole in response to presentation of a sound, and the number of false positive pole ascensions made in the absence of sound. After attaining criterion performance of 80% correct responses with presentation of the sound stimulus, tinnitus is induced.

Several aspects of this active avoidance paradigm are notable. First, rapid assessment of tinnitus was possible. Subjects treated with sodium salicylate demonstrated a decrease in performance accuracy reflected as an increase in false positive responses beginning 2 h after the initial salicylate injection. Second, the assessment was sensitive to changes in auditory perception. Correct responses to sound presentation resumed the day after discontinuing salicylate treatment. Third, the method may discriminate between hearing loss and tinnitus. Daily salicylate produced a decrease in the percentage of correct responses to sound presentation. The decrease corresponded to the elevation in auditory nerve compound action potential (CAP) threshold. Salicylate induced a temporary threshold shift that interfered with detection of the sound cue signaling a foot shock. When the intensity of the conditioning sound paired with the footshock was increased as a function of the CAP threshold, effectively maintaining an equivalent level of tone presentation throughout salicylate treatment, the percentage of correct responses remained at criterion performance and the false-positive rate still increased. This divergence between the two test conditions (silence and tone presentation) suggests that hearing acuity affects the accuracy of conditioned avoidance, whereas the presence of a phantom sound (tinnitus) affects the false-positive response rate. Potential limitations of this model derive from the dependent measure of pole climbing. Pole climbing is an effortful and coordinated motor activity that may be directly compromised by procedures used to induce tinnitus, to the extent that those procedures interfere with motor coordination or motivation, for example, high drug doses.

References

Abel MD, Levine RA (2004) Muscle contractions and auditory perception in tinnitus patients and nonclinical subjects. Cranio 22:181–191.

Allen PD, Virag TM, Ison JR (2002) Humans detect gaps in broadband noise according to effective gap duration without additional cues from abrupt envelope changes. J Acoust Soc Am 112:2967–2974.

Axelsson A, Sandh A (1985) Tinnitus in noise-induced hearing loss. Br J Audiol 19: 271–276.

Barnea G, Attias J, Gold S, Shahar A (1990) Tinnitus with normal hearing sensitivity: extended high-frequency audiometry and auditory-nerve brain stem-evoked responses. Audiology 29:36–45.

Bauer CA (2003). Animal models of tinnitus. In: Sismanis A (ed) Tinnitus: Advances in Evaluation and Management. Philadelphia: Elsevier, pp. 267–285

Bauer CA, Brozoski TJ (2001) Assessing tinnitus and prospective tinnitus therapeutics using a psychophysical animal model. J Assoc Res Otolaryngol 2:54–64.

Bauer CA, Brozoski TJ (2006) Effect of gabapentin on the sensation and impact of tinnitus. Laryngoscope 116:675–681.

Bauer CA, Brozoski TJ, Rojas R, Boley J, Wyder M (1999) Behavioral model of chronic tinnitus in rats. Otolaryngol Head Neck Surg 121:457–462.

Bauer CA, Meyers K, Brozoski T (2007) Primary afferent dendrite degeneration as a cause of tinnitus. J Neurosci Res 85(7):1489–1498.

Borchgrevink HM, Tambs K, Hoffman HJ (2001) The Nord-Trondelag Norway Audio-metric Survey 1996–98: unscreened adult high-frequency thresholds, normative thresholds and noise-related socio-acusis. In: Henderson D, Prasher D, Kopke R (eds) Noise Induced Hearing Loss: Basic Mechanisms, Prevention and Control. London: Noise Research Network Publications, pp. 377–385.

Brozoski TJ, Bauer CA (2005) The effect of dorsal cochlear nucleus ablation on tinnitus in rats. Hear Res 206:227–236.

Brozoski TJ, Bauer CA, Caspary DM (2002) Elevated fusiform cell activity in the dorsal cochlear nucleus of chinchillas with psychophysical evidence of tinnitus. J Neurosci 22:2383–2390.

Cacace AT (2003) Expanding the biological basis of tinnitus: crossmodal origins and the role of neuroplasticity. Hear Res 175:112–132.

Caspary DM, Pazara KE, Kossl M, Faingold CL (1987) Strychnine alters the fusiform cell output from the dorsal cochlear nucleus. Brain Res 417:273–282.

Caspary DM, Backoff PM, Finlayson PG, Palombi PS (1994) Inhibitory inputs modulate discharge rate within frequency receptive fields of anteroventral cochlear nucleus. J Neurophysiol 72:2124–2133.

Caspary DM, Milbrandt JC, Helfert RH (1995) Central auditory aging: GABA changes in the inferior colliculus. Exp Gerontol 30:349–360.

Cazals Y, Horner KC, Huang ZW (1998) Alterations in average spectrum of cochleoneural activity by long-term salicylate treatment in the guinea pig: a plausible index of tinnitus. J Neurophysiol 80:2113–2120.

Chen GD, Jastreboff PJ (1995) Salicylate-induced abnormal activity in the inferior colliculus of rats. Hear Res 82:158–178.

De Ridder D, De Mulder G, Verstraeten E, Van der Kelen K, Sunaert S, Smits M, Kovacs S, Verlooy J, Van de Heyning P, Moller AR (2006) Primary and secondary auditory cortex stimulation for intractable tinnitus. ORL J Otorhinolaryngol Relat Spec 68:48–54.

Dobie RA (1999) A review of randomized clinical trials in tinnitus. Laryngoscope 109:1202–1211.

Eggermont JJ (1990) On the pathophysiology of tinnitus; a review and a peripheral model. Hear Res 48:111–123.

Eggermont JJ (2005) Tinnitus: neurobiological substrates. Drug Discov Today 10: 1283–1290.

Evans EF, Borerwe TA (1982) Ototoxic effects of salicylates on the responses of single cochlear nerve fibres and on cochlear potentials. Br J Audiol 16:101–108.

Evered D, Lawrenson G (1981) Tinnitus. Summit, NJ: Ciba Pharmaceutical Co. Medical Education Administration.

Eybalin M (1993) Neurotransmitters and neuromodulators of the mammalian cochlea. Physiol Rev 73:309–373.

Feldmeier I, Lenarz T (1996) An electrophysiologic approach to the localization of tinnitus generators. (Abstract). Association for Research in Otolaryngology Midwinter Research Meeting, St. Petersburg Beach, FL.

Flor H, Hoffmann D, Struve M, Diesch E (2004) Auditory discrimination training for the treatment of tinnitus. Appl Psychophysiol Biofeedback 29:113–120.

Forrest TG, Green DM (1987) Detection of partially filled gaps in noise and the temporal modulation transfer function. J Acoust Soc Am 82:1933–1943.

Girardi E (1965) The Complete Works of Michelangelo. New York: Reynal.

Green DM, Forrest TG (1989) Temporal gaps in noise and sinusoids. J Acoust Soc Am 86:961–970.

Guitton MJ, Caston J, Ruel J, Johnson RM, Pujol R, Puel JL (2003) Salicylate induces tinnitus through activation of cochlear NMDA receptors. J Neurosci 23:3944–3952.

Guitton MJ, Pujol R, Puel JL (2005) m-Chlorophenylpiperazine exacerbates perception of salicylate-induced tinnitus in rats. Eur J Neurosci 22:2675–2678.

Heffner HE, Harrington IA (2002) Tinnitus in hamsters following exposure to intense sound. Hear Res 170:83–95.

Heffner HE, Koay G (2005) Tinnitus and hearing loss in hamsters (*Mesocricetus auratus*) exposed to loud sound. Behav Neurosci 119:734–742.

Hiller W, Haerkotter C (2005) Does sound stimulation have additive effects on cognitive-behavioral treatment of chronic tinnitus? Behav Res Ther 43:595–612.

Ison JR, Allen PD, Rivoli PJ, Moore JT (2005) The behavioral response of mice to gaps in noise depends on its spectral components and its bandwidth. J Acoust Soc Am 117:3944–3951.

Itoh K, Kamiya H, Mitani A, Yasui Y, Takada M, Mizuno N (1987) Direct projections from the dorsal column nuclei and the spinal trigeminal nuclei to the cochlear nuclei in the cat. Brain Res 400:145–150.

Jastreboff PJ, Brennan JF (1994) Evaluating the loudness of phantom auditory perception (tinnitus) in rats. Audiology 33:202–217.

Jastreboff PJ, Sasaki CT (1986) Salicylate-induced changes in spontaneous activity of single units in the inferior colliculus of the guinea pig. J Acoust Soc Am 80:1384–1391.

Jastreboff PJ, Brennan JF, Coleman JK, Sasaki CT (1988a) Phantom auditory sensation in rats: an animal model for tinnitus. Behav Neurosci 102:811–822.

Jastreboff PJ, Brennan JF, Sasaki CT (1988b) An animal model for tinnitus. Laryngoscope 98:280–286.

Johnson RM, Brummett R, Schleuning A (1993) Use of alprazolam for relief of tinnitus. A double-blind study. Arch Otolaryngol Head Neck Surg 119:842–845.

Kaltenbach JA (2006) The dorsal cochlear nucleus as a participant in the auditory, attentional and emotional components of tinnitus. Hear Res 216:224–234.

Kaltenbach JA, Afman CE (2000) Hyperactivity in the dorsal cochlear nucleus after intense sound exposure and its resemblance to tone-evoked activity: a physiological model for tinnitus. Hear Res 140:165–172.

Kaltenbach JA, Heffner HE (1999) Spontaneous activity in the dorsal cochlear nucleus of hamsters tested behaviorally for tinnitus. (Abstract). Association for Research in Otolaryngology Midwinter Research Meeting, St. Petersburg Beach, FL.

Kaltenbach JA, Czaja JM, Kaplan CR (1992) Changes in the tonotopic map of the dorsal cochlear nucleus following induction of cochlear lesions by exposure to intense sound. Hear Res 59:213–223.

Kaltenbach JA, Zhang J, Afman CE (2000) Plasticity of spontaneous neural activity in the dorsal cochlear nucleus after intense sound exposure. Hear Res 147:282–292.

Kaltenbach JA, Rachel JD, Mathog TA, Zhang J, Falzarano PR, Lewandowski M (2002) Cisplatin-induced hyperactivity in the dorsal cochlear nucleus and its relation to outer hair cell loss: relevance to tinnitus. J Neurophysiol 88:699–714.

Kaltenbach JA, Zhang J, Finlayson P (2005) Tinnitus as a plastic phenomenon and its possible neural underpinnings in the dorsal cochlear nucleus. Hear Res 206:200–226.

Kiang NY, Liberman MC, Levine RA (1976) Auditory-nerve activity in cats exposed to ototoxic drugs and high-intensity sounds. Ann Otol Rhinol Laryngol 85:752–768.

Kleinjung T, Eichhammer P, Langguth B, Jacob P, Marienhagen J, Hajak G, Wolf SR, Strutz J (2005a) Long-term effects of repetitive transcranial magnetic stimulation (rTMS) in patients with chronic tinnitus. Otolaryngol Head Neck Surg 132:566–569.

Kleinjung T, Steffens T, Langguth B, Eichhammer P, Marienhagen J, Hajak G, Strutz J (2005b) Treatment of chronic tinnitus with neuronavigated repetitive transcranial magnetic stimulation (rTMS). HNO 54:439–444.

Levine RA (1999) Somatic (craniocervical) tinnitus and the dorsal cochlear nucleus hypothesis. Am J Otolaryngol 20:351–362.

Levine RA, Abel M, Cheng H (2003) CNS somatosensory-auditory interactions elicit or modulate tinnitus. Exp Brain Res 153:643–648.

Lobarinas E, Sun W, Cushing R, Salvi R (2004) A novel behavioral paradigm for assessing tinnitus using schedule-induced polydipsia avoidance conditioning (SIP-AC). Hear Res 190:109–114.

Man A, Naggan L (1981) Characteristics of tinnitus in acoustic trauma. Audiology 20: 72–78.

Martin WH, Schwegler JW, Scheibelhoffer J, Ronis ML (1993) Salicylate-induced changes in cat auditory nerve activity. Laryngoscope 103:600–604.

Milbrandt JC, Holder TM, Wilson MC, Salvi RJ, Caspary DM (2000) GAD levels and muscimol binding in rat inferior colliculus following acoustic trauma. Hear Res 147:251–260.

Moller AR (1997) Similarities between chronic pain and tinnitus. Am J Otol 18:577–585.

Morgenstern L (2005) The bells are ringing: tinnitus in their own words. Perspect Biol Med 48:396–407.

Nondahl DM, Cruickshanks KJ, Wiley TL, Klein R, Klein BE, Tweed TS (2002) Prevalence and 5-year incidence of tinnitus among older adults: the epidemiology of hearing loss study. J Am Acad Audiol 13:323–331.

Norena AJ, Eggermont JJ (2003) Changes in spontaneous neural activity immediately after an acoustic trauma: implications for neural correlates of tinnitus. Hear Res 183: 137–153.

Ochi K, Eggermont JJ (1996) Effects of salicylate on neural activity in cat primary auditory cortex. Hear Res 95:63–76.

Oestreicher E, Arnold W, Ehrenberger K, Felix D (1998) Memantine suppresses the glutamatergic neurotransmission of mammalian inner hair cells. ORL J Otorhinolaryngol Relat Spec 60:18–21.

Penner MJ (1990) An estimate of the prevalence of tinnitus caused by spontaneous otoacoustic emissions. Arch Otolaryngol Head Neck Surg 116:418–423.

Plewnia C, Bartels M, Gerloff C (2003) Transient suppression of tinnitus by transcranial magnetic stimulation. Ann Neurol 53:263–266.

Rosenhall U, Karlsson AK (1991) Tinnitus in old age. Scand Audiol 20:165–171.

Ryugo DK, Haenggeli CA, Doucet JR (2003) Multimodal inputs to the granule cell domain of the cochlear nucleus. Exp Brain Res 153:477–485.

Salvi RJ, Hamernik RP, Henderson D (1978) Discharge patterns in the cochlear nucleus of the chinchilla following noise induced asymptotic threshold shift. Exp Brain Res 32:301–320.

Salvi RJ, Saunders SS, Gratton MA, Arehole S, Powers N (1990) Enhanced evoked response amplitudes in the inferior colliculus of the chinchilla following acoustic trauma. Hear Res 50:245–257.

Sanchez TG, Guerra GC, Lorenzi MC, Brandao AL, Bento RF (2002) The influence of voluntary muscle contractions upon the onset and modulation of tinnitus. Audiol Neurootol 7:370–375.

Shore SE (2005) Multisensory integration in the dorsal cochlear nucleus: unit responses to acoustic and trigeminal ganglion stimulation. Eur J Neurosci 21:3334–3348.

Shore SE, Zhou J (2006) Somatosensory influence on the cochlear nucleus and beyond. Hear Res 216:90–99.

Shore SE, El Kashlan H, Lu J (2003) Effects of trigeminal ganglion stimulation on unit activity of ventral cochlear nucleus neurons. Neuroscience 119:1085–1101.

Sindhusake D, Golding M, Newall P, Rubin G, Jakobsen K, Mitchell P (2003) Risk factors for tinnitus in a population of older adults: the blue mountains hearing study. Ear Hear 24:501–507.

Smith J (1970) Conditioned suppression as an animal psychophysical technique. In: Stebbins WC (ed) Animal Psychophysics. New York: Appleton-Century-Crofts, pp. 125–159.

Tonndorf J (1981) Tinnitus and physiological correlates of the cochleo-vestibular system: peripheral; central. J Laryngol Otol Suppl 4:18–20.

Tonndorf J (1987) The analogy between tinnitus and pain: a suggestion for a physiological basis of chronic tinnitus. Hear Res 28:271–275.

Turner JG, Brozoski TJ, Bauer CA, Parrish JL, Myers K, Hughes LF, Caspary DM (2006) Gap detection deficits in rats with tinnitus: a potential novel screening tool. Behav Neurosci 120:188–195.

Wang J, Salvi RJ, Powers N (1996) Plasticity of response properties of inferior colliculus neurons following acute cochlear damage. J Neurophysiol 75:171–183.

Wang J, Ding D, Salvi RJ (2002) Functional reorganization in chinchilla inferior colliculus associated with chronic and acute cochlear damage. Hear Res 168:238–249.

Willott JF, Lu S-M (1981) Noise-induced hearing loss can alter neural coding and increase excitability in the central nervous system. Science 216:1331–1332.

Wu JL, Chiu TW, Poon PW (2003) Differential changes in Fos-immunoreactivity at the auditory brainstem after chronic injections of salicylate in rats. Hear Res 176:80–93.

Young ED, Voigt HF(1982) Response properties of type II and type III units in dorsal cochlear nucleus. Hear Res 6:153–169.

Zhang JS, Kaltenbach JA (1998) Increases in spontaneous activity in the dorsal cochlear nucleus of the rat following exposure to high-intensity sound [published erratum appears in Neurosci Lett 1998 Aug 14;252(2):668]. Neurosci Lett 250:197–200.

5
Autoimmune Inner Ear Disease

QUINTON GOPEN AND JEFFREY P. HARRIS

1. Introduction

Substantial insights into our understanding of autoimmune inner ear disease (AIED) have occurred since McCabe first described this condition in 1979 (McCabe 1979). Harris proved that the inner ear is not an immunologically privileged site, as was once theorized (Harris 1983). Extensive evidence exists that the immune response of the inner ear can be extremely beneficial in protecting the auditory and balance systems from pathogens, but can also cause tremendous destruction and damage to the delicate inner ear. This chapter describes the basic immunology of the inner ear, the pathophysiology of AIED including experimental animal models, as well as the clinical presentation, diagnosis, and current treatment of this disorder.

2. History and Epidemiology

2.1 History

Although autoimmune damage to other organ systems has been acknowledged for centuries, the inner ear was not implicated as a possible autoimmune target until the 1950s. The first clinical report of autoimmune inner ear damage came in 1958 from Lehnhardt, who reported on 13 patients with progressive bilateral sensorineural hearing loss. He proposed anticochlear antibodies as the likely cause of the inner ear damage, as 9 of the 13 patients had hearing loss that involved the contralateral ear in a delayed fashion (Lehnhardt 1958).

Over the ensuing years, scattered reports of steroid-responsive sensorineural hearing loss appeared in the literature. Schiff and Brown in 1974 described the use of adrenocorticotropic hormone (ACTH), corticosteroids, and heparin for the treatment of sudden deafness (Schiff and Brown 1974). In a review article, Clemis (1975) stated "antigen-antibody reactions do occur within the inner ear and are associated with progressive sensorineural hearing losses." He goes on to discuss treatment for fluctuating hearing loss, including corticosteroids, histamine, antihistamines, and heparin. However, it was not until McCabe's review in 1979 that AIED became an established clinical entity (McCabe 1979).

He described 18 patients who had vestibular dysfunction as well as progressive bilateral hearing loss worsening over several weeks to months. In each case, an extensive workup for neoplastic and infectious etiologies was negative. Nearly all of the patients went on to demonstrate a dramatic response to immunosuppressive therapy, specifically dexamethasone concurrent with cyclophosphamide. This landmark publication began a new era where otolaryngologists became aware of AIED as one of the few treatable causes of sensorineural hearing loss.

2.2 Epidemiology

The true incidence of AIED is difficult to estimate. Confounding factors include similarity in presentation to Ménière's syndrome, a much more common audiovestibular disorder, as well as a lack of confirmatory laboratory tests or radiological imaging diagnostic for AIED. Current estimates of AIED's incidence place it behind idiopathic sudden sensorineural hearing loss (Rauch 1997), which has been estimated to involve 1 in 5,000 to 1 in 10,000 individuals per year (Ruckenstein 2004). AIED is therefore currently considered a rare disorder. Although AIED can affect patients at any age, it is more common in adults than in children. A recent review of 67 AIED patients found an age range from 18 to 70 years old with an equal male to female ratio (Harris et al. 2003).

AIED can be divided into two subtypes: patients with a known systemic autoimmune disease and patients without systemic autoimmune symptoms or diagnosis. The AIED patients without other systemic autoimmune disease constitute the majority of cases (70%), making the recognition of this illness difficult to separate from other progressive forms of hearing loss (genetic, viral, metabolic, toxic) that may have similar presentations (Hughes et al. 1988; Harris and Keithley 2002).

3. Immunology of the Inner Ear

Analogous to the central nervous system, the inner ear is separated from the systemic circulation by a blood–labyrinthine barrier that allows the inner ear to generate separate compartments containing high concentrations of both potassium in the endolymph and sodium in the perilymph, compartments that are critical for the normal functioning of the vestibular and auditory systems. Although the passage of immunoglobulins through the blood–labyrinthine barrier is restricted, immunoglobulins can be found within the inner ear at roughly 1/1000th the level present within the serum. IgG is the most abundant immunoglobulin found within the inner ear, with IgA and IgM also present (Palva and Raunio 1967, Mogi 1982).

Because of the existence of this blood–labyrinthine barrier, the inner ear has long been considered an immunologically privileged end organ incapable of participating in the immune response. However, research by Harris and coworkers has disproved this premise, and we now understand that the inner ear can participate as the afferent limb of an immune response, capable of local antigen processing, release of proinflammatory cytokines, and recruitment of immunocompetent cells from the systemic circulation (Harris 1983; Harris and Ryan 1984; Tomiyama and Harris 1986; Fukuda et al. 1992). Further, it has been demonstrated that antibodies can be produced locally within the inner ear, and the secondary immune response to an antigen previously processed by the inner ear leads to a much higher systemic antibody level than is seen with the primary response. This secondary response is also independent of cerebrospinal fluid or serum, proving the inner ear's participation in immune surveillance and processing (Harris 1989).

A critical component of the immune response is the release of proinflammatory cytokines when antigen arrives within the inner ear. These cytokines include tumor necrosis factor-α (TNF-α), interleukin-1β (IL-1β), and IL-6 (Satoh et al. 2003). The antigen must be processed and presented by immunocompetent cells just as in other parts of the body. Current research implicates the endolymphatic sac and the surrounding perisaccular tissues as the main antigen processing center of the inner ear. The endolymphatic sac is likely responsible for the generation of both the local response of the inner ear as well as the subsequent recruitment of the systemic immune response. Several facts substantiate this claim. The required cells for immunologic processing and presentation are present in the endolymphatic sac and perisaccular tissues. Specifically, lymphocytes including T-cells, macrophages, and B cells bearing IgM, IgG, and IgA have been identified within the endolymphatic sac but not within the cochlea (Takahashi and Harris 1988). Further, obliterating either the endolymphatic sac or the endolymphatic duct results in a much decreased immune response of the inner ear (Tomiyama and Harris 1986). In addition, horseradish peroxidase injected into the inner ear was found phagocytosed within hours of injection in a distribution consistent with uptake and processing by the endolymphatic sac and not the cochlea (Harris et al. 1997).

Once the antigen has been processed by these cells residing in and around the endolymphatic sac, proinflammatory cytokines are released, which results in the recruitment of a variety of cells and further elaboration of cytokines. The release of interleukin-2, platelet endothelial cell adhesion molecule 1, and other mediators from the endolymphatic sac are thought to cause an increased expression of intercellular adhesion molecule one (ICAM-1) as well as other surface molecules on the spiral modiolar vein (Suzuki et al. 1994; Takasu and Harris 1997). These steps are all critical for the diapedesis of specific systemic cells. Polymorphonuclear cells as well as macrophages enter the inner ear as early as 6 h after antigenic challenge. In the cochlea, the main conduit for entry has been identified as the spiral modiolar vein (Harris et al. 1990), with additional contributions from neighboring dilated bone marrow channels (Yamanobe et al. 1993).

Within the cochlea, immunoglobulin-bearing cells are also seen relatively early in the immune response (Ryan et al. 1997). IgG-bearing cells can be found as early as the first day after antigen challenge, with IgM-bearing cells found shortly thereafter. T-helper cells are noted in the endolymphatic sac approximately 24 h after antigenic challenge and gradually increase in their concentration with peak levels 2–3 weeks after antigen stimulation. T-suppressor cells can be detected in the cochlea and endolymphatic sac 3 weeks after antigenic challenge. IgA also appears roughly at 3 weeks post antigenic challenge. Antigen-specific antibodies within the inner ear are a relatively late event occurring at 4–6 weeks after a primary antigen challenge. When an animal has been systemically sensitized before receiving an inner ear challenge with the same antigen, the secondary immune response is faster and more vigorous, peaking at 2 weeks and 10-fold larger than the primary response described in the preceding text (Harris et al. 1997).

4. Clinical

4.1 Presentation

Autoimmune inner ear disease typically presents as a sensorineural hearing loss rapidly progressive over weeks to months. Fluctuations in hearing levels are common and asymmetry of the hearing loss between ears is typical. Although the initial presentation can be unilateral, both ears are ultimately affected in approximately 80% of cases. Vestibular symptoms including imbalance, vertigo, and ataxia have been found in 65% to 79% of cases (Hughes et al. 1988; Broughton 2004). Patients also commonly have aural fullness and tinnitus, both of which may fluctuate and are correlated with exacerbations in hearing loss and vestibular dysfunction.

Autoimmune inner ear disease can be divided into organ-specific disease (involves inner ear alone) or non-organ-specific disease (involves inner ear along with another systemic autoimmune illness). The majority of patients have autoimmune inner ear disease in the absence of other systemic autoimmune disease. However, approximately one third of patients have autoimmune inner ear disease along with other systemic autoimmune illness, such as those listed in Table 5.1.

4.2 Diagnosis

Autoimmune inner ear disease lacks any imaging modality or definitive laboratory test that allows for confirmation of the disorder. Currently, the diagnosis of AIED is most often made using a combination of clinical findings combined with a response to immunosuppressive therapy. A typical patient has bilateral fluctuating sensorineural hearing loss and a variable degree of vestibulopathy. A systemic autoimmune disorder is found in roughly 30% of patients and may aid in the diagnosis. A classification scheme is provided as Table 5.2 (Harris and Keithley 2002).

TABLE 5.1. Systemic diseases associated with immune-mediated hearing loss.

Polyarteritis nodosum
Cogan's syndrome
Wegener's granulomatosis
Behçet's disease
Relapsing polychondritis
Systemic lupus erythematosus
Sjögren's syndrome
Rheumatoid arthritis
Inflammatory bowel disease (ulcerative colitis, Crohn's disease)
Susac's syndrome
Idiopathic thrombocytic purpura
Scleroderma
Polymyositis
Dermatomyositis

4.2.1 Audiology

An audiogram may demonstrate either unilateral or bilateral sensorineural hearing loss. This hearing loss will often worsen over the ensuing weeks if the patient is left untreated. A recent review evaluated audiograms in 116 adults with AIED ranging from 18 to 70 years old. They found the mean six-frequency pure tone average was 52.4 dB in the better hearing ear and 65.4 dB in the poorer hearing ear. Similarly, the mean word recognition score was 71.4% in the better hearing ear and 44.7% in the poorer hearing ear (Niparko 2005). Although there is no specific audiologic pattern of hearing loss that is seen in AIED, the usual audiogram demonstrates a flat loss across all frequencies. Tympanometry is normal as the patients do not have middle ear disease unless they are affected by a systemic autoimmune disorder.

Electronystagmonography may reveal decreased responses on caloric testing. Rotational chair testing may show an asymmetrical gain with unilateral loss of vestibular function or reduced gain with phase lag consistent with bilateral vestibular loss.

4.2.2 Laboratory Tests

Using a Western blot against bovine inner ear antigens, a patient's serum can be screened for the presence of antibodies against a 68-kDa protein. These antibodies were first discovered using the guinea pig model for autoimmune inner ear disease (Harris 1987). Guinea pigs were injected with bovine cochlear tissue and 32% developed sensorineural hearing loss. Analysis of the serum from the guinea pigs that developed hearing loss revealed an antibody specific for an inner ear antigen with a molecular mass of 68 kDa. Subsequent analysis of human serum in patients with rapidly progressive bilateral sensorineural hearing loss demonstrated antibodies to the same 68-kDa antigen using Western blot analysis. A recent study of 72 patients found 89% of patients who had active AIED were also found to have antibodies to the 68-kDa antigen (Moscicki et al. 1994). None of the 25 patients with inactive AIED had antibodies to the 68-kDa antigen. Further, of those patients with antibodies to the 68-kDa antigen, 75% responded to corticosteroid treatment. In contrast, only 18% of patients who were negative for antibodies to the 68-kDa antigen responded to corticosteroid therapy. In a separate study, Western blot analysis against the 68-kDa antigen of 82 patients with AIED had a sensitivity of 42% and a specificity of 90% (Hirose et al. 1999). However, in recent years, there has been a move away from the anti-68-kDa antigen assay to an anti-heat shock protein 70 assay because of the reported cross reactivity between the two antigens (Shin et al. 1997). With this shift in an antigenic target, the specificity of the assay has declined and there is a need to return to inner ear antigen as the antigen source for AIED diagnostic testing. Currently, the Western blot assay for antibodies against the 68-kDa antigen remains the most specific test for AIED and serves to predict the likelihood of steroid responsive disease.

Tests for cellular immunity, such as the lymphocyte transformation test and the lymphocyte migration inhibition test, may assist in the diagnosis of AIED (Rocklin 1980; Oppenheim and Shecter 1980). Nonspecific antigen tests can be used to identify circulating immune complexes. These include cryoglobulins, C1q binding assay, and the hemolytic complement assay CH_{50}.

It is also important to screen for systemic autoimmune disorders that are correlated with AIED. Tests frequently used are the Antinuclear antibody (ANA), Erythrocyte sedimentation rate (ESR), Cytoplasmic antineutrophilic antibody (c-ANCA), Urinalysis (UA), Florescent absorption test for treponemal antibodies (FTA-ABS), and Venereal Disease Research Laboratory test for syphilis (VDRL). The FTA-ABS or VDRL are important tests used to exclude syphilis, which can also present with bilateral fluctuating hearing loss with vestibular symptoms.

4.3 Classification

TABLE 5.2. Classification of AIED (Harris and Keithley 2002).

Type	Conditions
1	Organ (ear)-specific
2	Rapidly progressive bilateral sensorineural hearing loss with systemic autoimmune disease - other autoimmune condition is present (see table 1)
3	Immune-mediated Ménière's disease
4	Rapidly progressive bilateral sensorineural hearing loss with associated inflammatory disease (chronic otitis media, Lyme Disease, otosyphilis, serum sickness)
5	Cogan's Syndrome

4.4 Treatment

4.4.1 Acute Therapy

When faced with a sudden drop in hearing threshold, the most effective initial treatment is the use of high-dose systemic steroids. The choice of oral prednisone (60 mg/day) or Solu-Medrol (1–2 g IV per day) is up to the practitioner. We recommend oral prednisone for 3–4 weeks to initially manage the acute hearing loss. If the hearing improves, we begin a slow taper. If once again the hearing falls with the reduction of steroid dosage, the prednisone at high doses is restarted and plans are made to move to the chronic treatment strategies as discussed in the following subsection. If hearing does not improve after the first course of high dose prednisone for 1 month, no further immunosuppressives are given as this indicates a lack of steroid sensitivity. Plasmapharesis has been used successfully in treating patients with AIED, and some investigators advocate its use in patients who have not responded to corticosteroid therapy (Luetje and Berliner 1997).

4.4.2 Chronic Therapy

The attempt here is to manage patients with steroid-sparing medications so that the lowest dose of prednisone possible is used. In this scenario, the patient's hearing has proven to be steroid dependent, and any reduction in dose results in abrupt hearing loss that often is preceded by fullness and loud tinnitus. Long-term corticosteroid therapy has substantial risks and side effects including peptic ulcers, mood disorders including suicidal ideation, insomnia, truncal obesity, moon facies, hyperglycemia, hypertension, immunosuppression, and avascular necrosis of the femoral head.

   Consequently, the patient is routinely seen by our rheumatologists and together we find a steroid sparing medication to institute before we begin a steroid taper. A recent multi-institutional trial evaluated the efficacy of methotrexate (Trexall) in the treatment of AIED and found no benefit (Harris et al. 2003). Cyclophosphamide (Cytoxan) is a potent immunosuppressant, but the side effects are prohibitive. Bladder toxicity with the potential of bladder carcinoma, sterility, and bone marrow suppression with leukocytopenia are significant risks when using this medication. Typical dosing is 2–5 mg/kg per day each morning with heavy fluid intake to minimize the bladder toxicity. A peripheral blood smear must be performed periodically to monitor for the development of leukocytopenia. Tumor necrosis factor antagonists such as Remicade (infliximab), Enbrel (etanercept), and Humira (adalimumab) as well as the anti-B cell agent Rituxan (rituximab) are undergoing active investigation for use in AIED. There certainly will be many more biological modifying drugs developed in the years ahead that may be found to suppress the immunological damage that this illness causes in the inner ear without side effects that are unpleasant or life-threatening.

4.4.3 Experimental Therapy

Here the notion is to apply sound logic to the treatment of AIED without specific data to support the regimens. It is always important to remember that these unproven methods need to be studied in a randomized clinical trial to prove their efficacy and safety. Currently, it is popular to recommend intratympanic steroids or even intratympanic biological modifying drugs to the middle ear with the hopes that they will cross the round window membrane and act directly on the inner ear. Of the steroid preparations, dexamethasone and methylprednisolone are the agents of choice. The dose and timing of these medications has not been established, nor has their efficacy been proven over and above systemic steroids; however, they do offer the theoretical benefit of not causing systemic side effects seen with the latter route. There is also the possibility that a TNF-α blocker could be instilled directly into the middle ear with the benefit of getting high dose local effects compared to the rather meager benefit recently reported using systemic Enbrel (Matteson 2005; van Wijk et al. 2006).

5. Animal Models for Autoimmune Inner Ear Disease

The first attempt to create an animal model for autoimmune inner ear disease was made by Bieckert in 1961. He immunized guinea pigs with isologous inner ear tissue and on sacrifice found lesions within the cochlea. However, he was not able to show any evidence of a cellular or humoral immune response to the inner ear tissues (Bieckert 1961). Shortly thereafter, Terayama immunized guinea pigs with isologous cochlear tissue in Freund's adjuvant and also documented lesions within the cochlea. He too could not demonstrate any cellular or humoral immune response to inner ear tissues (Yoshihiko 1963).

Yoo et al. (1983) developed an animal model for immune-mediated hearing loss based on type II collagen. Rats were immunized with native bovine type II collagen and developed sensorineural hearing loss as well as vestibular dysfunction; they also developed antibodies to the native bovine type II collagen. Histopathologic evaluation revealed perivascular lesions within the cochlear artery including a mononuclear cell infiltration around the artery with thickening of the endothelial cells of the vessel. Immunofluorescence demonstrated IgG deposition within the vessel wall, the perivascular fibrous tissue, and the bone surrounding the cochlear and vestibular arteries. There was also a mild fibrosis in these vessels. Spiral ganglion cell degeneration was found along with swelling of the ganglion cell bodies and pyknotic nuclei displaced towards the axonal poles of the cells. The organ of Corti, stria vascularis, and cochlear nerve showed no evidence of injury at 1 month with only mild atrophic changes at two months. No pathologic changes could be found within the vestibular system. Interestingly, rats immunized with type I collagen or with denatured type II collagen did not develop sensorineural hearing loss or vestibular dysfunction.

Subsequent animal experiments have also demonstrated autoimmune inner ear damage (Harris 1983). Specifically, when a foreign protein antigen, in this case keyhole limpet hemocyanin (KLH), was injected into the inner ear of naïve animals, a systemic immune response could be measured including elevation of inner ear and serum anti-KLH antibody levels. Importantly, no change in hearing levels occurred during this process. When the identical experiment was performed on animals that had been previously systemically exposed to KLH, a more vigorous systemic immune response was witnessed, with antibody levels within the perilymph elevated to levels far greater than in the naïve animals. More importantly, these animals developed hearing loss as determined by cochlear microphonic responses and also demonstrated histopathologic damage to the inner ear consisting of perilabyrinthine fibrosis, hemorrhage, loss of spiral ganglion cells, and degeneration of the organ of Corti. In these systemically exposed animals, inflammatory cells infiltrated the scala tympani and to a lesser extent the scala vestibule of the cochlea. The amount of damage was correlated with the degree of infiltration of the inflammatory cells within each turn of the cochlea. For example, the apical turns had fewer inflammatory cells and consequently displayed less damage and less hearing loss at the low frequencies. When the inflammation was severe, the cochlea went on to form a fibrous matrix followed by eventual calcification of the cochlea (Harris 1983).

Although these experiments demonstrated autoimmune inner ear damage via a specific antibody, the rise in the level of this specific antibody could have resulted from increased vascular permeability to serum immunoglobulins instead of local antibody production. To investigate this possibility, a study was designed that created a serum marker by systemically immunizing guinea pigs with bovine serum albumin (BSA). Guinea pigs once again had inner ear immunization with KLH, and an increase in anti-KLH perilymph antibody levels without a corresponding increase in anti-BSA levels was found. This proved that changes in vascular permeability was not the factor responsible for elevation in the increased perilymph antibody levels but in fact local production of the antibody within the inner ear was occurring (Harris 1984).

Harris went on to immunize guinea pigs with heterologous bovine inner ear antigen in Freund's adjuvant. He found hearing deficits and lesions within their inner ears as well as circulating antibody titers to inner ear antigens both in the serum and in the perilymph. Interestingly, however, not all animals developed hearing loss. In fact, only 32% of the ears tested had significant deficits compared to control animals. In addition, of those animals that did develop hearing loss, the contralateral ear was affected only 50% of the time. There was also very little correlation between the magnitude of hearing loss and antibody levels within the serum or other markers of cellular immunity. Spiral ganglion cell degeneration, perivascular infiltration by plasma cells, nonspecific damage to the organ of Corti, and diffuse edema with hemorrhage could all be identified in these ears (Harris 1987).

More recently, a mouse monoclonal antibody was developed that binds to inner ear supporting cells in guinea pigs. When this monoclonal antibody, termed KHRI-3, was infused into the cochlea of guinea pigs, it induced hearing loss in the 25- to 55-dB range in the majority of tested animals. Control animals had no hearing loss with infusion of placebo into the cochlea (Nair et al. 1997). This monoclonal antibody binds a 68- to 72-kDa protein antigen within the supporting cells of the guinea pig. This protein could very well be a choline transporter, most likely CTL-2 (Nair et al. 2004).

Finally, some of the animal models used to study other systemic autoimmune illnesses also display auditory and vestibular dysfunction. In particular, the MRL-Fas mouse model used for systemic lupus erythematosus develops significant degrees of sensorineural hearing loss. These MRL-Fas mice lose the ability to eliminate autoimmune T cells by apoptosis. As these autoimmune T cells accumulate, various endorgans are injured as occurs in systemic lupus erythematosus. Histopathologic studies of these mice reveal hydropic cellular degeneration isolated to the stria vascularis in the absence of any inflammatory infiltrate or immune complex deposition. However, significant antibody deposition within the capillaries of the stria vascularis has been identified. Whether these antibodies or other factors, such a uremia or genetic predisposition, lead to the sensorineural hearing loss found in these animals remains unclear (Ruckenstein 1999).

It is important to note that although the immune response of the inner ear can cause autoimmune damage, it more often serves a critical protective role against pathogens. Guinea pigs receiving live cytomegalovirus (CMV) injected into their inner ears developed progressive deafness within 1 week. If these guinea pigs were exposed systemically to CMV prior to having the live CMV injected into their inner ears, they quickly developed elevated levels of antibody within the perilymph against CMV. These antibodies prevented inner ear damage and the animals did not develop any degree of hearing loss (Woolf et al. 1985).

6. Mechanisms of Autoimmune Injury to the Inner Ear

Autoimmune injury in general can occur either via humoral or cell-mediated reactions. Injury to tissues can occur from direct damage by an antibody directed toward a specific antigen, by direct cell destruction from cytotoxic T cells, or by immune complex deposition into tissues. All three mechanisms of autoimmune injury can potentially damage inner ear tissues. However, confirmation of any of these mechanisms has proven difficult to obtain, and the exact mechanisms of autoimmune injury to the inner ear remain unknown.

It seems highly likely, however, that some subsets of patients with AIED do indeed develop organ-specific autoantibodies to a particular antigenic epitope within the inner ear. When inner ear proteins are injected into the inner ear of animals, hearing loss as well as inflammation within the cochlea develops along with antibodies specific to the inner ear tissues. Curiously, the hearing loss which is incurred by such measures is modest and transient. Also, control animals immunized with antigens that are not from the inner ear do not demonstrate any hearing loss or cochlear inflammation. The greatest degree of hearing loss and cochlear inflammation comes from a secondary immune response to purified antigen used on animals with a high degree of preexisting immunity (Ryan et al. 2002). In addition, the mouse monoclonal antibody KHRI-3 binds to the inner ear supporting cells in guinea pigs and causes hearing loss that is not seen in control guinea pigs. This gives further evidence for antibody-mediated hearing loss in humans (Nair et al. 1997).

Evidence of antibody–antigen immune complex deposition is found in patients with certain systemic autoimmune diseases who go on to develop inner ear damage. Examples are collagen vascular diseases such as systemic lupus erythematosus, Wegener's granulomatosis, and polyarteritis nodosum, all of which have immune complex deposition but only sporadically develop inner ear damage. Veldman et al. (1984) reported a case of a 14-year-old girl who presented with fatigue, arthralgias, and sensorineural hearing loss. She had circulating immune complexes and was diagnosed with juvenile arthritis. She was placed on prednisone and had a dramatic resolution of her symptoms including normalization of her hearing within 2 weeks. This case report is convincing for immune complex deposition as a possible etiology in some cases of autoimmune inner ear disease (Veldman et al. 1984). Immune complex disease has also been implicated as the mechanism of injury in patients who develop hearing loss, vertigo, and tinnitus in association with Cogan's syndrome after vaccinations for small pox (Harris 1984).

Evidence of cytotoxic T cell damage to inner ear tissues comes from the lymphocyte transformation test, a test designed to identify cytotoxic T cell injury. The lymphocyte transformation test involves taking lymphocytes and exposing them to an antigen with subsequent evaluation of their activity. In a study performed on 68 patients with bilateral progressive sensorineural hearing loss of unknown etiology using type II collagen as an antigen, Berger and coworkers found 34 out of 68 patients had a strong stimulation in the transformation test as compared to only 4 of 68 control patients (Berger 1991).

Although definitive evidence is lacking for either humoral or cell-mediated mechanisms of injury, one can gain some insight into the mechanism of damage from autoimmune inner ear disease through the histopathologic examination of inner ear tissue from affected patients. In temporal bones evaluated from patients suffering from autoimmune inner ear disease, damage to both the organ of Corti and the stria vascularis has been found. Similarly, animal models of autoimmune inner ear disease also demonstrate damage to both the organ of Corti and stria vascularis. Further, the degree of damage to the organ of Corti and stria vascularis is closely correlated to the severity of inflammation in secondary immune responses. The most damage occurs to the organ of Corti in the cochlear turns with the highest amount of inflammatory response. Typically, the amount of inflammation is most severe in the basal turns with the apical cochlear turns often displaying much less inflammation and damage to the organ of Corti and stria vascularis. In human tissue and animal models, both the inner and outer hair cells as well as supporting cells undergo apoptosis. The stria vascularis also demonstrates signs of injury with cell protrusion, hyperplasia, thinning, and even total degeneration. Although the inflammatory response does not involve Rosenthal's canal at any point, the spiral ganglion neurons demonstrate significant loss and degeneration at roughly 6 weeks after the initial immune response, suggesting their retrograde degeneration after hair cell loss. However, the mechanisms behind the spiral ganglion cell loss are not well understood. When the inflammatory response is vigorous and goes unchecked, the cochlea often develops an inflammatory matrix. This matrix is often quite difficult for the inner ear to clear, which ultimately leads to ossification within the cochlea (Ma et al. 2000).

The mechanisms of how the hair cells, the supporting cells, and the cells of the stria vascularis are damaged in the immune response remain unclear. Perhaps clues can be taken from other pathologies such as presbycusis, noise trauma, or ototoxicity in which oxidant stress may be a factor initiating apoptotic or necrotic cell death (see Chapters 6, 7, and 8 on these topics) but substantial research efforts will be required to elucidate these undoubtedly complex mechanisms.

7. Summary

The inner ear can participate as the afferent limb of the immune response with the endolymphatic sac and perisaccular tissue functioning as the primary antigen processing site of the inner ear. Systemic cells are then recruited into the inner ear via the spiral modiolar vein. It remains unclear whether autoimmune damage to the inner ear occurs via cell-mediated, humoral, or a combined mechanism. Patients typically present with fluctuating sensorineural hearing loss progressing over weeks to months. Although one third of patients suffer from a concurrent systemic autoimmune illness, two thirds of patients present with isolated autoimmune inner ear damage. High-dose corticosteroids remain the treatment of choice for patients with AIED, but more specific medications including TNF-α antagonists such as Remicade, Enbrel, and Humira as well as the anti-B cell agent Rituxan show promise in preliminary studies.

References

Berger P, Hillman M, Tabak M, Vollrath M (1991) The lymphocyte transformation test with type II collagen as a diagnostic tool of autoimmune sensorineural hearing loss. Laryngoscope 101:895–899.

Bieckert P (1961) On the problem of perception deafness and autoallergy. Z Laryngol Rhinol Otol 40:837–842.

Broughton SS, Meyerhoff WE, Cohen SB (2004) Immune-mediated inner ear disease: 10-year experience. Semin Arthritis Rheum 34:544–548.

Clemis JD (1975) Allergy as a cause of fluctuant hearing loss. Otol Clin N Am 8: 375–383.

Fukuda S, Harris JP, Keithley EM, Ishikawa K, Kucuk B, Inuyama Y (1992) Spiral modiolar vein: its importance in viral load of the inner ear. Ann Otol Rhinol Laryngol Suppl 157:67–71.

Harris JP (1983) Immunology of the inner ear: response of the inner ear to antigen challenge. Otolaryngol Head Neck Surg 91:18–32.

Harris JP (1984) Immunology of the inner ear: evidence of local antibody production. Ann Otol Rhinol Laryngol 93:157–162.

Harris JP (1987) Experimental autoimmune sensorineural hearing loss. Laryngoscope 97:63–76.

Harris JP (1989) Autoimmunity of the inner ear. Am J Otol 10:193–195.

Harris JP, Keithley EM (2002) Autoimmune inner ear disease. In Snow JB, Ballenger JJ (eds) Otorhinolaryngology Head and Neck Surgery. Baltimore: BC Decker, pp. 396–407.

Harris JP, Ryan AF (1984) Immunobiology of the inner ear. Am J Otolaryngol 5:418–425.

Harris JP, Fukuda S, Keithley EM (1990) Spiral modiolar vein: its importance in inner ear inflammation. Acta Otolaryngol 110:357–365.

Harris JP, Heydt J, Keithley EM, Chen MC (1997) Immunopathology of the inner ear: an update. Ann NY Acad Sci 830:166–178.

Harris JP, Weisman WH, Derebery JM, Espeland MA, Gantz BJ, Gulya AS, Hammerschlag PE, Hannley M, et al. (2003) Treatment of corticosteroid-responsive autoimmune inner ear disease with methotrexate: a randomized controlled trial. JAMA 290:1875–1883.

Hirose K, Wener MH, Duckert LG (1999) Utility of laboratory testing in autoimmune inner ear disease. Laryngoscope 109:1749–1754.

Hughes GB, Barna BP, Kinney SE, Calabrese LH, Nalepa NJ (1988) Clinical diagnosis of immune inner-ear disease. Laryngoscope 98:251–253.

Lehnhardt E (1958) Sudden hearing disorders occurring simultaneously or successively on both sides. Z Laryngol Rhinol Otol 37:1–16.

Luetje CM, Berliner KI (1997) Plasmapharesis in autoimmune inner ear disease: long-term follow-up. Am J Otol 18:572–576.

Ma C, Billings P, Harris JP, Keithley EM (2000) Characterization of an experimentally induced inner ear immune response Laryngoscope 110:451–456.

Matteson EL, Choi HK, Poe DS, Wise C, Lowe VJ, McDonald TJ, and Rahman MU (2005) Etanercept therapy for immune-mediated cochleovestibular disorders: a multi-center, open-label, pilot study. Arthritis Rheum 53:337–342.

McCabe BF (1979) Autoimmune sensorineural hearing loss. Ann Otol Rhinol Laryngol 88:585–589.

Mogi G, Lim DJ, Watanabe N (1982) Immunologic study on the inner ear. Immunoglobulins in perilymph. Arch Otolaryngol 108:270–275.

Moscicki RA, San Martin JE, Quintero CH, Rauch SD, Nadol JB Jr, and Bloch KJ. (1994) Serum antibody to inner ear proteins in patients with progressive hearing loss. JAMA 272:611–616.

Nair TS, Prieskorn DM, Miller JM, Mori A, Gray J and Carey TE. (1997) In vivo binding and hearing loss after intracochlear infusion of KHRI-3 antibody. Hear Res 107:93–101.

Nair TS, Kozma KE, Hoefling NL, Kommareddi PK, Ueda Y, Gong TW, Lomax MI, Lansford CD, Telian SA, Satar B, Arts HA, El-Kashlan HK, Berryhill WE, Raphael Y, Carey TE. (2004) Identification and characterization of choline transporter-like protein 2, an inner ear glycoprotein of 68 and 72 kDa that is the target of antibody-induced hearing loss. J Neurosci 24:1772–1779.

Niparko JK, Wang NY, Rauch SD, Russell GB, Espeland MA, Pierce JJ, Bowditch S, Masuda A, et al. (2005) Serial audiometry in a clinical trial of AIED treatment. Otol Neurotol 26:908–917.

Oppenheim JJ, Shecter B (1980) Lymphocyte transformation. In: Zachary AA, Braun WE (eds) Manual of Clinical Immunology, 2nd ed. Washington, DC: American Society Microbiology, pp. 233–245.

Palva T, Raunio V (1967) Disc electrophoretic studies of human perilymph. Ann Otol Rhinol Laryngol 76:23–36.

Rauch SD (1997) Clinical management of immune-mediated inner-ear disease. Ann NY Acad Sci 830:203–210.

Rocklin RE (1980) Production and assay of macrophage inhibitory factor. In: Zachary AA, Braun WE (eds) Manual of Clinical Immunology, 2nd ed. Washington, DC: American Society Microbiology, pp. 246–251.

Ruckenstein MJ (2004) Autoimmune inner ear disease. Curr Opin Otolaryngol Head Neck Surg 12:426–430.

Ryan AF, Gloddek B, Harris JP (1997) Lymphocyte trafficking to the inner ear. Ann NY Acad Sci 830:236–242.

Ryan AF, Harris JP, Keithley EM (2002) Immune-mediated hearing loss: basic mechanisms and options for therapy. Acta Otolaryngol Suppl 548:38–43.

Satoh H, Firestein GS, Billings PB, Harris JP, Keithley EM (2003) Proinflammatory cytokine expression in the endolymphatic sac during inner ear inflammation. J Assoc Res Otolaryngol 4:139–147.

Schiff M, Brown M (1974) Hormones and sudden deafness. Laryngoscope 84:1959–1981.

Shin S, Billings P, Keithley E, Harris JP (1997) Comparison of anti-heat shock protein 70 (anti-hsp70) and anti-68-kDa inner ear protein in the sera of patients with Ménière's disease. Laryngoscope 107:222–227.

Suzuki M, Mogi G, Harris JP (1994) Expression of intercellular adhesion molecule-1 during inner ear inflammation. Ann Otol Rhinol Laryngol 104:69–75.

Takahashi M, Harris JP (1988) Anatomic distribution and localization of immunocompetent cells in normal mouse endolymphatic sac. Acta Otolaryngol 106:409–416.

Takasu T, Harris JP (1997) Reduction of inner ear inflammation by treatment with anti-ICAM-1 antibody. Ann Otol Rhinol Laryngol 106:1070–1075.

Tomiyama S, Harris JP (1986) The endolymphatic sac: its importance in inner ear immune responses. Laryngoscope 96:685–691.

van Wijk F, Staecker H, Keithley E, Lefebvre P (2006) Local perfusion of the tumor necrosis factor alpha blocker infliximab to the inner ear improves autoimmune neurosensory hearing loss. Audiol Neurootol 11:357–365.

Veldman JE, Roord JJ, O'Connor AF, Shea JJ (1984) Autoimmunity and inner ear disorders: an immune-complex mediated sensorineural hearing loss. Laryngoscope 94:501–507.

Woolf NK, Harris JP, Ryan AF, Butler DM, and Richman DD. (1985) Hearing loss in experimental cytomegalovirus infection of the guinea pig inner ear: prevention by systemic immunity. Ann Otol Rhinol Laryngol 94:350–356.

Yamanobe S, Harris JP, Keithley EM (1993) Evidence of direct communication of bone marrow cells with the endolymphatic sac in experimental autoimmune labyrinthitis. Acta Otolaryngol 113:166–170.

Yoo TJ, Tomoda K, Stuart JM, Cremer MA et al. (1983) Type II collagen-induced autoimmune sensorineural hearing loss and vestibular dysfunction in rats. Ann Otol Rhinol Laryngol 92:267–271.

Yoshihiko T, Yukihiro S (1964) Studies on experimental allergic (isoimmune) labyrinthitis in guinea pigs. Acta Otolaryngol 58:49–64.

6
Age-Related Hearing Loss and Its Cellular and Molecular Bases

Kevin K. Ohlemiller and Robert D. Frisina

1. Introduction

Age-related hearing loss (ARHL, or presbycusis) affects most people age 65 and older and represents the predominant neurodegenerative disease of aging. Despite decades of research, there remains disagreement as to how many forms of ARHL exist and what the environmental and genetic risk factors are. Nor are there any widely accepted prevention or treatment strategies. Age-related changes in hearing reflect alterations in both the peripheral and central auditory systems. Changes in the periphery are typically degenerative, and include cell loss in the organ of Corti, spiral ganglion, and stria vascularis. Perceptual correlates of these alterations include reduced sensitivity, reduced sharpness of tuning, loss of compression, and reduced signal-to-noise ratios. Some age-related pathology of the central auditory system appears secondary to degenerative changes in the periphery. Other central pathology occurs independently, and includes reduced connectivity and alterations in synaptic chemistry. (Central consequences of cochlear trauma are detailed in Chapter 9 by Morest and Potashner.) Central physiological effects of ARHL that have no significant peripheral origin include reduced temporal fidelity and weakened suppressive effects of olivocochlear efferent activation on cochlear processing. Perceptually, these may result in degraded signal-to-noise ratios and impaired speech perception.

Application of mouse models to the study of ARHL is promoting rapid advancement in this area. Compared to other models, mice offer a shorter natural lifespan, genetic homogeneity, and the potential for genetic engineering approaches. Nevertheless, their value for understanding human ARHL must follow establishment of common pathology and common genetic and environmental bases for disease. This chapter addresses the history, epidemiology, and current status of understanding of both peripheral and central ARHL, and attempts to identify major unanswered questions as a basis for future research. Recent findings in mice are emphasized, yet not to the exclusion of other models, and only insofar as they complement clinical findings.

2. History and Epidemiology

Historical accounts of research in ARHL are given by Gilad (1979a,b), by Schuknecht (1993), and more recently by Schacht and Hawkins (2005). The first presentation of ARHL as a distinct pathology at a scientific meeting may have been by S.J. Roosa to the American Otological Society in 1885, where he introduced the term *presbycusis*. It was not until the 1930s that Guild, and also Crowe and colleagues, associated hearing loss in aging with specific cochlear structures. von Fieandt and Saxen (1937), (cited in Schacht and Hawkins [2005]) coined the terms *senile atrophy* of auditory neurons and *angiosclerotic degeneration*, the latter term foreshadowing decades of speculation about the role of microvascular pathology in ARHL. Crowe (1934) and Saxen (1937) are credited with the initial division of ARHL into forms emphasizing pathology of organ of Corti versus afferent neurons. Glorig and Davis (1961) and later Nixon and Glorig (1962) argued for nondegenerative pathology of the mechanical structures of the inner ear, later supported by Schuknecht as inner ear conductive ARHL. Schuknecht, in a series of papers beginning in 1953, expanded the number of posited forms to include ARHL characterized by pathology of stria vascularis, and began evaluating temporal bones in terms of putatively independent pathology of organ of Corti, afferent neurons, stria, and spiral ligament. To the present day, there is no clearly superior competing paradigm for understanding ARHL.

Figure 6.1 shows the average progression of ARHL across the span of life (Glorig et al. 1957). The common pattern is one of progressive loss of sensitivity to high frequencies, with relatively little change below 2 kHz to about age 40. Ages 40–60 are characterized by further erosion of high-frequency hearing, typically accompanied by flat losses at low frequencies. After age 60, the entire audiogram may indicate increased hearing loss, such that more than 40% show a clinically significant hearing loss (generally taken to be greater than 25 dB at any frequency) by age 60. Between ages 70 and 80, the prevalence of ARHL exceeds 50%. Nevertheless, the tendencies graphed in Fig. 6.1 obscure a highly skewed character of the underlying distributions, particularly at advanced ages. The rate of hearing decline is quite individualized, so that among those 60 years of age or older will be people with near-normal hearing, sometimes referred to as "golden ears." Debilitating hearing loss is not an inevitable part of aging, and if understood well enough, may be avoidable. Interindividual differences must reflect different environments and experiences, as well as differences in genetic makeup, predisposing some people to cellular damage from health habits or events that would be benign to others. Also not evident in Fig. 6.1 is the fact that different cochlear cells and structures are affected in different people, so that an unknown number of distinct conditions have been combined.

The pattern of sound frequencies affected by ARHL depends on gender, albeit in a complex manner. Figure 6.2, based on a summary of USPHS surveys by Jerger et al. (1993), shows that men and women have comparable hearing thresholds at 0.5 kHz up to about age 50. After that, thresholds in women deteriorate more rapidly than in men at this frequency. By contrast, above

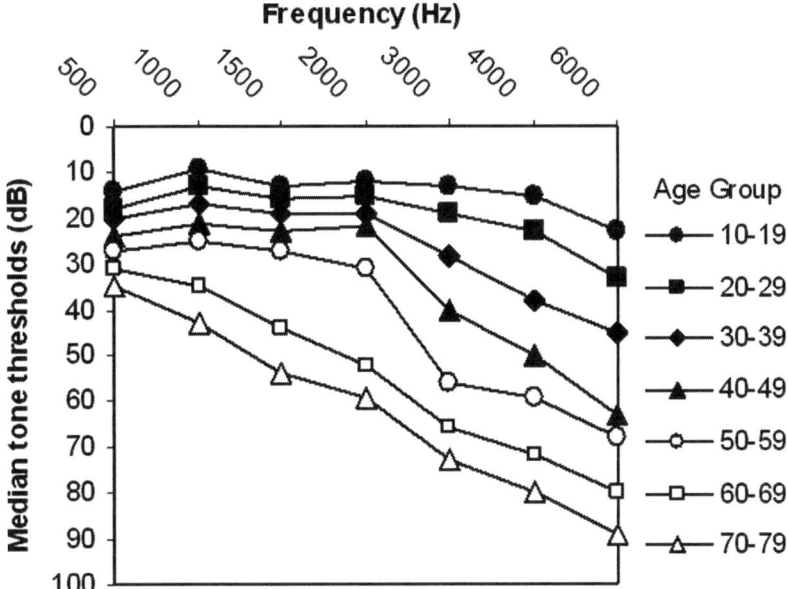

FIGURE 6.1. Median thresholds in dB SPL for men by age group. (Adapted from Glorig et al. 1957.)

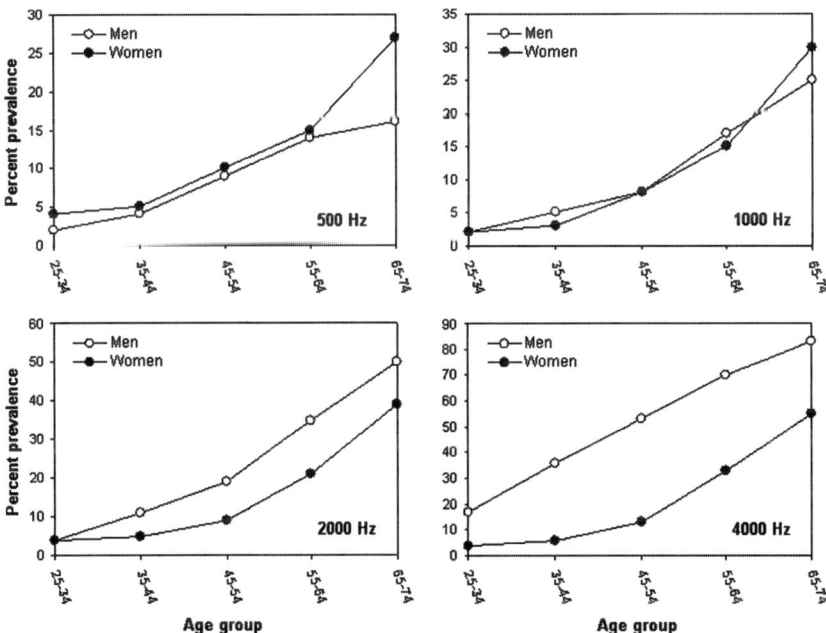

FIGURE 6.2. Prevalence of clinically significant hearing loss (>25 dB above norms) versus age for men and women at 500, 1000, 2000, and 4000 Hz. (Data are taken from a 1975 USPHS National Health Survey, and adapted from Jerger et al. 1993.)

2–4 kHz women hear better than men after about age 35. If subjects are separated into groups with a known history of noise exposure and those without, threshold disparities at high frequencies between men and women are revealed to largely reflect permanent noise-induced hearing loss (NIHL) in men (Fig. 6.3). However, the overall frequency pattern by gender is retained. The role of noise injury in ARHL and possible reasons for gender effects are considered in later sections.

Subdividing ARHL in a way that is diagnostically and prognostically meaningful has remained elusive. While the most dramatic impact of ARHL on hearing ability typically reflects auditory peripheral pathology, the auditory central nervous system (ACNS) ages in characteristic ways and differentially

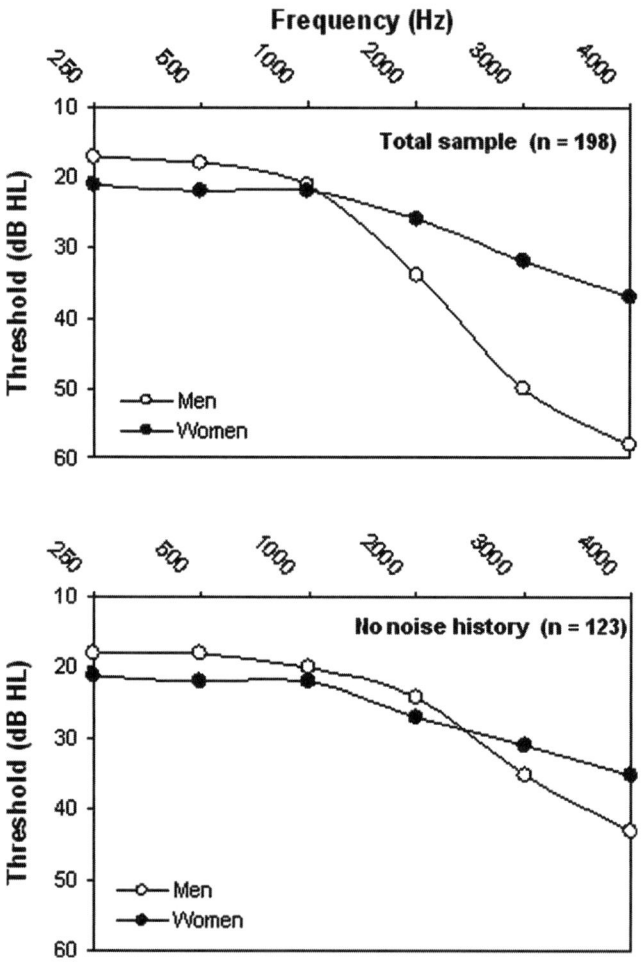

FIGURE 6.3. Hearing thresholds at six test frequencies in men and women 50–89 years of age, separated by noise exposure history. (Adapted from Jerger et al. 1993.)

impacts hearing. Separating peripheral and central pathology is therefore appropriate, and one of the organizing principles of this chapter.

3. Peripheral Aspects of ARHL in Humans

In attempting to impose some kind of typology on ARHL, one might take one of the following approaches: classifying ARHL by its clinical characteristics, by cause, or by the cell type(s) affected. Clinical measures of age-at-onset, severity, laterality, stability, and audiogram shape (see later) are not reliably diagnostic of any particular ARHL type. They are, of course, quite useful for distinguishing other forms of inner ear dysfunction from ARHL, as it is generally symmetrical and nonfluctuating. An elevated "notch" in the audiogram at 4–6 kHz may be useful in separating hearing loss that is principally noise-related from ARHL (e.g., McBride and Williams 2001). Absent any clear indications of causes other than aging, the "cause" of ARHL is the very goal of much current basic research, not a clinical tool. The single current diagnostic for a particular type of ARHL (noted later) is that speech perception in noise may be impacted, even when hearing sensitivity remains normal. To approach the cell biology of ARHL, it has proven more useful to classify cases according to the pattern of cells affected. Classifications made this way unfortunately arrive too late to help the donors of temporal bones, but thus far so have any real cures.

3.1 Classification by Affected Cochlear Structure and Cell Type

The presently dominant framework for classifying of ARHL using histopathology is one championed by Schuknecht in a series of papers written over 40 years, and in his classic book *Pathology of the Ear* (Schuknecht 1953, 1964, 1993; Schuknecht et al. 1974; Schuknecht and Gacek 1993). From his studies of a modest collection of human temporal bones, Schuknecht was struck by cases of relatively isolated degeneration of organ of Corti, afferent neurons, and stria vascularis. He proposed that these represent distinct types of ARHL: *sensory* ARHL (hearing loss due principally to organ of Corti pathology), *neural* ARHL (hearing loss reflecting primary loss of neurons despite the presence of inner hair cells), and *strial* ARHL (hearing loss due mainly to strial degeneration and reduction of the endocochlear potential). Other forms proposed included *mixed* ARHL (multiple apparent contributors), *cochlear conductive* ARHL (possible nondegenerative changes in passive mechanics), and *indeterminate*. Isolated occurrences of delimited pathology of organ of Corti, afferent neurons, and stria were taken to demonstrate the potential for independent degeneration of these structures, and for the existence of environmental and genetic risk factors specific to each. Dysfunction of any one structure/population was further posited to impact independently hearing ability and the shape of the audiogram.

Completely isolated pathology of any cochlear structure is the exception rather than the rule, and most cochleae will show a mixture of pathologies encompassing many cell types. The prevalence of mixed pathology is not really surprising, given that hair cell, neural, and fibrocyte loss increases steadily with age in human temporal bones (Fig. 6.4) (Bredberg 1968; Wright and Schuknecht 1972; Otte et al. 1978). Recognizing this, Schuknecht sought in each case to identify the major contributing degeneration(s) to ARHL. Such a process may seem arbitrary, but is less so than one might expect. Hearing sensitivity is well correlated with outer hair cell (OHC) loss (e.g., Hamernik et al. 1989), but appears surprisingly resistant to loss of afferent neurons and reduction of functional strial epithelial area (Pauler et al. 1986, 1988). In principle, it should be possible to separate the contributions of organ of Corti, afferent neurons, and stria to hearing loss, assuming the surviving cells are functional. Against a backdrop of gradual loss of many cochlear cell types with age, the specific cell types that show accelerated loss and become limiting for hearing will determine the form of ARHL diagnosed. So-called *mixed* ARHL may occur by coincidence of independent causes, or because of genetic or environmental factors that promote broad cochlear degeneration. Both of these conditions probably apply.

Consideration of Figs. 6.1 to 6.4 may raise questions about the distinctness of ARHL from other problems of aging. If one lives long enough, odds are that he or she will experience clinically significant hearing loss as part of a broad spectrum of age-related dysfunctions, one of which ultimately becomes fatal. The goal of responsible aging research is not to overthrow this ultimate limitation. Rather, it is "healthy aging"—to maximize the quality of life up to the immutable limits of the lifespan. Unless one steps in front of an unheard bus, ARHL is not often fatal. Hard limits on longevity will typically be imposed by age-related dysfunction of cardiovascular, pulmonary, or neuroendocrine systems. One's "true age" must be that of the organ system(s) whose age-related failure is life-ending. Accordingly, aging researchers have sought "biomarkers" for aging, metrics that predict longevity (Harper et al. 2003, 2004). Several good candidates have emerged spanning multiple organ systems. The possibility of such markers permits—at least conceptually—refinement of when ARHL is accumulating more rapidly than other age-related pathologies, and when it is just one facet of healthy aging. Assuming an appropriate biomarker were to be sampled in a patient, one may ask, "Is hearing loss progressing at the rate predicted from the marker, or more rapidly?" If the former applies, the ARHL may reflect broad aging processes applicable to a host of tissues, and good preventive strategies might be those shown to delay aging of the whole organism (e.g., caloric restriction). Genetic or environmental factors that promote this "biological-age-synchronous" ARHL would be expected to accelerate a host of age-related pathologies. In the latter case, the ARHL may reflect progressive failure of cochlea-specific cells, or vulnerabilities unique to the cochlea based on its mechanics or metabolic demand. Genetic and environmental risk factors that promote this "accelerated" ARHL might be expected to exert their effects principally on the cochlea, and not other organs. Optimal therapies might be aimed at replacement of lost cells, compensating for missing repair mechanisms,

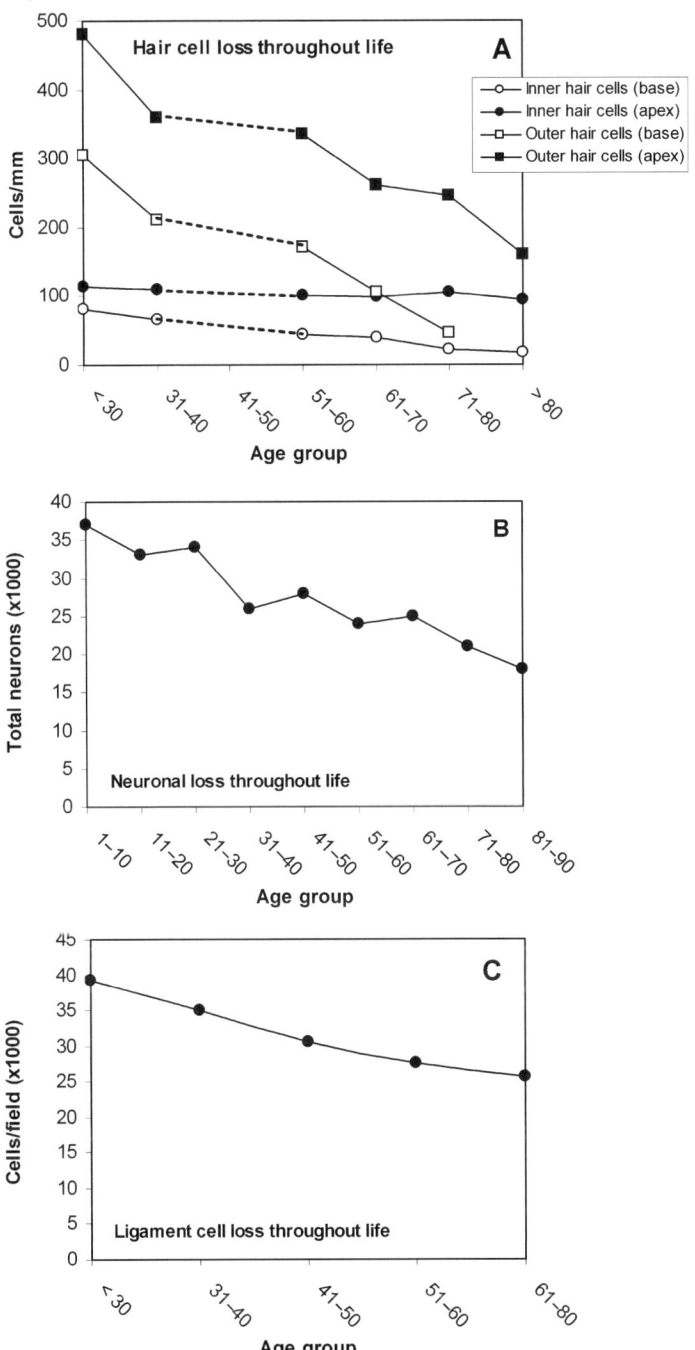

FIGURE 6.4. (**A**) Inner and outer hair cell density versus age in the basal and mid-to-apical cochlea. Data were pooled across subjects of varying hearing ability. (Adapted

or gene therapy to replace nonfunctional genes. Note that the pattern of cell loss or mechanisms of cell death need not differ between these general forms of ARHL. All ARHL is, of course, worth understanding and treating, yet the perceived seriousness of the problem will be greater if the rate of ARHL greatly outpaces overall biological age and singularly impinges on quality of life.

3.1.1 Sensory ARHL

As defined for humans, sensory ARHL refers to degeneration of the organ of Corti that extends at least 10 mm from the cochlear base, that is, into the region serving speech reception. Although most studies have emphasized hair cell loss, Schuknecht included other types of changes in the organ of Corti. Pathology principally of the basal organ of Corti would result in an audiogram that is abnormal only at the highest frequencies. Schuknecht felt that sensory ARHL was the least common form, with an incidence of less than 10% of all cases, although this seems likely to be a considerable underestimate. As hair cells degenerate, secondary neuronal degeneration eventually follows, so that some have recommended the term "sensorineural" ARHL. Among primates at least (including humans), secondary neuronal degeneration may be delayed for years after loss of hair cell targets (with favorable implications for cochlear implants). Sensory ARHL is particularly difficult to distinguish from injury due to noise or ototoxins, as these also often mainly affect the basal organ of Corti. Accordingly, it may be the form most closely associated with injury, and may in fact be mechanistically and anatomically indistinguishable from certain forms of noise or ototoxic injury.

3.1.2 Neural ARHL

Neural ARHL refers to loss of radial afferent neurons that project to inner hair cells (IHCs). Schuknecht estimated the incidence of neural ARHL at 15% to 30% of all cases. As shown in Fig. 6.4, loss of afferent neurons progresses slowly with age in most people, so that clinically significant neural ARHL means unusually rapid loss that becomes limiting for hearing, and is not secondary to loss of IHCs. In Schuknecht's sample, neuronal loss typically appeared evenly distributed along the cochlear spiral, although some studies have indicated greater loss in the extreme cochlear base and apex (Felder and Schrott-Fischer 1995). The severity criterion for this classification is taken to be a total loss of 50% or more of nerve fibers, based on the extent that may lead to impaired word discrimination. Losses up to 50% may produce few clinical signs (Pauler et al. 1986). Even

FIGURE 6.4. (continued) from Bredberg 1968.) (**B**)Total spiral ganglion cells versus age. (Adapted from Otte et al. 1978.) (**C**) Density of fibrocytes in the cochlear spiral ligament versus age. Subjects had no known hearing impairment. (Adapted from Wright 1972.)

more remarkably, significant changes in the audiogram may not occur before nearly 90% of neurons are lost. The fact that speech discrimination and threshold sensitivity are very different in their imperviousness to neuronal loss suggests that redundancy of neurons is far more important for perception of complex stimuli than for simple detection. It is highly fortunate for the operation of cochlear implants that loss of even half of all neurons may have modest consequences.

The existence of neural ARHL is controversial. Some (e.g., Starr et al. 2001) have suggested it merely represents delayed auditory neuropathy, not an explicitly age-related condition. One complication for its diagnosis and study is heterogeneity of form (e.g., Zimmermann et al. 1995; Felix et al. 2002; Chen et al. 2006). In some cases, the most obvious anomaly is loss of dendritic processes from the osseous spiral lamina, while the density of cell bodies in Rosenthal's canal may appear near normal. In other cases dendrites, cell bodies, and presumably their central projections are lost. Most likely, both observations reflect a "dying back" process, whereby dendrites are lost first, followed by cell bodies and axons. The surprising retention of cell bodies that have lost their distal processes has important implications for the success of cochlear implants, as these cells appear "drivable" by electrical stimulation. It may also have physiological implications, given evidence that cell bodies in the spiral ganglion may become commonly ensheathed in myelin and electrically coupled (Felder et al. 1997). This may still allow neurons that have lost their inner hair cell inputs to be excited acoustically.

Neither noise nor ototoxins typically produce a cochlear state resembling neural ARHL, although evidence from animals (see later) suggests that noise exposure at a young age may cause primary neural loss. One proposed mechanism (Pujol et al. 1991, 1993) assigns a significant causal role to excitotoxicity, a process whereby excessive glutamate release from IHCs promotes injury to afferent dendrites. Although excitotoxic injury after noise exposure and pharmacologic manipulations appears reversible, prolonged excitotoxic stress could lead to permanent loss of dendrites, and eventually entire neurons. Some observations indicate that some neuronal loss may follow subtle pathology of pillar cells and other supporting cells within the organ of Corti (Suzuka and Schuknecht 1988). Thus, some loss that would be superficially considered "primary" (since the hair cell targets are still present), may actually be secondary to subtle degeneration in the organ of Corti. Because there is no physiological test that is specific for IHCs and no assay for their trophic influence on neurons, apparent neural ARHL may often have its origin within the inner hair cell.

3.1.3 Strial ARHL

Strial ARHL denotes hearing loss caused by degeneration of the stria vascularis, usually in the mid-cochlear to apical regions, with presumed reduction in the endocochlear potential (EP) that is taken to be the proximate cause of hearing loss. The stria-generated EP provides a significant portion of the driving force for hair cell receptor currents, and reduction of this driving force would be expected to cause elevated hearing thresholds. The hearing loss may be "flat" at lower frequencies, with sloping threshold elevation at higher frequencies. Because differences in the EP from base to apex are typically modest, larger sensitivity losses at high frequencies probably reflect greater dependence of the cochlear base on active mechanical processes that are "fueled" by the EP. Prevalence of strial ARHL was estimated by Schuknecht at 20% to 35% of cases.

The cochlea appears surprisingly tolerant to degenerative changes in the stria. Up to 30% to 40% of this structure may degenerate all along the cochlear spiral before changes in the audiogram occur (Pauler et al. 1988). Schuknecht noted that strial ARHL tends to occur earlier in life than other forms, and shows a stronger familial component. This has received support from inheritance studies Gates et al. (1999). Studies by Jerger et al. (1993), showing flat hearing loss at low frequencies and differences by gender (Figs. 6.2 and 6.3), have led to the interpretation by some that strial ARHL disproportionately affects women, and may reflect gender differences in microvascular pathology (Gates and Mills 2005). A causal role for microvascular disease in strial ARHL has been proposed (Hawkins et al. 1972; Johnsson and Hawkins 1972), but was not supported by Schuknecht, based on his own observations. The most subtle pathology he noted more often involved marginal cells; moreover, strial capillary anomalies he observed did not appear associated with abnormalities of strial cells. No causal link between strial degeneration and microvascular disease has been demonstrated in humans, nor have any genes that predispose individuals to strial ARHL been identified. Noise and ototoxins can cause strial injury (e.g., Ulehlova, 1983; Garetz and Schacht 1996; Hirose and Liberman 2003), but they rarely promote permanent EP reduction or produce strial pathology resembling effects of aging. Although human and animal studies have tied strial pathology to that of adjacent spiral ligament, Schuknecht saw little connection, and did not include spiral ligament in the hallmarks of strial ARHL.

3.1.4 Cochlear Conductive ARHL

Cochlear conductive ARHL remains unproven. This classification was created to cover cases showing linearly descending audiograms (greater than 50 dB decline overall) that appeared unexplained by obvious degeneration of any cochlear cells or structures (about 15% to 22% of cases). Schuknecht suggested that in some individuals subtle changes in the passive mechanical properties of the organ of Corti, basilar membrane, or adjacent spiral ligament may interfere with transduction, even without notable cell loss. Although this form of ARHL seems plausible, some have questioned its validity, pointing out that detailed cellular/molecular analysis would probably reveal pathology to sensory cells (Gates and Mills 2005).

3.1.5 Mixed ARHL

Mixed ARHL was taken by Schuknecht to be present when multiple forms of degeneration were found, each of which seemed likely to contribute to hearing loss. The shape of the audiogram was suggested to be consistent with the sum of their effects. Mixed ARHL might arise by coincidence, because a single process common to multiple cell types is impaired, or because of the interdependency for survival among cell types. Schuknecht classified about 25% of cases as mixed, and guessed that the pathologies were often linked.

3.2 The Status of Schuknecht's Framework

Although Schuknecht was not the first to propose all the above categories, his synthesis remains today the most comprehensive. The framework has many limitations. First, its formulation was subject to problems often posed by human temporal bones. These are best interpreted in light of life history, yet health records often paint an incomplete picture of recreational and occupational noise, and ototoxin exposure. Frequently mediocre preservation of samples forced an emphasis on cell numbers rather than cell appearance for evaluation and classification. The framework also has an ad hoc character, perhaps attempting to "shoehorn" cases into too few categories. Up to 25% of cases were considered indeterminate, showing no apparent relationship between histopathology and the appearance of the audiogram. Many have expressed doubt regarding the diagnostic value of the shape of the audiogram (e.g., Chisolm et al. 2003). Even if pathologies of different structures or cells independently contribute to the audiogram, many possible combinations may yield a particular shape. The framework is also incomplete. Details that did not seem to fit anywhere included occasional atrophy of Reissner's membrane, degeneration of spiral limbus, and variable patterns in the loss of fibrocytes from the spiral ligament.

Despite its deficiencies, the core assertion of Schuknecht's framework—independent pathology of organ of Corti, afferent neurons, and stria vascularis—has appealing clarifying power and testability. If each of these structures/cell types possesses distinct environmental and genetic risk factors for age-related degeneration, then identifying these risk factors is an important research goal. Although Schuknecht's scheme has been criticized, it has not been replaced or greatly refined. Observations in humans and animals have largely provided general support, yet few studies have been directed at testing its foundations. The alternative is a murkier and unproductive notion of ARHL, whereby multiple pathologies with multiple causes invariably coincide, and may be inseparable.

4. Central Auditory Aspects of Human ARHL

Relative to the burgeoning reports on the neural and molecular bases of presbycusis from animal models, little is known about anatomical and structural changes in the human ACNS. In considering age-related pathological changes taking place in the brain, two broad etiologies present themselves (Frisina et al. 2001). Some changes, particularly those at the level of the cochlear nucleus, for example, are driven by the declines in peripheral cochlear inputs that occur with age, typically starting with the high frequencies (Frisina and Walton 2006). These cochlear-driven changes with age are sometimes referred to as "peripherally induced central effects." In contrast, some anatomical or functional changes in the ACNS of old humans and animals appear to occur somewhat independently of the periphery, and may reflect "true aging" neurodegenerative changes in the brain. These may share some similarities or common mechanisms with other central nervous system conditions of the aged such as Alzheimer's and Parkinson's diseases.

4.1 Structural Changes in the Central Auditory System

Several noteworthy studies of structural changes with age have been performed, and what is known is generally consistent with animal model results presented below. Konigsmark and Murphy (1970, 1972) found age-dependent declines in the volume of the human ventral cochlear nucleus (VCN), while Seldon and Clark (1991) observed VCN neuron size reductions. In neither case was a change in the number of VCN neurons with age noted, but increased lipofuscin deposits and declines in capillary density were found.

Higher in the human brain stem at the level of the lateral lemniscus, which is the major ascending input tract for the inferior colliculus (IC), Ferraro and Minckler (1977) examined the anatomy of 15 brains, from persons ranging in age from birth to 97 years. They reported a significant decline in the number of lemniscal nerve fibers with age. In one of the first neurochemical studies of the ACNS, foreshadowing later animal work presented later in this chapter, McGeer and McGeer (1975) reported that the postmortem presence of glutamic acid decarboxylase (GAD), the primary synthetic enzyme for the main inhibitory neurotransmitter of the inferior colliculus, γ-aminobutyric acid (GABA), decreased with age. At the level of the auditory cortex, Brody (1955) reported a striking negative correlation ($r = -0.99$) for age and the number of neurons in the human superior temporal gyrus. This reduction was much greater than for neighboring cortical regions such as the inferior temporal cortex, striate (visual) cortex, and pre- and postcentral gyri (sensorimotor).

4.2 Functional Declines with Age

One tactic for teasing out age-related brain-specific changes from peripherally induced central effects is to perform behavioral or physiological experiments that assess some aspect of central auditory processing in human or animal subjects that have relatively good hearing. For instance, Frisina and Frisina performed the SPIN test (Speech Perception In Noise) on young adult and old human subjects with good peripheral hearing to assess speech perception problems in background noise that might have central auditory brainstem or cortical components (Frisina and Frisina 1997). Old subjects having audiograms in the normal range nevertheless performed worse on the SPIN test, requiring a stronger speech signal and higher signal-to-noise ratio for a particular speech recognition criterion. SPIN tests performed while simultaneously assessing brain activity in a positron emission technology (PET) scanner (Frisina 2001; Salvi et al. 2002) revealed changes in brain activity of old subjects with normal audiograms and 50% correct speech recognition performance level. Specifically, the old subjects had less brain activity in the auditory midbrain/thalamus regions, and in some of the auditory/visual processing areas.

Event-related potentials have also been utilized to assess central auditory functional changes with age. For example, the amplitude and latency of the P3, or P300, an event-related potential that marks an updating of our current sensory environment in working memory, has been employed as a probe to the aging brain. Polich's lab found that the auditory P3 amplitude declines, and the latency increases with age (Polich et al. 1985). Frisina, Walton and co-workers confirmed this basic finding, and extended it to musical stimuli (Swartz et al. 1994). Interestingly, when early Alzheimer's patients were compared to young adult and old controls on these musical P3 auditory tasks, even Alzheimer's patients who could not make an overt behavioral response to the musical stimuli displayed significant processing of music as indexed by the P3 (Swartz et al. 1992).

It has been known for some time from psychoacoustic experiments on young adults that there is a link between auditory temporal processing and speech perception capabilities. Consistent with this, Fitzgibbons and Gordon-Salant (1996) and Snell and Frisina (2000; Snell et al. 2002) discovered an age-related decline in temporal processing related to speech perception. It is interesting to note that even subjects with good peripheral hearing, i.e., audiograms in the normal range, experience problems with gap detection and speech recognition in background noise as they proceed through middle age into old age.

Jerger and his colleagues pioneered investigation of central auditory processing problems of presbycusis while taking into account cognitive factors (for a review see Martin and Jerger 2005). Interactions between cognitive slowing and auditory processing deficits of aging are still controversial given the difficulty in equating for age-related hearing loss in assessing language processing abilities in the elderly (Pichora-Fuller 2003). Human neurophysiological studies shed some light on auditory temporal dysfunction at cortical levels. For example, Pekkonen (2000) utilized the mismatch negativity to demonstrate that aged subjects exhibited deficits in processing sound duration, but not frequency, in the auditory cortex.

5. Peripheral Aspects of ARHL in Animal Models

Age-related cochlear pathology has been examined in aging chinchillas, guinea pigs, primates, dogs, and rodents (e.g., Covell and Rogers 1957; Keithley and Feldman 1979; Bohne et al. 1990; Tarnowski et al. 1991; Shimada et al. 1998). These models show the range of cochlear pathologies noted for humans (organ of Corti, neural, strial), but they have not typically been analyzed in terms of whether they would meet human criteria for sensory, neural, or strial ARHL.

Whether the relationship between measures such as neural loss and threshold elevation, or between percentage of strial impairment and EP reduction, is the same as for humans has not been examined in all models, but has received support (Schulte and Schmiedt 1992). Particularly valuable—albeit rare—animal models are those that exhibit relatively isolated forms of ARHL, allowing these to be studied with minimal confounds. Currently the best example is the gerbil, which compellingly models strial ARHL (Schulte and Schmiedt 1992; Gratton and Schulte 1995; Gratton et al. 1996, 1997; Schmiedt et al. 2002; Spicer and Schulte 2002, 2005). For the majority of animal models showing a complex mix of cochlear pathologies, there is unfortunately no way to separate these (or at least see if they can be separated). Such is possible only if there exist multiple varieties of highly inbred subpopulations of a species, allowing comparison of their aging characteristics and the possibility of segregating traits in genetic crosses. So is revealed the value of mouse models. One can raise an "old" mouse in 2 years and examine multiple genetic variants. As will be expanded upon, there are many strains of mice that show ARHL. Most are not well characterized, and only some of these will ultimately be found to be useful models of human ARHL. Like other animals, the mouse ARHL models show a mix of pathologies, and attempts to separate these are not well advanced. Nevertheless, mouse models have emerged that mirror the defining characteristics of human ARHL.

5.1 Sensory ARHL in Animals

Nearly all animal models characterized to date most resemble sensory ARHL, and studies of these have used hair cell loss as their primary metric. The best characterized mouse ARHL models, including C57BL/6 (B6), BALB/c (BALB), CD-1, 129S6/SvEv, and SAMP-1, show degeneration of the organ of Corti, and also variably include some degeneration of afferent neurons, stria vascularis, and spiral ligament (Mikaelian et al. 1974; Henry and Chole 1980; Saitoh et al. 1995; Willott et al. 1998; Hequembourg and Liberman 2001; Wu et al. 2001; Ohlemiller 2002; Ohlemiller and Gagnon 2004b). For ages up to which hearing loss is pronounced, the EP appears normal in these models, and changes in the organ of Corti can account for most hearing loss. A rapidly expanding collection of genes, collectively termed *ahl* genes (e.g., Johnson et al. 2006), have been identified that account for the hearing loss in some of these strains. The *Cdh23^{ahl}* allele, which is common to B6, BALB, and several other strains (Johnson et al. 2000), is the most intensively studied and most used to extract principles of cochlear aging. This locus codes for cadherin 23 (or otocadherin), believed to be a component of stereocilia (Siemens et al. 2004). Most strains that carry this allele show a mix of pathologies that bear no obvious relation to stereocilia function (e.g., Figs. 6.5 and 6.6), and understanding whether and how these are related may be important for how mixed ARHL occurs. The contribution of *Cdh23^{ahl}* to age-related pathology in B6 mice can be isolated by examination of the congenic B6.CAST-*Cdh23^{CAST}* line. These mice show organ of Corti degeneration in the cochlear base and high frequency hearing loss

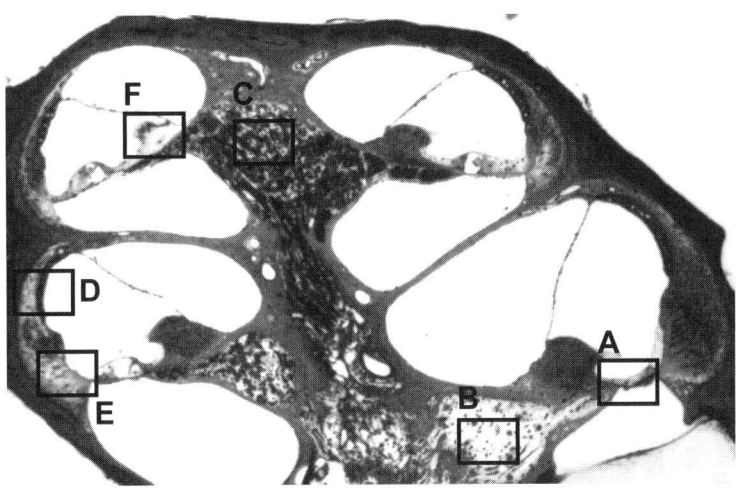

FIGURE 6.5. Mid-modiolar section from an 18-month-old female C57BL/6 mouse cochlea illustrating many types of cochlear changes found in aging. Cochlear pathology includes hair cell loss, organ of Corti anomalies, neuronal loss (probably both primary and secondary), strial degeneration, and loss of fibrocytes in spiral ligament and limbus. The animal had almost no hearing, but a normal endocochlear potential (110 mV). C57BL/6 mice carry an allele ($Cdh23^{ahl}$) that promotes sensory ARHL-like pathology. Other pathology may be related to additional unknown genes.

beginning after 1 year of age (Keithley et al. 2004), presumably reflecting the presence of additional alleles that promote ARHL. As might be expected from its gene product, $Cdh23^{ahl}$ promotes cochlear pathology that appears most similar to sensory ARHL. Notably, this allele also promotes NIHL (Erway et al. 1996; Davis et al. 2001), suggesting a connection between noise injury and this ARHL form. Although the role of otocadherin is not known, it interacts with known components such as the plasma membrane Ca^{2+}-ATPase, and thus may impact processes that regulate hair bundle integrity (Davis et al. 2003). There are many processes that, if impaired, could render the organ of Corti more vulnerable to injury, so that a general link between injury and sensory ARHL seems plausible. At present, at least six loci with alleles known to promote sensory ARHL-like pathology also promote NIHL (Ohlemiller 2006). A cautionary note is in order here in that the $Cdh23^{ahl}$ allele might be a confounding factor, i.e., it is a modified gene, and these mice may not reflect "true" age-related hearing loss in most human clinical cases.

5.2 Neural ARHL in Animals

Although primary loss of afferent neurons is frequently observed in animal models, whether such observations usefully model neural ARHL has remained unclear. At least three quite different "knockout" mutations (KOs, engineered

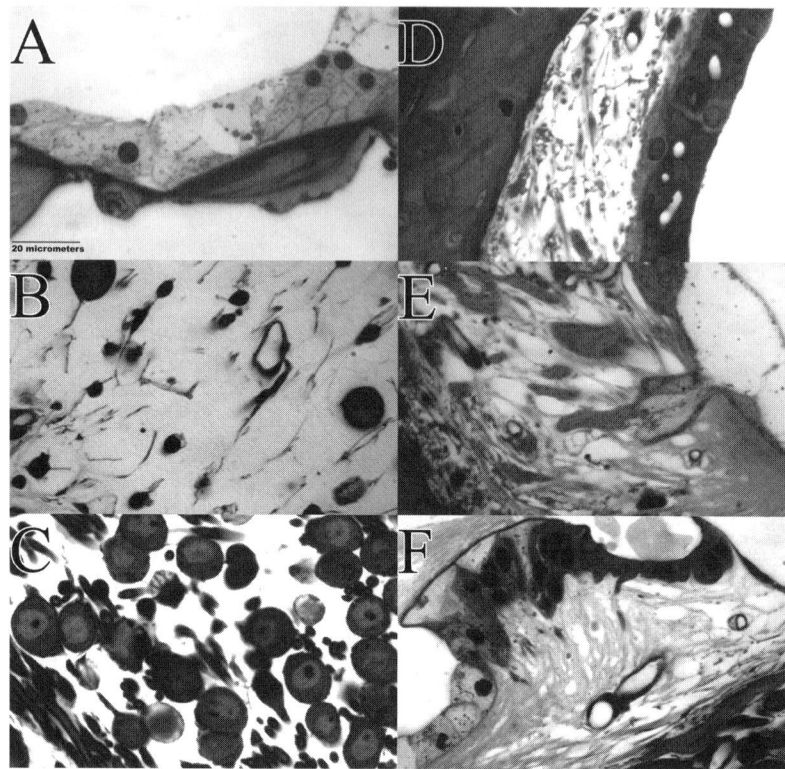

FIGURE 6.6. Enlarged view of boxed regions from Fig. 6.5. (**A**) Loss of hair cells and most recognizable cell types in organ of Corti of the lower basal turn. (**B**) Loss of spiral ganglion cells and their projections to hair cells in the lower base. Much of this loss may be secondary to hair cell loss. (**C**) Apparent primary loss of spiral ganglion cells in the upper apical turn. (**D**) Loss of type I fibrocytes in spiral ligament behind stria vascularis in the upper basal turn, along with thinning and disorganization of the stria. Overall strial degeneration is modest and may not be indicative of the kinds of cellular changes that occur in strial ARHL (see text). (**E**) Loss of type II and IV fibrocytes in the lower spiral ligament of the upper basal turn. (**F**) Loss of fibrocytes from the spiral limbus of the upper apical turn.

inactivating mutations of specific genes to test their role) promote accelerated neuronal loss, and thus point to specific genes and pathways as causes of this condition. The first is a knockout of the gene locus encoding Cu/Zn-superoxide dismutase (SOD1), a key antioxidant enzyme (Keithley et al. 2005). This suggests a role for oxidative stress and possible gene–environment interactions in neuronal survival. The second is a KO of the β_2 subunit of the nicotinic acetylcholine receptor (Bao et al. 2005), suggesting a trophic influence of lateral efferent neurons, which form synapses with afferent dendrites. In the third KO model, nuclear factor κB (NF-κB), a stress-activated transcription factor, was inactivated by elimination of the p50 subunit (Lang et al. 2006a). Transcription factors trigger the "reading" of whole families of genes related to particular functions. NF-κB is activated by stress-related increases in intracellular calcium, and may be important for preventing excitotoxic injury to afferent neural dendrites. Compared to wild-type controls, NF-κB knockout mice exhibit greatly increased neuronal dendrite and perikaryal loss with age, plus increased signs of excito-toxic injury. They may also be more vulnerable to NIHL. These mice support the proposal that some neural ARHL results from interplay between the environment and genes that regulate afferent synaptic function or glutamate transport. In a potentially related finding, Kujawa and Liberman (2006) reported in mice an interaction between the age at which noise exposure occurs and apparent primary age-related neuronal loss. (See Borg 1983 for potentially related findings in rats.) When CBA/CaJ mice were exposed to noise at 1.5 months of age (young but sexually mature), the resulting hearing loss increased with age along with neuronal loss, despite a stable inner hair cell population. Taken together, findings in NF-κB knockout mice and young noise-exposed mice raise the possibility that early noise exposure alters the developmental program for afferent synaptic function or calcium homeostasis.

The presence of IHCs does not ensure that they function properly, or exert necessary trophic influences on afferent dendrites. Apparent primary neural loss could result from abnormalities of hair cells, neurons, efferent innervation, or some more global factor such as perilymph oxygenation, pH, or ion content. Aging B6 mice show widespread neuronal loss that has been generally assumed to be secondary to loss of IHCs due to $Cdh23^{ahl}$. However, quantitative ultra-structural studies have shown that the withdrawal of dendrites from inner hair cells in B6 mice precedes any obvious anomalies of the hair cells, so that a general process not involving $Cdh23^{ahl}$ may be at work (Stamataki et al. 2006). Extension of this approach to other strains or the B6.CAST-$Cdh23^{CAST}$ line may help place this finding into a broader context. Schuknecht suggested that some apparent primary cochlear neural loss was actually secondary to subtle pathology of IHCs or supporting cells. Ohlemiller and Gagnon (2004a) demonstrated that primary loss of afferent neurons in the cochlear apex of several mouse strains (B6, BALB, CBA/J, and 129S6) was correlated with degenerative changes in pillar cells and (puzzlingly) Reissner's membrane. They proposed that a common factor local to the apex underlies these changes, and that the causes of primary neuronal loss may depend on basal–apical location.

5.3 Strial ARHL in Animals

Age-related strial degeneration has been described in chinchillas, guinea pigs, and rodents. Common strial changes include thinning of the strial epithelium and capillary loss and occlusion. Because EP decline is taken to be the most significant result of strial degeneration, it is important to understand which cellular changes and what degree of strial involvement indicate that the EP is likely to be reduced. Only in mice and Mongolian gerbils have anatomical changes and EP recordings been measured as a function of age. The gerbil appears to model strial ARHL in several respects. Age-related strial degeneration and EP reduction may drive most of the observed hearing loss in these animals (Schulte and Schmiedt 1992; Gratton and Schulte 1995; Gratton et al. 1997; Schmiedt et al. 2002; Spicer and Schulte 2002, 2005). The strial pathology is manifested as strial thinning and degeneration, followed by ligament degeneration. While the pathology in gerbils was initially interpreted as having a microvascular origin, more recent observations attribute the initial pathology to strial marginal cells. With time, other cells of stria, then spiral ligament, become affected.

Because marginal cells house ion transport systems that play a central role in EP generation, a marginal cell origin for strial ARHL makes sense from the standpoint of metabolic rate and oxidative stress. But what might be the basis of the suggested heritability of strial ARHL in humans? Because gerbils cannot be drawn from many different inbred strains, they do not readily facilitate genetic analysis. Use of mice could solve this problem, but application of mouse models to strial ARHL has been slowed by the near absence of mouse models that show significant late-onset EP reduction. B6 mice show many age-related features of the cochlear lateral wall that have been associated with strial ARHL, such as strial thinning, ligament thinning, loss of capillaries, and loss of fibrocytes (Figs. 6.5 and 6.6) (Ichimiya et al. 2000; Di Girolamo et al. 2001; Hequembourg and Liberman 2001). Yet these mice lack the key hallmark of this condition, showing little or no EP decline by 2 years of age (Lang et al. 2002). Cable et al. (1993) showed that about half of mice carrying the $Tyrp1^{B-lt}$ allele, which affects melanocyte function, undergo EP reduction by 22 months. No clear morphological correlate was identified. A mouse lupus model, MRL-Fas^{lpr}, shares characteristics with strial ARHL that merit mention. These mice hear as well as controls in their first months, but thereafter show progressive threshold elevation, strial degeneration, and EP reduction (Ruckenstein et al. 1999a,b). Strial pathology in MRL-Fas^{lpr} initially impacts strial vasculature and neighboring intermediate cells. As claimed for human strial ARHL, the pathology is more pronounced in females than in males (Trune and Kempton 2002). In general, autoimmune disease impacts principally females, so that some strial ARHL may possess an autoimmune component. The MRL-Fas^{lpr} model may be found to have implications for how some human strial ARHL arises.

Ohlemiller (2006) found moderate EP reduction beginning at 19 months in BALBs, even though the stria and adjacent ligament showed only modest changes. The contrast between BALB mice and B6 mice (which undergo more striking changes in the spiral ligament than do BALBs) presented an opportunity to isolate the major contributors to age-related EP decline in BALBs. In each strain, EP was compared with a host of factors previously associated with strial ARHL including fibrocyte density in spiral ligament, strial cell density (basal, intermediate, and marginal cells), strial thickness, ligament thickness, plus strial capillary density, diameter, and basement membrane thickness. Among all measures, only marginal cell density and ligament thickness were correlated with the EP in BALBs. B6 mice showed little age-related loss of marginal cells and little reduction in ligament thickness, even though there was significant fibrocyte loss. Neither strain revealed any predictive value of changes in strial microvasculature. Observations in BALB mice and gerbils therefore support a marginal cell origin for some strial ARHL, specifically the form described in human temporal bones. A mouse strial ARHL model may promote identification of the underlying genes. Consideration of alleles common to B6s and BALBs, plus additional strain comparisons in the Ohlemiller et al. study permitted elimination of the $Cdh23^{ahl}$ locus and loci involved in melanin synthesis as a genetic basis for the findings in BALBs. Both BALB mice and gerbils share an important feature in that only about half of animals show EP decline. Because inbred mice are essentially genetically identical and gerbils used for research are highly inbred, this suggests substantial environmental modulation of the genetic tendency toward age-related EP decline.

Accumulating evidence from animals undermines an obligate role for microvascular disease in strial ARHL. Studies using the mitotic tracer bromodeoxyuridine (BrdU, commonly used to detect new cell division) indicate that some cells in spiral ligament and stria vascularis are replaced over time (Conlee et al. 1994; Yamashita et al. 1999; Lang et al. 2003; Hirose et al. 2005). Net loss of marginal cells in BALB stria could therefore reflect either an abnormally high rate of cell death, or impaired replacement.

Findings in mice have also helped clarify the relationship between age-related degeneration of stria vascularis and spiral ligament. Schuknecht and colleagues (Wright and Schuknecht 1972; Schuknecht and Gacek 1993) treated these structures as independent, and did not include ligament pathology in the hallmarks of strial ARHL. Spicer and Schulte (2002) proposed a sequence in gerbil whereby strial pathology spreads to ligament. Certainly, the potential exists for dependence of ligament on the stria. In keeping with the notion of K^+ recycling from the organ of Corti, through ligament, and back to the stria (Wangemann 2002; see also Wangemann, Chapter 3), strial dysfunction could promote toxic K^+ accumulation in ligament. This relationship may not work in reverse, however. Mice carrying a single functional copy of $Brn4$, an X-linked transcription factor, and homolog of human $DFN3$ (Minowa et al. 1999; Xiu et al. 2002) develop pathology of spiral ligament and Reissner's membrane, and EP reduction. Similarly, mice deficient in otospiralin, a protein normally present in fibrocytes of the ligament, show ligament pathology (Delprat et al. 2005). Notably, neither of these models shows strial degeneration, supporting relative insulation of the stria from moderate ligament pathology.

6. Central Auditory Aspects of ARHL in Animal Models

In the above consideration of age-related changes in the human ACNS, a distinction was made—insofar as possible—between those intrinsic to the ACNS and those that result from peripheral degeneration. Animal models permit better separation of these, especially mice and rats, wherein strains with differing degrees of age-related peripheral pathology can be compared.

6.1 Cochlear Nucleus

6.1.1 Structural and Neurochemical Alterations in Cochlear Nucleus

Studies led by Willott were among the first to exploit differences in age-related cochlear pathology to isolate peripheral influences on the aging ACNS (Willott 1991). Unlike B6 and some of the other strains introduced in the preceding text, CBA mice (both CBA/J and CBA/CaJ) lose hearing sensitivity slowly, at a rate comparable to the slowest progression rates of human ARHL, normalizing for lifespan. Willott's group examined the neuroanatomical aspects of these peripherally induced central effects. In B6 mice, neuron size, number, and packing density decline in the VCN, in concert with the loss of high-frequency inputs from the cochlea. Changes of this nature rarely occur in very old CBAs (Lambert and Schwartz 1982; Willott et al. 1987). Neurons of the VCN in aging B6 mice also show an increase in lipofuscin deposits, nucleoplasm pathologies, and nuclear invaginations (Briner 1989). These were most pronounced in high-frequency (dorsal) regions of the VCN, and more in multipolar cells than in bushy cells. In the dorsal cochlear nucleus (DCN), only layer III, which receives direct inputs from the auditory nerve, showed significant aging declines in B6, while the CBA mice DCN was relatively stable (Willott et al. 1992).

The DBA mouse strain shows even faster age-related peripheral hearing loss than B6, presumably due to a greater number of progressive deafness alleles (Erway et al. 1993). As one might predict, neuronal declines in the anteroventral cochlear nucleus (AVCN) occur faster in DBA than in B6 (Willott and Bross 1996). In a light microscopic investigation of octopus cell region of the VCN in CBAs and B6s, Willott and Bross (1990) observed significant age-related declines in octopus cell number, number of primary dendrites, and octopus cell volume. Increases in glial cell packing density were also noted.

Age-related neuronal changes do not always take the form of degradation. In a neuroanatomical investigation of the VCN in aging rats, Keithley and Croskrey (1990) observed that axonal terminations may become larger and more complex in nature, perhaps in an attempt by the system to compensate for the decline in neuron numbers with age.

In an ultrastructural examination of age-related changes in the AVCN of Fischer-344 rats (which show good preservation of hearing with age), Helfert and coworkers (2003) observed that the synaptic terminal specializations of distal dendrites of both excitatory and inhibitory neurons (likely glycinergic) decreased in both size and length. Age-related declines in glycine receptors therefore appear associated with reduced size of synapses. In contrast, the number of dendrites and density of synapse decline with age in the IC, but not in the AVCN. In experiments involving expression of α- and β-glycine receptor subunits in the AVCN of young adult and old rats, Helfert's group uncovered age-related changes in the subunit gene expression. The α_1 and β subunits declined with age, while α_2 increased, thus altering the overall glycine receptor functionality (Krenning et al. 1998).

6.1.2 Physiological Alterations in Cochlear Nucleus

For animals with fairly good hearing late into life such as the CBA mouse and Fischer-344 rat, neural coding of sounds does not appear to change drastically with age. By contrast, B6 mice show age-dependent tonotopic map plasticity in the central nucleus of the IC and auditory cortex. Willott et al. (1991) compared aging physiological responses of neurons in the cochlear nucleus for B6s and CBAs. The slow, progressive hearing loss in CBAs was associated with minimal change in ventral cochlear nucleus thresholds for simple sounds such as pure tones. In contrast, AVCN neuron responses in B6s showed major changes in regions representing high frequencies (peripherally induced central effect) due to the loss of inputs from the cochlear base with age. In B6 mice, DCN cells showed much less drastic changes with age, in response properties such as tuning curves and response areas, most likely due to the significantly higher proportion of their inputs that come from noncochlear sources.

Age-dependent declines in cochlear outputs can also drive aging changes at the synaptic level in the AVCN. Using brain stem slice electrophysiology, Wang and Manis (2005) measured pre- and postsynaptic potentials for the AVCN end-bulb synapses in young adult and aging DBA and CBA mice. Both pre- and postsynaptic mechanisms were altered in aging DBAs showing severe peripheral hearing loss. Presynaptic changes included reduced transmitter release probability. Postsynaptic deficits included declines in mEPSC frequency, speed and amplitude (Fig. 6.7). Apparent synaptic abnormalities were not observed in young adult mice of either strain, or in old age CBAs with relatively good hearing.

Milbrandt and Caspary (1995) performed biochemical investigations of the glycine inhibitory system in the cochlear nucleus of Fischer-344 rats. They found evidence of age-related impairment of this inhibitory system, including reductions in binding properties of glycine receptors in both the AVCN and the DCN. The posteroventral cochlear nucleus (PVCN) did not manifest age declines, as the levels of glycine receptors in young adult rats were already low. Willott et al. (1997) performed similar studies in mice, utilizing immunocyto-chemical and biochemical techniques, and also found reductions in glycine-based inhibitory synaptic transmission in the DCN. For example, in 18 month old B6s with severe high-frequency hearing loss the number of glycine immunoreactive neurons and strychnine-sensitive glycine receptors declined significantly relative to younger B6s and old CBAs with good hearing. Caspary et al. (2005) delineated the functional effects of age-related reductions in the glycinergic inhibitory system in rats. Responses attributed to fusiform principal neurons were altered, such that rate-intensity functions grew at a faster rate in old rats. This finding is consistent with an age-related deficit in vertical cell on-best frequency inhibitory inputs to fusiform cells mediated by glycine.

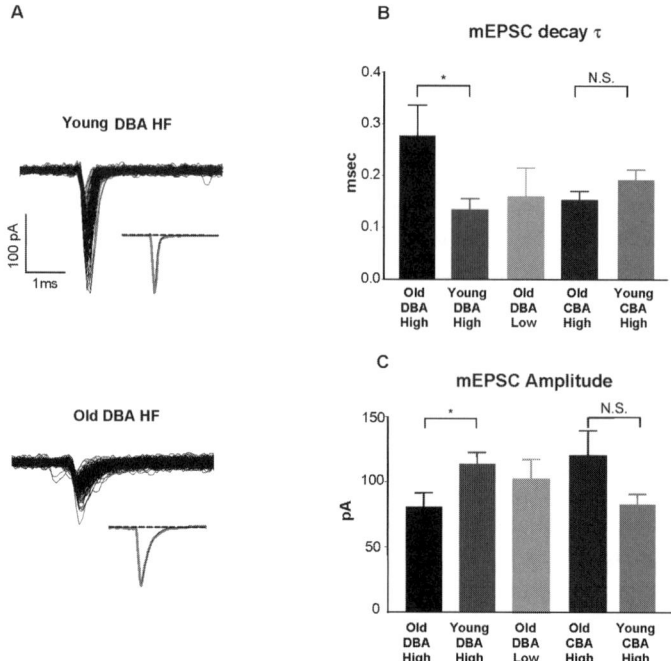

FIGURE 6.7. Spontaneous mEPSCs had slower decay time constant in hearing-impaired DBA bushy cells. (**A**) sample mEPSCs recorded from two bushy cells in HF regions from a young and an old DBA mouse. All detected mEPSCs were aligned to their onset. Insets: normalized average of the mEPSCs superimposed with the first-order exponential decay *(dark trace)*. (**B**) decay time constants for young and old bushy cells in HF regions of DBA mice were significantly different. Decay time constants for all cells from normal hearing regions of the AVCN were comparable between old DBA and young CBA HF as well as old CBA HF cells. (**C**) spontaneous mEPSC amplitude was significantly different between hearing-impaired old DBA mice and young DBA mice. There were no statistical differences between the normal hearing low-frequency old DBA and the high-frequency young DBA cells, nor between the old CBA HF and the young CBA HF cells. HF = high frequency. (From Wang and Manis [2005], Fig. 6. Reprinted with permission.)

6.2 Inferior Colliculus

6.2.1 Structural and Neurochemical Alterations in Inferior Colliculus

Anatomical and neurochemical changes with age that have functional implications also occur at the level of the auditory midbrain, the inferior colliculus. Again utilizing the B6 and CBA strains, Willott et al. (1991) discovered that a major reorganization of the tonotopic map occurs in the IC of aging B6 mice. Regions expected to contain neurons tuned to high frequencies instead showed preferences for low frequencies, suggesting "rewiring" of these neurons to receive inputs from neurons in low frequency regions, where outputs from the cochlear apex were still available. The IC in CBA mice showed no evidence of such dramatic age-related plasticity of neuronal responses.

In a series of biochemical and neuroanatomical investigations, Caspary et al. (1990) investigated age-related changes in the IC inhibitory system using the Fischer rat. The primary inhibitory neurotransmitter here that shapes complex sound responses in the ACNS is GABA. Using an antibody for GABA, it was found that the number of immunolabeled neurons in the ventrolateral central nucleus of the IC (high-frequency region) decreased with age. Basal levels and potassium ion-evoked effluxes of GABA from preparations of the central nucleus of the IC also diminished with age. By contrast, release of glutamate and aspartate (the main excitatory neurotransmitters in the IC), acetylcholine (another inhibitory transmitter), and the amino acid tyrosine, were stable with age.

Milbrandt et al. (1994) uncovered an age-related deficit in $GABA_B$ receptor binding using quantitative receptor binding assays. In this case, reductions were noted in the central nucleus, dorsal cortex and external nucleus, whereas nearby cerebellar tissue showed no such age-related changes and the rat IC volume was age stable. Perhaps as a compensation mechanism, $GABA_A$ receptors in the IC showed an upregulation with age (Milbrandt et al. 1996).

Utilizing quantitative immunogold electron microscopy procedures similar to their cochlear nucleus synaptic structure studies, Helfert's group explored the synaptic age-related changes in the IC (Helfert et al. 1999). Unlike the cochlear nucleus, the number of synaptic specializations for both excitatory and inhibitory terminals in the rat IC declined with age. These decreases were correlated with declines in dendritic size, and with synaptic density remaining relatively stable on the larger, surviving proximal dendrites. In a related investigation, Milbrandt et al. (1997) examined $GABA_A$ receptor subunit composition in the IC. Evidence was found for compensatory changes that enhanced responses to GABA in the subunits, despite an age-related decline in the number of synapses. In particular, the γ_1 protein subunit increased with age, while the α_1 declined (Caspary et al. 1999). Also noted was an age-linked upregulation of a GABA-mediated chloride influx that is likely a result of the age-related receptor subunit composition change. These enhancements may help compensate for the $GABA_B$ age deficits. A summary of IC changes in the GABA inhibitory system is given in Caspary et al. (1995) and Frisina (2001). It is still not clear how generalizable these findings in the rat are to other mammals, and it is enlightening to note that there are no age-related changes in GABA at the level of the cochlear nucleus (Banay-Schwartz et al. 1889; Raza et al. 1994).

6.2.2 Physiological Alterations in Inferior Colliculus

As discussed in the preceding text, starting in middle age humans typically experience deficits in auditory temporal processing that are manifested in declines in speech understanding. Auditory midbrain neurons of unanesthetized CBA mice show gap encoding properties very similar to auditory gap coding as measured behaviorally using inhibition of acoustic startle paradigms (Walton et al. 1997).

This gap coding at the level of the IC appears to decline with age. Specifically, the number of neurons having short gap thresholds is reduced. In addition, there is a strong tendency for IC single neurons and near-field evoked potentials to have longer gap recovery functions in the old CBAs, especially at moderate sound levels (Allen et al 2003). Tract-tracing studies utilizing horseradish peroxidase (HRP), demonstrated a significant age-related decline in contralateral inputs from all three divisions of the cochlear nucleus to the IC region shown to have the age-related neural temporal processing deficit (Frisina and Walton 2001).

The brain is highly plastic, in that intercellular connections are readily modified by life experiences. Although central plasticity is not generally associated with the birth of new cells, the capacity for forming and eliminating synapses presumably enhances the brain's adaptability, even into advanced age. Higher sensory centers do not become isolated and inactive following peripheral pathology, but instead undergo a shift in the balance of excitatory and inhibitory inputs to become retuned. This retuning leads to the overrepresentation of frequencies with a cochlear "drive" that remains somewhat intact, does not confer any clear advantage, and may induce tinnitus. Nevertheless, it is possible that some accompanying features of this plasticity assist the brain in reducing the impact of cochlear degeneration and the resulting loss of information.

7. Perceptual Effects of Peripheral Auditory Pathology

Age-related changes in the peripheral and ACNS take different forms and exert different effects on auditory perception. In animals as in humans, it is therefore important to distinguish between *direct* effects of aging on the auditory periphery and ACNS respectively, and the effects of peripheral pathology alone on the function of the ACNS. The latter is considered first. The most dramatic coding effects of peripheral pathology are expected to be elevated thresholds, reduced dynamic range (through loss of nonlinear compression) and reduced frequency resolution. Both sensory and strial ARHL would be expected to exert all three effects, through their impact on OHC-mediated active processes. Elevated thresholds will, of course, impair detection. In addition, broadening of tuning and reduced dynamic range will distort the representation of the stimulus spectrum. Sound localization may also be impaired (McFadden and Willott 1994). Neural ARHL presents a different set of predictions. Since the organ of Corti may not be directly affected, threshold sensitivity, dynamic range, and frequency tuning of individual surviving afferent neurons may be normal (depending on whether the IHC/afferent synapse is functioning normally). Central auditory activity associated with detection tasks may be little altered. However, peripheral neural redundancy (many neurons having a broad range of sensitivities and dynamic ranges innervating any given hair cell) may be important for preservation of the stimulus spectrum. Neural ARHL would reduce this useful redundancy, altering representation of the stimulus spectrum, and probably, detection of signals in noise.

8. Perceptual Effects of Central Auditory Pathology

8.1 Temporal Processing and Speech Reception

As introduced in Section 4.2, examination of human subjects with good auditory sensitivity (suggesting a relatively healthy cochlea) is a principal method for teasing out peripheral vs. central etiologies. Using gap detection methodologies, Gordon-Salant and Fitzgibbons (1993) and Schneider et al. (1994) found that aged subjects with reasonably good peripheral sensitivity nevertheless exhibited temporal processing problems. These problems became worse as the temporal processing task became more complex. Subsequent work by Frisina and co-workers implicated temporal processing deficits in speech-in-noise perceptual problems that can start in middle age (Frisina and Frisina 1997; Snell et al. 2002). Using a speech-in-noise perception task, they demonstrated that aged subjects required a higher signal-to-noise ratio for suprathreshold speech perception. When subject groups with different degrees of peripheral hearing loss were compared in terms of temporal processing or speech perception in background noise, it became clear that peripheral loss exacerbated the perceptual deficits in a manner correlated with the degree of hearing loss. In cases where the high frequency portion of the hearing loss exceeded 50–60 dB, the peripheral loss dominated the perceptual temporal- or speech-processing deficit.

8.2 Changes in Auditory Efferent Feedback

Using distortion product otoacoustic emissions (DPOAEs), Frisina and colleagues have shown that efferent feedback from the brain stem to the cochlea declines with age, starting in middle age (Kim et al. 2002). DPOAEs are sounds measured in the ear canal that reflect mechanical activity of outer hair cells. Because normal hearing sensitivity depends on nonlinear mechanical amplification by the OHCs, delivery of a two-tone stimulus (containing frequencies f_1 and f_2, with $f_2 > f_1$) to the normal cochlea will lead to the generation of a recordable complex tone. For diagnostic purposes, it is standard to isolate the cubic distortion product, $2f_1 - f_2$. Frisina's group measured the DPOAE amplitudes in quiet and in the presence of moderate intensity wideband noise presented to the contralateral ear. In healthy cochleae, such contralateral stimuli suppress the level of the recorded DPOAE through a process involving medial olivocochlear (MOC) efferent control of OHC responses. Comparison of DPOAE amplitudes with and without contralateral stimulation thus permits assessment of the strength of MOC feedback. A significant difference in the strength of the MOC effect was noted between young adults and middle-aged subjects at all frequencies tested. Lesser declines were observed between the middle-aged and old subjects.

It is useful, on discovering a clinical decline in humans, to assess whether the same phenomenon exists in animal models. Frisina's group performed parallel experiments assessing MOC function in CBA mice (Jacobson et al. 2003). The mice showed a time course for age-linked MOC deterioration analogous to that in humans. Middle-aged animals showed a significant decline relative to the young adults, and further deficits in the efferent system were evident in old mice. As has been observed for humans (Varghese et al. 2005), wideband noise is more effective as a suppressing stimulus than narrowband signals such as pure tones.

8.3 Right Ear Advantage

Frisina's group also examined the effects of age on the peripheral "right ear advantage" (Tadros et al. 2005a). In most young adult listeners, the right ear shows a lower audiometric threshold and higher amplitude DPOAEs. Tadros et al. compared these measures in "golden ear" old subjects (audiograms in the normal range) to those in subjects with typical, sloping high-frequency hearing loss characteristic of presbycusis. The golden ear subjects tended to have lower thresholds and higher otoacoustic emission amplitudes in the right ear, whereas this situation was reversed in the presbycusis subject group (Fig. 6.8). These findings suggest that the peripheral right ear advantage is not lost with age per se, but rather is lost as a part of presbycusis hearing loss.

Jerger and colleagues conducted an elegant series of dichotic listening experiments to shed light on hemispheric changes in central auditory processing with age. Young adult observers with normal hearing typically perceive auditory information more accurately when presented to the right ear (Jerger and Martin 2004). The opposite is true for nonlinguistic materials. Jerger and Jordan (1992) and Jerger et al. (1994) provided convincing evidence that asymmetric cortical processing of speech materials increases with age, i.e., there is an increased right ear advantage in subjects with presbycusis. This robust finding was apparent for

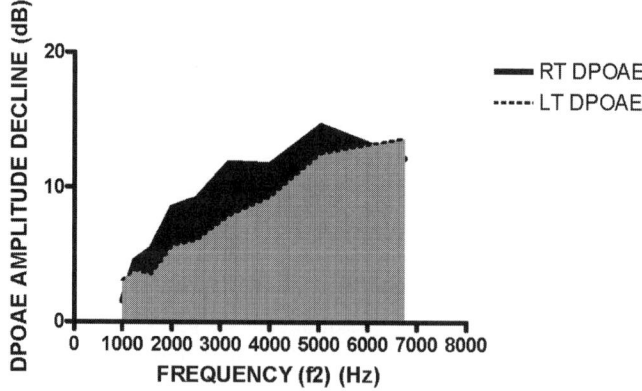

FIGURE 6.8. A significant difference in DPOAE amplitude decline, i.e., the normal hearing group relative to the presbycusis group. The right ear DPOAE decrement is more than the left ear decline especially in the f_2 = 2–5 kHz region. (From Tadros et al. [2005a], Fig. 3, with permission.)

measures of correct responses using speech material, as well as for reaction-time measurements of auditory performance (Jerger et al. 1995).

9. Cellular Aging Mechanisms in ARHL

Interfacing with the environment exposes all sensory epithelia to injury risk, because sensory stimuli contain energy that cannot all be usefully transduced. Just as excessive light injures the retina, excessive sound injures the organ of Corti. Unlike the retina, the organ of Corti is subject to both biochemical and mechanical injury in the course of its normal function. A lifetime of operating as an intermediary between the acoustic world and the brain inevitably yields injury that cannot be distinguished from ostensibly "pure" aging processes, so that aging-as-injury is a theme of this chapter.

Aging research in humans and animals has sought "longevity genes" and longevity-promoting environments and practices (Karasik et al. 2005; Sinclair 2005; Geesaman 2006; Harper et al. 2006). Alleles and environments that promote healthy aging and longevity—or the opposite—will probably often impact the apparent rate of ARHL. As in all tissues, however, the uniqueness of the cochlea arises from the activity of a unique set of genes. Some mutations in genes that govern hearing-specific structures and processes may therefore emerge as candidate "ARHL genes," actually meaning that certain alleles at particular loci can promote ARHL Most likely, all pro-ARHL alleles will be subject to strong environmental influences, as well as modulation by several other loci. Although no clear candidates have yet been identified in humans (Ates et al. 2005; Carlsson et al. 2005; Unal et al. 2005), possible ARHL-promoting genes and gene functions have been identified in mice. The following sections attempt to place ARHL into a wider context of cellular aging and the prevailing theories in this area

9.1 Theories of Cellular Aging

Models of cellular aging can generally be placed into two categories: (1) Aging as a regulated "program" and (2) aging as dysregulation. The first category supposes that aging evolved as an adaptive process, and has largely fallen out of favor. It is far more likely that longevity is simply not selected *for* after peak reproductive age. Nevertheless, one type of "built-in" limitation, a limit on the total number of cell divisions in mitotically active cell populations, may yet be adaptive. Each time chromosomes are duplicated as part of mitosis, end segments called *telomeres* are shortened (von Zglinicki and Martin-Ruiz 2005). At some point, the telomeres become too short for duplication to occur. This mechanism of aging appears most relevant to tissues that emphasize cell replacement over repair, and may have evolved to help protect against cancer. While it may have some applicability to cochlear aging, its meaning for actual age-related *hearing loss* is less clear, as is considered further.

The notion of aging as dysregulation and loss of homeostasis forms the basis of several aging hypotheses having relevance to ARHL. Cellular changes with aging typically include crippling modifications to DNA, important housekeeping proteins, and membrane lipids (e.g., Fenech 1998; Squier and Bigelow 2000; Leutner et al. 2001). Why should this occur given that cells have a host of DNA repair enzymes, and are constantly making or importing new proteins and lipids? The most significant and debilitating changes may permanently alter the DNA "blueprints" themselves. Proteins made from corrupted genes may have reduced function and be subject to improper folding and aggregation (Squier 2001). Proteins whose job it is to promote proper folding (chaperones) and to degrade damaged proteins (proteasomes) are also subject to modification, so that nonfunctional proteins may accumulate.

9.2 Progressive Cell Injury by Oxidative Stress

If much of aging can be equated with injury, the question arises regarding the kinds of injury to which the cells are subject. Most likely is oxidative stress (see also Wangemann, Chapter 3). The evolution of aerobic (oxygen-based) metabolism made it possible for cells to increase their energy output, and their range of activities. Oxygen is useful precisely because of its ability to break down carbon–carbon and carbon–hydrogen chemical bonds, the types of bonds that form biomolecules. This ability, however, might also present a problem. Cells use oxygen to synthesize metabolic intermediates and to fuel energy production, but must avoid being oxidized themselves. Inevitably, some oxidative attack on cellular DNA, proteins, and lipids does occur and is exacerbated by nearly any type of environmental stress. The observation that most injury to cells appears to be oxidative led to the Free Radical Theory of Aging, first proposed by Harman (1956), which asserts that aging is basically progressive oxidation. Aerobic cells have evolved to fend off oxidative attack in several ways. First, key reactions involving oxygen are quarantined to the mitochondrion. Inevitably, however, reactive oxygen-containing molecules (also known as reactive oxygen species or ROS) "escape" the intended reactions and boundaries. Cells then reduce oxidative injury by antioxidants, which either catalyze reactions that remove ROS or attenuate the activity of ROS. The Free Radical Theory has found much support, and currently provides the major framework for aging research (Fenech 1998; Squier and Bigelow 2000; Leutner et al. 2001; Barda 2002; Sinclair 2005; Hulbert et al. 2006). Oxidative modifications to cell constituents have detectable biochemical "signatures" for localization and semiquantitation, and studies have shown age-related increases in such changes in a host of tissues. Consequently, dietary antioxidants both decrease age-related infirmity and increase lifespan in animals. Moreover, treatments that increase lifespan, such as caloric restriction, also bolster antioxidant defenses and reduce oxidative tissue injury.

Experiments in humans and animals support the contention that the Free Radical Theory is applicable to ARHL. Cochlear injury caused by noise, ototoxins, and ischemia involves oxidative stress (see Henderson, Hu, and Bielefeld, Chapter 7 and Rybak, Talaska, and Schacht, Chapter 8 on noise-induced and drug-induced hearing loss, respectively). Oxidative modification to DNA, proteins, and lipids of cochlear sensory cells is increased during aging (Jiang et al. 2006). The inactivation of genes encoding antioxidant enzymes SOD1 and glutathione peroxidase (GPx1) exacerbates apparent age-related cochlear pathology such as loss of hair cells and neurons, as well as thinning of stria vascularis (Ohlemiller et al. 1999, 2000; McFadden et al. 1999, 2001; Keithley et al. 2005). Interestingly, deficiency of SOD1 and GPx1 does not appear to shorten lifespan. Impairment of these critical and widely expressed antioxidant enzymes might be expected to promote broad pro-aging effects, and to decrease longevity. The narrower effects that are observed may testify to the special susceptibility of sensory epithelia to oxidative injury. Consistent with an involvement of oxidative stress in these pathologies, applications of antioxidants such as glutathione, D-methionine, and N-acetylcysteine reduce noise and ototoxic injury (reviews: Forge and Schacht 2000; Le Prell et al. 2006), and dietary application of vitamin E, vitamin C, or L-carnitine slow the progression of cochlear degeneration and hearing loss (Seidman 2000; Derin et al. 2004; Takumida and Anniko 2005; Le and Keithley 2006).

9.3 Mitochondrial Viability as a Key Factor in Aging

Mitochondria are both key targets of age-associated oxidative injury, and key mediators of aging effects on cells (Sastre et al. 2000; Kujothet al. 2005). Uniquely among intracellular organelles, mitochondria house their own DNA. This DNA codes for many essential components of energy production, but not for protective or repair components as nuclear DNA does. Damage to mitochondrial DNA also means that new mitochondria will carry the same errors, as new mitochondria come from replication of the old. Finally, as stated earlier, reactions carried out in mitochondria create most of the cell's ROS, and ROS production may be exacerbated in compromised mitochondria. All of these factors may combine to promote the accumulation of DNA errors within individual mitochondria, so that overall energy production is reduced, and the function of the entire cell is impaired. A related view of aging, known as the Mitochondrial Clock Theory (see Seidman 2000), is perhaps a corollary to the Free Radical Theory. Accumulation of mitochondrial DNA mutations with age has been observed in essentially all tissues including the cochlea (Bai et al. 1997; Seidman et al. 2002; Pickles 2004;) and can be reduced both by antioxidants and by caloric restriction (Seidman 2000). Medical conditions that impact cochlear blood flow and possibly ARHL (see later), also promote mitochondrial DNA mutations (Dai et al. 2004).

9.4 Calcium Dysregulation

Calcium is a critical regulator of cellular events (see Wangemann, Chapter 3). Its levels normally remain very low in cytoplasm and in extracellular fluids, so that minute changes can modulate specialized functions such as transduction in stereocilia and neurotransmitter release (e.g., Chan and Hudspeth 2005; Keen and Hudspeth 2006), as well as fundamental functions such as growth, division, and death (Lu and Means 1993; Krebs 1998). Major cytoplasmic proteins that buffer calcium or bind calcium for signaling include calmodulin, calbindin, parvalbumin, and calretinin (Lu and Means 1993; Schwaller et al. 2002). Because of the prominence of Ca^{2+} in many vital processes, it is not surprising that dysregulation of calcium may contribute to cellular aging (Squier and Bigelow 2000; Crompton 2004; Toescu 2005). Disruption of calcium homeostasis is part of the mechanism of excitotoxicity, and may be among the causes of neural ARHL (Lang et al. 2006a). In addition, the *Cdh23* gene locus modifies other loci related to calcium, such as the one encoding plasma membrane Ca^{2+}-ATPase 2 (*Pmca2*) (Davis et al. 2003). Thus, the sensory ARHL-like pathology associated with the *Cdh23*^{ahl} allele may in part reflect calcium dysregulation.

Age-related changes in ACNS function also involve changes in calcium regulation. Zettel et al. (1997) examined changes in intracellular calcium binding proteins in the region of the IC shown by Walton's group to undergo age-related decline in temporal processing. In both CBA and B6 strains, calbindin levels declined with age. However, calretinin exhibited upregulation with age in CBA mice. To test whether this upregulation was strain-dependent or activity-dependent, Zettel et al. (2001) deafened young adult CBA mice by cochlear ablation and examined changes in calretinin immunochemistry in the IC with aging. Control CBAs showed upregulation of calretinin with aging, but the deafened CBAs did not, supporting the idea that maintenance of neuronal activity in IC (through preservation of cochlear function) is important for calretinin regulation. Subsequent work in B6 mice by the same group (O'Neill et al. 1997) also revealed an age-related decline in calbindin-labeled neurons in the medial nucleus of the trapezoid body.

Canlon and colleagues (Idrizbegovic et al. 2001a,b) examined age-related changes in calcium-binding proteins in the cochlear nucleus of CBA mice. The percentage of neurons in the DCN staining for calbindin, calretinin, and parvalbumin was *upregulated* with age for ages up to 29 months. The age-related upregulation of calretinin and parvalbumin was correlated with the degree of peripheral hearing loss, as measured by inner and outer hair cell loss and spiral ganglion cell loss, suggesting peripherally induced central effects. Quantitative stereology using optical fractionation to obtain total neuron counts in PVCN and DCN revealed that the total number of PVCN neurons remained constant with age, whereas DCN cell numbers declined. Only parvalbumin showed an age-related upregulation in the PVCN.

Subsequent investigations by the same group contrasted the changes in the CBA cochlear nucleus with those in B6. Like CBA mice, B6 showed age-related *upregulation* of calbindin and parvalbumin in the DCN and PVCN for ages up to 30 months (Idrizbegovic et al. 2003). Also as in CBA, calbindin and parvalbumin were correlated with peripheral hair cell and neurons loss in DCN and PVCN. The upregulation was accompanied by age-related declines in the absolute number of neurons in the DCN and PVCN (Idrizbegovic et al. 2004).

9.5 Limitations on Cell Repair and Replacement

Tissues that have sustained some degree of injury might compensate for cell loss by replacement of constituent cells by mitosis. Only a few cell types are normally replaced in the auditory system of adult mammals. Neurons are not replaced, nor are cochlear hair cells, nor most cells of the organ of Corti. This limitation places a premium on protective and repair capabilities. Clumping of hair cell stereocilia and deformation of the cuticular plate can be seen in aged humans (Scholtz et al. 2001), suggesting that normal bundle renewal is impaired in old hair cells, yet these nonfunctional hair cells may survive for some time. Supporting cells with pyknotic nuclei and dark cytoplasm are frequently observed in the organ of Corti of old mice (Ohlemiller and Gagnon 2004b). It is not clear whether cells showing these signs die and are replaced, or whether supporting cell pathology can also promote hair cell loss.

Within the cochlea, cell replacement appears limited to fibrocytes of the lateral wall and intermediate cells and marginal cells of the stria vascularis (Conlee et al. 1994; Yamashita et al. 1999; Dunaway et al. 2003; Hirose et al. 2005). Still these decrease in number with age (Wright and Schuknecht 1972; Ichimiya et al. 2000; Ohlemiller 2006), perhaps as a result of limits on proliferative capacity such as telomere status. Some new fibrocytes in ligament may derive from bone marrow (Lang et al. 2006b), so that it is not clear why there is net loss of fibrocytes with age. Both strial cells and fibrocytes in spiral ligament and limbus serve a distributed function, and are initially present in excess, as indicated by the fact that substantial numbers of these can be lost without any effect on hearing. While limits on strial cell replacement may play a role in strial ARHL (Ohlemiller 2006), it is at present not clear that ligament pathology is a significant factor in ARHL.

10. Risk Factors Affecting ARHL

The prevalence of clinically defined ARHL in the very old, while high, is not 100%. Although it appears in all societies, it occurs more frequently in industrial cultures than in nonindustrial cultures (Rosen et al. 1962). Such evidence argues that much of ARHL is influenced by the interplay between pro-ARHL alleles (or pro-aging alleles) and environment. No behavior, event, or environment carries a fixed degree of ARHL risk. Rather, the risk will depend on unknown alleles carried by the individual. Until such alleles are identified, the best strategy is to minimize environmental risks, which take many forms.

10.1 Noise and Ototoxins

Environmental risk factors for apparent ARHL include acute or chronic exposure to noise, ototoxic medications, industrial solvents, or combinations of these (Gilad and Glorig 1979b; Rosenhall et al. 1993; Rosenhall and Pedersen 1995; Toppila et al. 2001; Fransen et al. 2003; Fechter 2004). Assaults by these agents appear to promote largely oxidative injury that primarily injures hair cells (see Henderson, Hu, and Bielefeld, Chapter 7; Rybak, Talaska, and Schacht, Chapter 8). Note that the intention here is not to equate cochlear noise and ototoxic injury with ARHL, or to suggest that the cellular pattern of injury is exactly the same in all three cases although there are intriguing similarities. Both noise and ototoxin exposure can, for example, cause permanent strial injury (e.g., Ulehlova, 1983; Garetz and Schacht 1996; Hirose and Liberman 2003). They mostly do not, however, cause permanent EP reduction, and thus would not be expected to draw a diagnosis of strial ARHL. There is likewise no compelling evidence that ototoxins promote primary neural loss sufficient to bring a diagnosis of neural ARHL. By contrast, early noise exposure may yield neural ARHL (Kujawa and Liberman 2006). However, because the principal targets of noise and ototoxins will be hair cells, the diagnosis can often be confused with sensory ARHL.

10.2 Lifestyle and Risk of Vascular Pathology

Proper function of the cochlea, particularly the lateral wall, is energy intensive, and likely to be vulnerable to any restriction of blood flow. Accordingly, the role of vascular insufficiency has long been a prominent topic in ARHL research (reviews: Gilad and Glorig 1979b; Nakashima et al. 2003). Obesity and conditions to which it may lead (hyperlipidemia, hypercholesterolemia, hypertension, hyperhomocysteinemia, hyperlipoproteinemia, and cardiovascular disease) have all been implicated in ARHL (Rosen et al. 1970; Spencer 1973; Drettner et al. 1975; Tachibana et al. 1984; Axellson and Lindgren 1985; Pillsbury 1986; Saito et al. 1986; Sikora et al. 1986; Suzuki et al. 2000; Satar et al. 2001; Fransen et al. 2003). Poor health habits with regard to exercise, smoking, and diet may also be risk factors for ARHL insofar as they impact vascular health, tissue oxygenation, and diabetes risk (see later) (Rosenhall et al. 1993; Cruickshanks et al. 1998; Torre et al. 2005; Uchida et al. 2005), although probably only as part of a spectrum of conditions of aging. While it has been suggested that the most immediate cochlear target of vascular pathology is likely to be the stria, limited observations of affected human and animal cochleae suggest broad tissue degeneration, and no special relationship to strial ARHL.

10.3 Early Exposure to Stress

Environment extends to the prenatal environment. It has been hypothesized that prenatal stress can "program" individuals to pathology that resembles accelerated aging, or to possible risk factors for age-related pathology such as hypertension and cardiovascular disease (Barrenas et al. 2003). Suggested mechanisms involve "redeployment" of resources by the fetus to favor some tissues and organs, leaving others with fewer stem cells or poorly vascularized, and permanent alterations in endocrine function (Barker 1998). One result of such events may be sensorineural hearing loss associated with shortened stature in adulthood (Barrenäs et al. 2005). A potentially related finding is that exposure of prenatal rats to glucocorticoid stress hormones increases susceptibility in the young adult to NIHL possibly due to an overall increased vulnerability to oxidative stress (Canlon et al. 2003).

10.4 Mineralocorticoid Levels

Aging is often accompanied by medical comorbidities such as decreases in hormonal levels. A decline in levels of aldosterone, a mineralocorticoid produced by the adrenal cortex, may affect ionic balance, partly by its actions on Na^+, K^+-ATPase and the K^+, Na^+, Cl^- cotransporter. Because these enzymes are highly expressed in the cochlear lateral wall and are critical to ion regulation in the cochlea, aldosterone may participate in the regulation of the EP. Alternatively, it may operate at a systemic level by averting hypertension or reducing inflammation. Blood aldosterone titer probably reflects both genetic and environmental influences. Frisina's group examined the relation between aldosterone levels and age-related hearing loss in aged human subjects who had good hearing and showed good health overall (Tadros et al. 2005b). Based on standard audiometric criteria, the subjects were classified into three groups: golden ear (audiometric thresholds in the normal range), mild/moderate hearing loss, and severe loss. Serum aldosterone levels were significantly different between the groups, with the golden ears showing the highest aldosterone, the mild/moderate group next, and the subjects with severe hearing loss the least amount of aldosterone (Fig. 6.9). Interestingly, all aldosterone levels were within normal clinical limits. Regression analyses showed significant correlations between aldosterone levels, pure tone thresholds, and hearing-in-noise test (HINT) scores. These results suggest that aldosterone may be protective against presbycusis, as has been found for autoimmune hearing loss (Trune et al. 2006). At present there is no direct evidence to indicate which cochlear structures are preserved or affected.

10.5 Diabetes Mellitus

Non-insulin-dependent (type 2, adult onset) diabetes mellitus often appears as a condition of aging, frequently as a complication of obesity. In middle age, diabetes also produces multisystemic pathology that mimics aspects of aging (Geesaman 2006). Chronically elevated plasma glucose promotes malconformation and aggregation of proteins in all tissues, yet with particularly deleterious

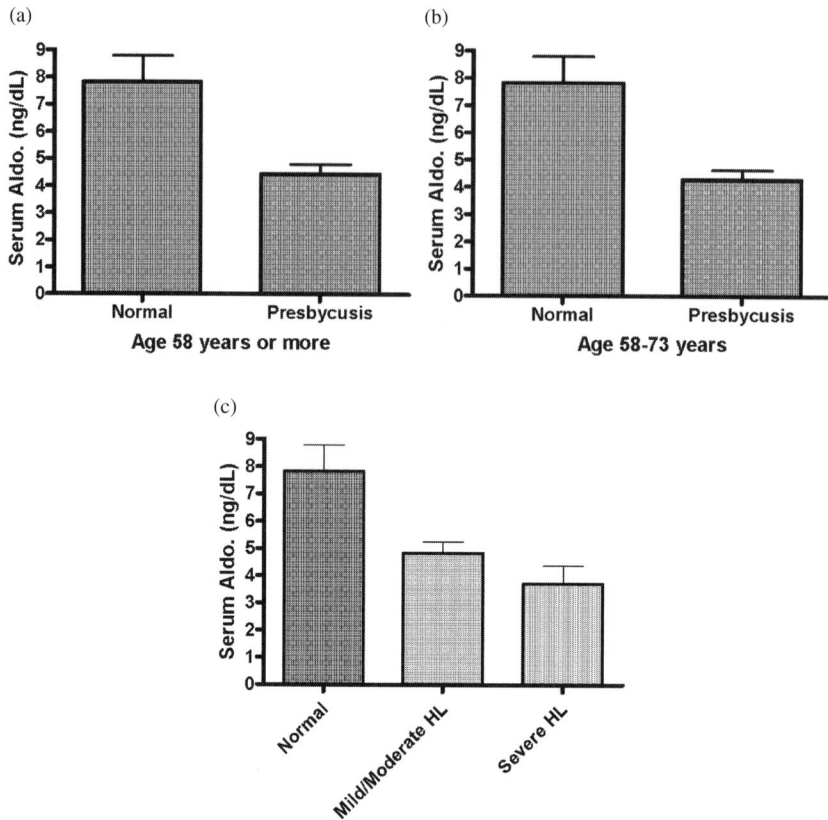

FIGURE 6.9. (**A**) A significant difference in serum aldosterone concentration was found between normal hearing and presbycusic groups, with a higher concentration in the normal hearing group. (**B**) A significant difference in serum aldosterone concentrations was found for the 58- to 73-year-old groups of normal hearing subjects and presbycusic subjects, with higher concentrations in the normal hearing group. This analysis eliminated the age factor. (**C**) A significant difference in serum aldosterone concentrations was found between normal hearing and both mild/moderate and severe presbycusic groups.(From Tadros et al. [2005b], Fig. 1. Reprinted with permission.)

consequences for the microvasculature. The most wide-ranging pathologies of diabetes therefore appear mediated by microangiopathy. Both type 1 (juvenile) and type 2 diabetes promote hearing loss and cochlear pathology in humans (Wackym and Linthicum 1986; Fukushima et al. 2005) and animals (Rust et al. 1992; Raynor et al. 1995; Ishikawa et al. 1995). Type 2 diabetes has been proposed as a cause of ARHL, but the evidence for this is mixed (Malpas et al. 1989; Ma et al. 1998; Ologe et al. 2005; Vaughan et al. 2005). To clarify whether presbycusis is accelerated in aged type 2 diabetics, a group of type 2 diabetics older than the age of 60 years were compared with a group of age- and sex-matched controls (Frisina et al. 2006). Both groups were otherwise healthy and had no history of major health or hearing problems. Audiometric thresholds, otoacoustic emission levels, and speech thresholds revealed deficits in the diabetic group, with the right ear showing a more severe loss relative to the left (Fig. 6.10). Tests involving the ACNS, such as suprathreshold gap detection and HINT scores, also exposed relative deficits in the diabetic group. These findings support a causal link between type 2 diabetes and both peripheral and central aspects of ARHL.

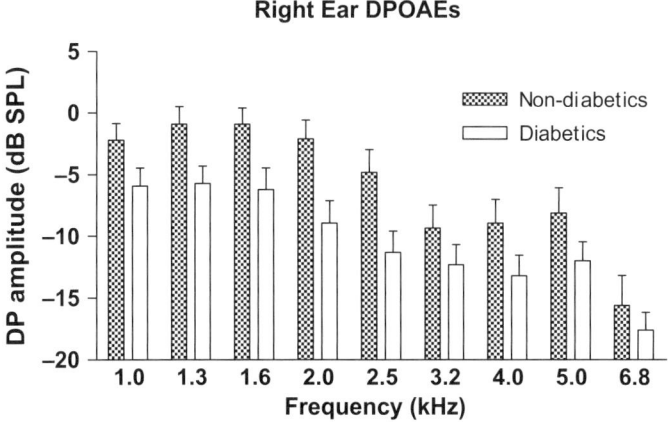

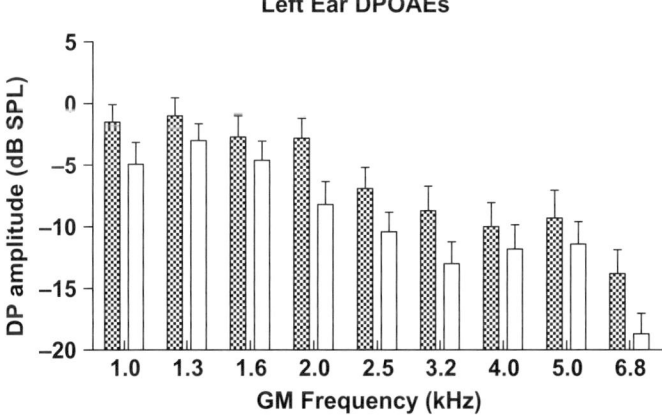

FIGURE 6.10. At all frequencies, DPOAEs were smaller for diabetics relative to non-diabetics. Like the threshold measures presented in the previous figures, the right ear was more affected than the left. ANOVA showed significant main effects of subject group (Right: $p < 0.0001$, $F = 31.1$, df= 1; Left: $p < 0.0001$, $F = 15.2$, df = 1). Interactions and subject group Bonferroni post-hoc analyses were not significant, except for the right ear at 2 kHz: $p < 0.05$, t = 2.82, df = 1. GM = geometric mean of f_1 and f_2. Error bars are SEM. (From Frisina et al. [2006], Fig. 3. Reprinted with permission.)

10.6 Caloric Restriction

By far the single best supported anti-aging regimen is caloric restriction (CR), chronic reduction of normal caloric intake by 10% to 20% (review: Sinclair 2005). In species ranging from flies and worms to humans, CR extends lifespan and reduces age-related pathology. Studies seeking the mechanism(s) by which CR delivers its impressive benefits have yielded many suspects, including slowing of metabolism, enhanced immune responses, decreased ROS production, enhanced ROS defenses, increased overall stress resistance, decreased circulating insulin levels, increased respiration (with decreased glycolysis), and reduced circulating thyroid hormones. Of more than 350 genes whose activity is significantly altered by CR in mice, at least 29 were also upregulated in the long-lived Snell dwarf mouse strain (Miller et al. 2002). However, there is little clear overlap among these 29 genes and genes shown to be upregulated in long-lived humans (Karasik et al. 2005). Moreover, different long-lived mouse strains show different subsets of the characteristics mentioned above (Harper et al. 2005). The most common gene profiles and characteristics shared by calorically restricted and long-lived organisms have led to the "Hormesis" hypothesis (Sinclair 2005), which proposes that enhanced stress resistance is the key to healthy aging. This key to longevity complements the Free Radical Theory of aging. Caloric restriction can slow the progression of ARHL in mice (Sweet et al. 1988; Park et al. 1990; Someya et al. 2006), presumably as part of an overall slowing of the aging process. The principles underlying CR are therefore clearly relevant to ARHL.

11. Prevention and Treatment of ARHL

11.1 Altering Behavior

As outlined in the preceding text, there are several behavioral/environmental factors whose association with added ARHL risk is plausible. Excessive noise, ototoxins, and smoking are clearly to be avoided. Conversely, behaviors that preserve overall health against aging (appropriate diet and exercise) very likely serve to preserve hearing as well.

11.2 Pharmacologic Approaches

Calcium channel blockers may be protective against ARHL (Mills et al. 1999), and dietary antioxidants have proven partially effective against age-related cochlear changes (Seidman 2000; Derin et al. 2004 ; Takumida and Anniko 2005; Le and Keithley 2006). A possible limitation to the ultimate efficacy of antioxidant therapy is that redox homeostasis comprises a complex web of checks and balances (see Wangemann, Chapter 3). When present at low concentrations, ROS perform important signaling functions. Exogenous agents, be they pro- or antioxidant, may disrupt this balance (Ohlemiller 2003).

Several drugs reproduce some of the positive effects of caloric restriction, including 2-deoxyglucose, metformin (and its analog phenformin, both used to treat diabetes), and resveratrol (Sinclair 2005). The first two compounds present their own health risks, and are not advocated as an anti-aging therapy. Resveratrol increases levels of SIRT1, a key longevity-promoting protein in mammals. It is one of the classes of sirtuin-activating compounds (STACS), which show tremendous promise in alleviating age-related pathology.

Another approach to protection is external stress applied in a controlled and noninjurious manner. This phenomenon, known as "preconditioning," has been demonstrated in brain, heart, and retina (Dirnagl et al. 2003; Ran et al. 2005; Whitlock et al. 2005). The types of stresses that may be protective include mild ischemia, hypoxia, and heat shock. Protection against cochlear noise injury has been linked to preconditioning by noise exposure, restraint (Wang and Liberman 2002), heat shock (Yoshida et al. 1999), and hypoxia (Gagnon et al. 2006). Protection by prior noise exposure includes both "noise conditioning" in which the initial exposure is noninjurious by itself (Niu and Canlon 2002), as well as "toughening," in which there is some permanent injury from the initial exposure (Hamernik et al. 2003). Protection against some ARHL as caused by the $Cdh23^{ahl}$ allele in mice is also provided by "acoustic augmentation," wherein mice are raised in moderate background noise (Willott and Turner 1999). These protective regimens may be impractical to apply clinically, but the innate processes they engage may be amenable to pharmacologic manipulation. Mediators of preconditioning in brain and retina include transcription factors such as hypoxia-inducible factor 1α (HIF-1α), heat shock factor 1 (HSF-1), and NF-κB. Their gene targets may include vascular endothelial growth factor and erythropoietin, which may promote vascular remodeling and exert trophic effects (Prass et al. 2003; Brimes and Cerami 2005). HIF-1α can be upregulated pharmacologically, and erythropoietin has been applied with therapeutic effects (Brimes and Cerami 2005). The effects of protective manipulations may also be genetically modulated, as shown for hypoxic preconditioning against NIHL in mice (Gagnon et al. 2006). People who show weak preconditioning effects often may also adapt poorly to environments that pose chronic stress to the cochlea, and have higher risk for NIHL and apparent ARHL.

11.3 Restoration of Lost Hearing

The best strategy against ARHL is clearly to prevent it. Until that is possible, restoration of hearing will remain the principal intervention, and this currently means hearing aids and cochlear implants. Digital hearing aids are far advanced over their predecessors and present a wide range of user options, tailored to specific acoustic environments. Cochlear implants are increasingly recommended to older adults, and appear to promote the survival of afferent neurons after loss of their hair cell targets.

True restoration of lost cells, however, poses a tremendous challenge. Loss of any cell population in the cochlea may trigger irreversible changes in other cell types. For example, sensory ARHL may begin with hair cell loss, but ultimately may be associated with replacement of the entire organ of Corti with a single undifferentiated cell layer. Currently most strategies for restoration are aimed at specific cell types, typically hair cells and neurons (see Raphael and Heller, Chapter 11). There are, however, many cell types in the organ of Corti and lateral wall whose functions and interdependence for survival are incompletely understood, and in some forms of ARHL the primary defect may lie in these cells. Successful gene therapies may require reprogramming of many types of the cells that make up the cochlear environment.

12. Questions for Future Research

Given the recent advances in areas of neuroregeneration and stem cell therapy, the future lies in biomedical interventions against ARHL. Generally, it would be very beneficial to start coordinating aging research across modalities, to focus in on a dietary regimen, including supplements as appropriate, to optimize sensory functioning in the elderly. However, interventions to the benefit of hearing must be scrutinized for their effect on vision, balance, touch, or taste. Agents to counteract the effects of the declining GABA (inferior colliculus) and glycine (cochlear nucleus) inhibitory systems in the auditory brain stem might embody such an example, where the generality of this phenomenon needs to be verified for the other senses.

Capitalizing on the presence of stem cells that are present in the inner ear and brain will require the development of gene therapy and/or pharmacological triggers to stimulate the differentiation into specialized cells of the cochlea or ACNS. The repair process may be more preventative or aimed at slowing down age-related changes. In contrast, restoration and regeneration are more important for full-fledged presbycusis, both peripheral, high-frequency hearing loss and central-understanding speech-in-background noise at suprathreshold, conversational levels.

Acknowledgments. R.D.F. was supported by NIH grants NIA P01 AG09524 from the National Institute on Aging, NIDCD P30 DC05409 from the National Institute on Deafness & Communication Disorders, and the Int. Ctr. Hearing Speech Res., Rochester NY. K.K.O. was supported by WU Medical School Dept. Otolaryngology, NIH P30 DC004665, and R01 DC08321. Thanks to P.M. Gagnon for assistance with figures.

References

Allen PD, Burkard RF, Ison JR, Walton JP (2003) Impaired gap encoding in aged mouse inferior colliculus at moderate but not high stimulus levels. Hear Res 186:17–29.
Ates NA, Unal M, Tamer L, Derici E, Karakas S, Ercan B, Camdevirin H (2005) Glutathione *S*-transferase gene polymorphisms in presbycusis. Otol Neurotol 26: 392–397.

Axellson A, Lindgren F (1985) Is there a relationship between hypercholesterolemia and noise-induced hearing loss? Acta Otolaryngol 100:379–386.

Bai U, Seidman MD, Hinojosa R, Quirk WS (1997) Mitochondrial DNA deletions associated with aging and possibly presbycusis: A human archival temporal bone study. Am J Otol 18:449–453.

Banay-Schwartz, M, Lajtha A, Palkovits M.(1889) Changes with aging in the levels of amino acids in rat CNS structural elements II. Taurine and small neutral amino acids. Neurochem Res 14:563–570.

Bao J, Lei D, Du Y, Ohlemiller KK, Beaudet AL, Role LW (2005) Requirement of nicotinic acetylcholine receptor subunit β2 in the maintenance of spiral ganglion neurons during aging. J Neurosci 25:3041–3045.

Barda G (2002) Rate of generation of oxidative stress–related damage and animal longevity. Free Radic Biol Med 33:1167–1172.

Barker DJP (1998) In utero programming of chronic disease. Clin Sci 95:115–128.

Barrenäs M-L, Bratthall A, Dahlgren J (2003) The thrifty phenotype hypothesis and hearing problems. Br Med J 327:1199–1200.

Barrenäs M-L, Bratthall A, Dahlgren J (2005) The association between short stature and sensorineural hearing loss. Hear Res 205:123–130.

Bohne BA, Gruner MM, Harding GW (1990) Morphological correlates of aging in the chinchilla cochlea. Hear Res 48:79–91.

Borg E (1983) Delayed effects of noise on the ear. Hear Res 9:247–254.

Bredberg G (1968) Cellular pattern and nerve supply of the human organ of Corti. Acta Otolaryngol (Suppl)236:1–135.

Brimes M, Cerami A (2005) Emerging biological roles for erythropoietin in the nervous system. Nat Neurosci 6:484–494.

Briner W, Willott JF (1989) Ultrastructural features of neurons in the CB57BL/6J mouse anteroventral cochlear nucleus: young mice versus old mice with chronic presbycusis. Neurobiol Aging 10:295–303.

Brody H (1955) Organization of the cerebral cortex: III. A study of aging in the human cerebral cortex. J Comp Neurol 102:511–556.

Cable J, Jackson IJ, Steel KP (1993) Light (B^ll), a mutation that causes melanocyte death, affects stria vascularis function, in the mouse inner ear. Pigment Cell Res 6:215–225.

Canlon B, Erichsen S, Nemlander E,Chen M, Hossain A, Celsi G, Ceccatelli S (2003) Alterations in intrauterine environment by glucocorticoids modifies the developmental program of the auditory system. Eur J Neurosci 17:2035–2041.

Carlsson P-I,VanLaer L, Borg E, Bondeson M-L, Thys M, Fransen E, Van Camp G (2005) The influence of genetic variation in oxidative stress genes on human noise susceptibility. Hear Res 202:87–96.

Caspary DM, Raza A, Armour BAL, Pippin J, Arneric SP (1990) Immunocytochemical and neurochemical evidence for age-related loss of GABA in the inferior colliculus: implications for neural presbycusis. J Neurosci 10:2363–2372.

Caspary DM, Milbrandt, JC, Helfert RH (1995) Central auditory aging: GABA changes in the inferior colliculus. Exp Gerontol 30:349–360.

Caspary DM, Holder TM, Hughes LF, Milbrandt JC, McKernan RM, Naritoku DK (1999) Age-related changes in GABA a receptor subunit composition and function in rat auditory system. Neuroscience 93:307–312.

Caspary DM, Schatteman TA, Hughes, LF (2005) Age–related changes in the inhibitory response properties of dorsal cochlear nucleus output neurons: role of inhibitory inputs. J Neurosci 47:10952–10959.

Chan DK, Hudspeth AJ (2005) Ca^{2+} current-driven nonlinear amplification by the mammalian cochlea in vitro. Nat Neurosci 8:149–155.

Chen MA, Webster P, Yang Y, Linthicum FH (2006) Presbycusic neuritic degeneration within the osseous spiral lamina. Otol Neurotol 27:316–322.

Chisolm TH, Willott JF, Lister JJ (2003) The aging auditory system: anatomic and physiologic changes and implications for rehabilitation. Int J Audiol 42:2S3–2S10.

Conlee JW, Gerrity LC, Bennett ML (1994) Ongoing proliferation of melanocytes in the stria vascularis of adult guinea pigs. Hear Res 79:115–122.

Covell WP, Rogers JB (1957) Pathologic changes in the inner ear of senile guinea pigs. Laryngoscope 67:118–129.

Crompton M (2004) Mitochondria and aging: a role for the permeability transition? Aging Cell 3:3–6.

Cruickshanks KJ, Klein R, Klein BEK, T.L. W, Nondahl DM, Tweed TS (1998) Cigarette smoking and hearing loss. JAMA 279:1715–1719.

Dai P, Yang W, Jiang S, Gu R, Yuan H, Han D, Guo W, Cao J (2004) Correlation of blood supply with mitochondrial DNA common deletion in presbycusis. Acta Otolaryngol 124:130–136.

Davis RR, Newlander JK, Ling X–B, Cortopassi GA, Kreig EF, Erway LC (2001) Genetic basis for susceptibility to noise–induced hearing loss in mice. Hear Res 155:82–90.

Davis RR, Kozel P, Erway LC (2003) Genetic influences in individual susceptibility to noise: a review. Noise Health 5:19–28.

Delprat B, Ruel J, Guitton MJ, Hamard G, Lenoir M, Pujol R, Puel JL, Brabet P, Hamel CPJA (2005) Deafness and cochlear fibrocyte alterations in mice deficient for the inner ear protein otospiralin. Mol Cell Biol 25:847–853.

Derin A, Agirdir B, Derin N, Dinc O, Guney K, Ozcaglar H, Kilincarslan S (2004) The effects of L–carnitine on presbycusis in the rat model. Clin Otolaryngol 29:238–241.

Di Girolamo S, Quaranta N, Picciotti P, Torsello A, Wolf F (2001) Age-related histopathological changes of the stria vascularis: an experimental model. Audiology 40:322–326.

Dirnagl U, Simon RP, Hallenbeck JM (2003) Ischemic tolerance and endogenous neuroprotection. Trends Neurosci 26:248–254.

Drettner B, Hedstrand H, Klockhoff I, Svedberg A (1975) Cardiovascular risk factors and hearing loss. Acta Otolaryngol 79:366–371.

Dunaway G, Mhaskar Y, Armour G, Whitworth C, Rybak LP (2003) Migration of cochlear lateral wall cells. Hear Res 177:1–11.

Erway Lc, Willott JF, Archer JR, Harrison DE (1993) Genetics of age–related hearing loss in mice: I. Inbred and F1 hybrid strains. Hear Res 65:125–132.

Erway LC, Shiau Y–W, Davis RR, Kreig EF (1996) Genetics of age-related hearing loss in mice. III. Susceptibility of inbred and F1 hybrid strains to noise-induced hearing loss. Hear Res 93:181–187.

Fechter LD (2004) Promotion of noise–induced hearing loss by chemical contaminants. J Toxicol Appl Environ Health A 67:727–740.

Felder E, Schrott-Fischer A (1995) Quantitative evaluation of myelinated nerve fibers in cochlea of humans with age-related high-tone hearing loss. Hear Res 91:19–32.

Felder E, Kanonier G, Scholtz A, Rask-Andersen H, Schrott-Fischer A (1997) Quantitative evaluation of cochlear neurons and computer-aided three-dimensional reconstruction of spiral ganglion cells in humans with a peripheral loss of nerve fibers. Hear Res 105:183–190.

Felix H, Pollak A, Gleeson MJ, Johnsson L–G (2002) Degeneration pattern of human first-order cochlear neurons. Adv Otorhinolaryngol 59:116–123.

Fenech M (1998) Chromosomal damage rate, aging, and diet. Ann NY Acad Sci 854: 23–26.

Ferraro JA, Minckler J (1977) The human lateral lemniscus and its nuclei. Brain Lang 4:156–164.

Fitzgibbons PJ, Gordon-Salant S (1996) Auditory temporal processing in elderly listeners. J Am Acad Audiol 7:183–189.

Forge A, Schacht J (2000) Aminoglycoside antibiotics. Audiol Neuro-Otol 5:3–22.

Fransen E, Lemkens N, Van Laer L, Van Camp G (2003) Age-related hearing impairment (ARHI): environmental risk factors and genetic prospects. Exp Gerontol 38:353–359.

Frisina RD (2001) Possible neurochemical and neuroanatomical bases of age-related hearing loss—presbycusis. Semin Hear 22:213–225.

Frisina DR, Frisina RD (1997) Speech recognition in noise and presbycusis: relations to possible neural mechanisms. Hear Res 106:95–104.

Frisina RD, Walton JP (2001) Aging of the mouse central auditory system. In: Willott JP (ed) From Behavior to Molecular Biology. New York: CRC Press, pp. 339–379.

Frisina RD, Walton JP (2006) Age-related structural and functional changes in the cochlear nucleus. Hear Res. 217:216–233.

Frisina DR, Frisina RD, Snell KB, Burkard R, Walton JP, Ison JR (2001) Auditory temporal processing during aging. In: Hof PR, Mobbs CV (eds) Functional Neurobiology of Aging. San Diego: Academic Press, pp. 565–579.

Frisina ST, Mapes F, Kim S-H, Frisina DR, Frisina RD (2006) Characterization of hearing loss in aged type II diabetics. Hear Res 211:103–113.

Fukushima H, Cureoglu S, Schachern PA, Kusunoki T, Oktay MF, Fukushima N, Paparella MM, Harada T (2005) Cochlear changes in patients with type I diabetes mellitus. Otolaryngol Head Neck Surg 133:100–106.

Gagnon PM, Simmons DD, Bao J, Lei D, Ortmann A, J., Ohlemiller KK (2007) Temporal and genetic influences on protection against noise-induced hearing loss by hypoxic preconditioning in mice. Hear Res 226: 79–91.

Garetz SL, Schacht J (1996) Ototoxicity: of mice and men. In: Van De Water TR, Popper AN, Fay RR (eds) Clinical Aspects of HearingNew York: Springer-Verlag, pp. 116–154.

Gates GA, Mills JH (2005) Presbycusis. Lancet 366:1111–1120.

Gates GA, Couropmitree NN, Myers RH (1999) Genetic associations in age–related hearing thresholds. Arch Otolaryngol Head Neck Surg 125:654–659.

Geesaman BJ (2006) Genetics of aging: implications for drug discovery and development. Am J Clin Nutr 83:466S–469S.

Gilad O, Glorig A (1979a) Presbycusis: The aging inner ear. Part I. J Am Audit Soc 4:195–206.

Gilad O, Glorig A (1979b) Presbycusis: The aging ear. Part II. J Am Audit Soc 4:207–217.

Glorig A, Wheeler D, Quiggle R, Grings W, Summerfeld A (1957) 1954 Wisconsin state fair hearing survey—statistical treatment of clinical and audiometric data. Am Acad Ophthalmol Otolaryngol (Monograph).

Gordon-Salant S, Fitzgibbons PJ (1993) Temporal factors and speech recognition performance in young and elderly listeners. J Speech Hear Res 36:1272–1285.

Gratton MA, Schulte BA (1995) Alterations in microvasculature are associated with atrophy of the stria vascularis in quiet-aged gerbils. Hear Res 82:44–52.

Gratton MA, Schmiedt RA, Schulte BA (1996) Age-related decreases in endocochlear potential are associated with vascular abnormalities in the stria vascularis. Hear Res 102:181–190.

Gratton MA, Smyth BJ, Lam CF, Boettcher FA, Schmiedt RA (1997) Decline in the endocochlear potential corresponds to decreased Na,K-ATPase activity in the lateral wall of quiet-aged gerbils. Hear Res 108:9–16.

Hamernik RP, Patterson JH, Turrentine GA, Ahroon WA (1989) The quantitative relation between sensory cell loss and hearing thresholds. Hear Res 38:199–212.

Hamernik RP, Qiu W, Davis B (2003) Cochlear toughening, protection, and potentiation of noise–induced hearing loss by non-Gaussian noise. J Acoust Soc Am 113:969–976.

Harman D (1956) Aging: a theory based on free radical and radiation chemistry. J Gerontol 11:98–300.

Harper JM, Wolf N, Galecki AT, Pinkosky SL, Miller RA (2003) Hormone levels and cataract scores as sex-specific, mid-life predictors of longevity in genetically heterogeneous mice. Mech Ageing Dev 124:801–810.

Harper JM, Galecki AT, Burke DT, Miller RA (2004) Body weight, hormones and T cell subsets as predictors of life span in genetically heterogeneous mice. Mech Ageing Dev 125:381–390.

Harper JM, Durkee SJ, Smith-Wheelock M, Miller RA (2005) Hyperglycemia, impaired glucose tolerance and elevated glycated hemoglobin in a long-lived mouse stock. Exp Gerontol 40:303–314.

Harper JM, Salmon AB, Chang Y, Bonkowski M, Bartke A, Miller RA (2006) Stress resistance and aging: influence of genes and nutrition. Mech Ageing Dev 127:687–694.

Hawkins JE, Johnsson L-G, Preston RE (1972) Cochlear microvasculature in normal and damaged ears. Laryngoscope 82:1091–1104.

Helfert RD, Sommer TJ, Meeks J, Hofstetter P, Hughes L F (1999) Age-related synaptic changes in the central nucleus of the inferior colliculus of the Fischer-344 rat. J Comp Neurol 406:285–298.

Helfert RD, Krenning J, Wilson TS, Hughes LF (2003) Age-related synaptic changes in the anteroventral cochlear nucleus of Fischer-344 rats. Hear Res 183:18–28.

Henry KR, Chole RA (1980) Genotypic differences in behavioral, physiological and anatomical expressions of age-related hearing loss in the laboratory mouse. Audiology 19:369–383.

Hequembourg S, Liberman MC (2001) Spiral ligament pathology: a major aspect of age-related cochlear degeneration in C57BL/6 mice. J Assoc Res Otolaryngol 2:118–129.

Hirose K, Liberman MC (2003) Lateral wall histopathology and endocochlear potential in the noise-damaged mouse cochlea. J Assoc Res Otolaryngol 4:339–352.

Hirose K, Discolo CM, Keasler JR, Ransohoff R (2005) Mononuclear phagocytes migrate into the murine cochlea after acoustic trauma. J Comp Neurol 489:180–194.

Hulbert AJ, Faulks SC, Harper JM, Miller RA, Buffenstein R (2006) Extended longevity of wild-derived mice is associated with peroxidation-resistant membranes. Mech Ageing Dev 127:653–657.

Ichimiya I, Suzuki M, Goro M (2000) Age-related changes in the murine cochlear lateral wall. Hear Res 139:116–122.

Idrizbegovic E. Canlon B. Bross LS. Willott JF. Bogdanovic N (2001a) The total number of neurons and calcium binding protein positive neurons during aging in the cochlear nucleus of CBA/CaJ mice: a quantitative study. Hear Res 158:102–115.

Idrizbegovic E, Viberg A, Bogdanovic N, Canlon B (2001b) Peripheral cell loss related to calcium binding protein immunocytochemistry in the dorsal cochlear nucleus in CBA/CaJ mice during aging. Audiol Neuro-Otol 6:132–139.

Idrizbegovic E, Bogdanovic N, Viberg A, Canlon B (2003) Auditory peripheral influences on calcium binding protein immunoreactivity in the cochlear nucleus during aging in the C57BL/6J mouse. Hear Res 179:33–42.

Idrizbegovic E. Bogdanovic N. Willott JF. Canlon B (2004) Age-related increases in calcium-binding protein immunoreactivity in the cochlear nucleus of hearing impaired C57BL/6J mice. Neurobiol Aging 25:1085–1093.

Ishikawa T, Naito Y, Taniguchi K (1995) Hearing impairment in WBN/Kob rats with spontaneous diabetes mellitus. Diabetologia 38:649–655

Jacobson M, KimS-H, Romney J, Zhu XX, Frisina RD (2003) Contralateral suppression of distortion-product otoacoustic emissions declines with age: a comparison of findings in CBA mice with human listeners. Laryngoscope 113:1707–1713.

Jerger J, Jordan C (1992) Age-related asymmetry on a cued-listening task. Ear Hear 13:272–277.

Jerger J, Martin J (2004) Hemispheric asymmetry of the right ear advantage in dichotic listening. Hear Res 198:125–136.

Jerger J, Chmiel R, Stach B, Spretnjak M (1993) Gender affects audiometric shape in presbycusis. J Am Acad Audiol 4:42–49.

Jerger J, Chmiel R, Allen J, Wilson A (1994) Effects of age and gender on dichotic sentence identification. Ear Hear 15:274–286.

Jerger J, Alford B, Lew H, Rivera V, Chmiel R (1995) Dichotic listening, event-related potentials, and interhemispheric transfer in the elderly. Ear Hear 16:482–498.

Jiang H, Talaska AE, Schacht J, Sha S-H (2006) Oxidative imbalance in the aging inner ear. Neurobiol Aging 28: 1605–1612.

Johnson KR, Zheng QY, Erway LC (2000) A major gene affecting age-related hearing loss is common to at least 10 inbred strains of mice. Genomics 70:171–180.

Johnson KR, Zheng QY, Noben-Trauth K (2006) Strain background effects and genetic modifiers of hearing in mice. Brain Res 1091:79–88.

Johnsson L-G, Hawkins JE (1972) Strial atrophy in clinical and experimental deafness. Laryngoscope 82:1105–1125.

Karasik D, Demissie S, Cupples AL, Kiel DP (2005) Disentangling the genetic determinants of human aging: biological age as an alternative to the use of survival measures. J Gerontol Biol Sci 60A:574–587.

Keen EC, Hudspeth AJ (2006) Transfer characteristics of the hair cell's afferent synapse. Proc Natl Acad Sci USA 103:5537–5542.

Keithley, EM, Croskrey KL (1990) Spiral ganglion cell endings in the cochlear nucleus of young and old rats. Hear Res 49:169–177.

Keithley EM, Feldman ML (1979) Spiral ganglion cell counts in an age-graded series of rat cochleas. J Comp Neurol 188:429–444.

Keithley EM, Canto C, Zheng QY, Fischel-Ghodsian N, Johnson KR (2004) Age-related hearing loss and the ahl locus in mice. Hear Res 188:21–28.

Keithley EM, Canto C, Zheng QY, Wang X, Fischel-Ghodsian N, Johnson KR (2005) Cu/Zn superoxide dismutase and age-related hearing loss. Hear Res 209:76–85.

Kim S-H, Frisina DR, Frisina RD (2002) Effects of age on contralateral suppression of distortion-product otoacoustic emissions in human listeners with normal hearing. Audiol Neuro-Otol 7:348–357.

Konigsmark BW, Murphy EA (1970) Neuronal populations in the human brain. Nature 228:1335–1336.

Konigsmark BW, Murphy EA (1972) Volume of the ventral cochlear nucleus in man: Its relationship to neuronal population and age. J Neuropathol Exp Neurol 31:304–316.

Krebs J (1998) The role of calcium in apoptosis. Biometals 11:375–382.

Krenning J, Hughes L, Caspary D, Helfert, RH (1998) Age–related glycine receptor subunit changes in the cochlear nucleus of Fischer-344 rats. Laryngoscope 108:26–31.

Kujawa SG, Liberman MC (2006) Acceleration of age–related hearing loss by early noise: Evidence of a misspent youth. J Neurosci 26:2115–2123.

Kujoth GC, Hiona A, Pugh TD, Someya S, Panzer K, Wohlgemuth SE, Hofer T, Seo AY, Sullivan R, Jobling WA, Morrow JD, Van Remmen H, Sedivy JM, Yamasoba T, Tanokura M, Weindruch R, Leeuwenburgh C, Prolla TA (2005) Mitochondrial DNA mutations, oxidative stress, and apoptosisin mammalian aging. Science 309:481–484.

Lambert PR, Schwartz IR (1982) A longitudinal study of changes in the cochlear nucleus in the CBA mouse. Otolaryngol Head Neck Surg 90:787–794.

Lang H, Schulte BA, Schmiedt RA (2002) Endocochlear potentials and compound action potential recovery: functions in the C57BL/6J mouse. Hear Res 172:118–126.

Lang H, Schulte BA, Schmiedt RA (2003) Effects of chronic furosemide treatment and age on cell division in the adult gerbil inner ear. J Assoc Res Otolaryngol 4:164–175.

Lang H, Schulte BA, Zhou D, Smythe NM, Spicer SS, Schmiedt RA (2006a) Nuclear factor κB deficiency is associated with auditory nerve degeneration and increased noise-induced hearing loss. J Neurosci 26:3541–3550.

Lang H, Ebihara Y, Schmiedt RA, Minamiguchi H, Zhou D, Smythe N, M., Liu L, Ogawa M, Schulte BA (2006b) Contribution of bone marrow hematopoietic stem cells to adult mouse inner ear: mesenchymal cells and fibrocytes. J Comp Neurol 496: 187–201.

Le T, Keithley EM (2006) Effects of antioxidants on the aging inner ear. Hear Res 226: 194–202.

Le Prell CG, Yamashita D, Minami SB, Yamasoba T, Miller JM (2006) Mechanisms of noise-induced hearing loss indicate multiple methods of prevention. Hear Res 226: 22–43.

Leutner S, Eckert A, Muller WE (2001) ROS generation, lipid peroxidation and antioxidant enzyme activities in the aging brain. J Neural Transm 108:955–967.

Lu KP, Means AR (1993) Regulation of the cell cycle by calcium and calmodulin. Endocr Rev 14:40–58.

Ma F, Gomez-Martin O, Lee DJ, Balkany T (1998) Diabetes and hearing impairment in Mexican American adults: A population-based study. J Laryngol Otol 112:835–839.

Malpas S, Blake P, Bishop R, Robinson B, Johnson R (1989) Does autonomic neuropathy in diabetes cause hearing deficits? N Zeal Med J 102:434–435.

Martin JS, Jerger JF (2005) Some effects of aging on central auditory processing. J Rehab Res Dev 42:25–44.

McBride DJ, Williams S (2001) Audiometric notch as a sign of noise-induced hearing loss. Occup Environ Med 58:46–51.

McFadden SL, Willott JF (1994) Responses of inferior colliculus neurons in C57BL/6J mice with and without sensorineural hearing loss: effects of changing the azimuthal location of an unmasked pure-tone stimulus. Hear Res 78:115–131.

McFadden SL, Ding D, Reaume AG, Flood DG, Salvi RJ (1999) Age-related cochlear hair cell loss is enhanced in mice lacking copper/zinc superoxide dismutase. Neurobiol Aging 20:1–8.

McFadden SL, Ding D-L, Ohlemiller KK, Salvi RJ (2001) The role of superoxide dismutase in age-related and noise-induced hearing loss: clues from Sod1 knockout mice. In: Willott JF (ed) From Behavior to Molecular Biology. New York: CRC Press, pp. 489–504.

McGeer EGG, McGeer PL (1975) Age changes in the human for some enzymes associated with metabolism of catecholamines, GABA and acetylcholine. In: Ordy JM, Brizzee KR (eds) Neurobiology of Aging. New York: Plenum Press, pp. 287–305.

Mikaelian DO, Warfield D, Norris O (1974) Genetic progressive hearing loss in the C57b16 mouse. Acta Otolaryngol 77:327–334.

Milbrandt JC, Albin RL, Caspary DM (1994) Age-related decrease in GABAb receptor binding in the Fischer 344 rat inferior colliculus. Neurobiol Aging 15:699–703.

Milbrandt JC, Caspary DM (1995) Age-related reduction of [3H]strychnine binding sites in the cochlear nucleus of the Fischer 344 rat. Neuroscience 67(3): 713–719.

Milbrandt JC, Albin RL, Turgeon SM, Caspary DM (1996) $GABA_A$ receptor binding in the aging rat inferior colliculus. Neuroscience 73:449–458.

Milbrandt JC, Hunter C, Caspary DM (1997) Alterations of $GABA_A$ receptor subunit mRNA levels in the aging Fischer rat inferior colliculus. J Comp Neurol 379:455–465.

Miller RA, Chang Y, Galecki AT, Al-Regaiey K, Kopchick JJ, Bartke A (2002) Gene expression patterns in calorically restricted mice: partial overlap with long-lived mutant mice. Mol Endocrinol 16:2657–2666.

Mills JH, Matthews LJ, Lee FS, Dubno JR, Schulte BA, Weber PC (1999) Gender-specific effects of drugs on hearing levels of older persons. Ann NY Acad Sci 884:381–388.

Minowa O, Ikeda K, Sugitani Y, Oshima T, Nakai S, Katori Y, Suzuki M, Furukawa M, Kawase T, Zheng Y, Ogura M, Asada Y, Watanabe K, Yamanaka H, Gotoh S, Nishi-Takeshima M, Sugimoto T, Kikuchi T, Takasaka T, Noda T (1999) Altered cochlear fibrocytes in a mouse model of DFN3 nonsyndromic deafness. Science 285:1408–1411.

Nakashima T, Naganawa S, Sone M, Tominaga M, Hayashi H, Yamamoto H, Liu X, Nuttall AL (2003) Disorders of cochlear blood flow. Brain Res Rev 43:17–28.

Niu X, Canlon B (2002) Protective mechanisms of sound conditioning. Adv Otorhinolaryngol 59:96–105.

Nixon JC, Glorig A (1962) Changes in air and bone conduction threshoulds as a function of age. J Laryngol 76:288–298.

Ohlemiller KK (2002) Reduction in sharpness of frequency tuning but not endocochlear potential in aging and noise-exposed BALB/cJ mice. J Assoc Res Otolaryngol 3: 444–456.

Ohlemiller KK (2003) Oxidative cochlear injury and the limitations of antioxidant therapy. Semin Hear 24:123–133.

Ohlemiller KK (2006) Contributions of mouse models to understanding of age- and noise-related hearing loss. Brain Res 1091:89–102.

Ohlemiller KK, Gagnon PM (2004a) Apical-to-basal gradients in age-related cochlear degeneration and their relationship to 'primary' loss of cochlear neurons. J Comp Neurol 479:103–116.

Ohlemiller KK, Gagnon PM (2004b) Cellular correlates of progressive hearing loss in 129S6/SvEv mice. J Comp Neurol 469:377–390.

Ohlemiller KK, McFadden SL, Ding D-L, Reaume AG, Hoffman EK, Scott RW, Wright JS, Putcha GV, Salvi RJ (1999) Targeted deletion of the cytosolic Cu/Zn-superoxide dismutase gene (SOD1) increases susceptibility to noise-induced hearing loss. Audiol Neuro-Otol 4:237–246.

Ohlemiller KK, McFadden SL, Ding D-L, Lear PM, Ho Y-S (2000) Targeted mutation of the gene for cellular glutathione peroxidase (Gpx1) increases noise-induced hearing loss in mice. J Assoc Res Otolaryngol 1:243–254.

Ohlemiller KK, Lett JM, Gagnon PM (2006) Cellular correlates of age-related endocochlear potential reduction in a mouse model. Hear Res 220:10–26.

Ologe FE, Okoro EO, Oyejola BA (2005) Hearing function in Nigerian children with a family history of type II diabetes. Int J Pediatr Otorhinolaryngol 69:387–391.

O'Neill WE, Zettel ML, Whittemore KR, Frisina RD (1997). Calbindin D-28k immunoreactivity in the medial nucleus of the trapezoid body declines with age in C57BL/6, but not CBA/CaJ, mice. Hear Res 112:158–166.

Otte J, Schuknecht HF, Kerr AG (1978) Ganglion cell populations in normal and pathological human cochleae: Implications for cochlear implantation. Laryngoscope 38:1231–1246.

Park JC, Cook KC, Verde EA (1990) Dietary restriction slows the abnormally rapid loss of spiral ganglion neurons in C57BL/6 mice. Hearing Res 48:275–280.

Pauler M, Schuknecht HF, Thornton AR (1986) Correlative studies of cochlear neuronal loss with speech discrimination and pure-tone thresholds. Arch Otolaryngol 243: 200–206.

Pauler M, Schuknecht HF, White JA (1988) Atrophy of the stria vascularis as a cause of sensorineural hearing loss. Laryngoscope 98:754–759.

Pekkonen E (2000) Mismatch negativity in aging and in Alzheimer's and Parkinson's diseases. Audiol Neuro-Otol 5:216–224.

Pichora-Fuller MK (2003) Processing speed and timing in aging adults: psychoacoustics, speech perception, and comprehension. Int J Audiol (Suppl 1)42:S59–67.

Pickles JO (2004) Mutation in mitochondrial DNA as a cause of presbyacusis. Audiol Neuro-Otol 9:23–33.

Pillsbury HC (1986) Hypertension, hyperlipoproteinemia, chronic noise exposure: Is there synergism in cochlear pathology? Laryngoscope 96:1112–1138.

Polich J, Howard L, Starr A (1985) Effects of age on the P300 component of the event–related potential from auditory stimuli: peak definition, variation, and measurement. J Gerontol 40:721–726.

Prass K, Scharff A, Ruscher K, Lowl D, Muselmann C, Victorov I, Kapinya K, Dirnagl U, Meisel A (2003) Hypoxia-induced stroke tolerance in the mouse in mediated by erythropoietin. Stroke 34:1981–1986.

Pujol R, Rebillard G, Puel J-L, Lenoir M, Eybalin M, Recasens M (1991) Glutamate neurotoxicity in the cochlea: a possible consequence of ischaemic or anoxic conditions occurring in ageing. Acta Otolaryngol (Stockh) 476:32–36.

Pujol R, Puel J-L, D'Aldin CG, Eybalin M (1993) Pathophysiology of the glutamate synapses of the cochlea. Acta Otolaryngol 113:330–334.

Ran R, Xu H, Lu A, Bernaudin M, Sharp FR (2005) Hypoxic preconditioning in the brain. Dev Neurosci 27:87–92.

Raynor EM, Carrasco VN, Prazma J, Pillsbury HC (1995) An assessment of cochlear hair-cell loss in insulin-dependent diabetes mellitus diabetic and noise-exposed rats. Arch Otolaryngol Head Neck Surg 121:452–456.

Raza A, Milbrandt JC, Arneric SP, Caspary DM (1994) Age-related changes in brainstem auditory neurotransmitters: measures of GABA and acetylcholine functions. Hear Res 77:221–230.

Rosen S, Olin P, Rosen HV (1970) Dietary prevention of hearing loss. Acta Otolaryngol 70:242–247.

Rosen S, Bergman M, Plester D, El Mofti A, Satti M (1962) Presbycusis study of a relatively noise-free population in the Sudan. Ann Otol Rhinol Laryngol 71:727–742.

Rosenhall U, Pedersen KE (1995) Presbycusis and occupational hearing loss. Occup Med 10:593–607.

Rosenhall U, Sixt E, Sundh V, Svanborg A (1993) Correlations between presbycusis and extrinsic noxious factors. Audiology 32:234–243.

Ruckenstein MJ, Milburn M, Hu L (1999a) Strial dysfunction in the MRL-Fas[lpr] mouse. Otolaryngol Head Neck Surg 121:452–456.

Ruckenstein MJ, Keithley EM, Bennett T, Powell HC, Baird S, Harris JP (1999b) Ultrastructural pathology in the stria vascularis of the MRL-Fas[lpr] mouse. Hear Res 131:22–28.

Rust KR, Prazma J, Triana RJ, Michaelis OEt, Pillsbury HC (1992) Inner ear damage secondary to diabetes mellitus. II. Changes in aging SHR/N-cp rats. Arch Otolaryngol Head Neck Surg 118:397–400.

Saito T, Sato K, Saito H (1986) An experimental study of auditory dysfunction associated with hyperlipoproteinemia. Archives of Otorhinolaryngol 243:242–245.

Saitoh Y, Hosokawa M, Shimada A, Watanabe Y, Yasuda N, Murakami Y, Takeda T (1995) Age-related cochlear degeneration in senescence-accelerated mouse. Neurobiol Aging 16:129–136.

Salvi RJ, Lockwood AH, Frisina RD, Coad ML, Wack DS, Frisina DR (2002) PET imaging of the normal human auditory system: responses to speech in quiet and in background noise. Hear Res 170:96–106.

Sastre J, Pallardo FV, De La Asuncion JG, Vina J (2000) Mitochondria, oxidative stress and aging. Free Radic Res 32:189–198.

Satar B, Ozkaptan Y, Surucu HS, Ozturk H (2001) Ultrastructural effects of hypercholesterolemia on the cochlea. Otol Neurotol 22:786–789.

Schacht J, Hawkins JE (2005) Sketches of otohistory. Part 9: Presby[a]cusis. Audiol Neuro–Otol 10:243–247.

Schmiedt RA, Lang H, Okamura H, Schulte BA (2002) Effects of furosemide applied chronically to the round window: A model of metabolic presbycusis. J Neuroscience 22:9643–9650.

Schneider BA, Pichora-Fuller MK, Kowalchuk D, Lamb M (1994) Gap detection and the precedence effect in young and old adults. J Acoust Soc Am 95:980–991.

Scholtz AW, Kammen-Jolly K, Felder E, Hussl B, Rask-Andersen H, Schrott-Fischer A (2001) Selective aspects of human pathology in high-tone hearing loss of the aging inner ear. Hear Res 157:77–86.

Schuknecht HF (1953) Lesions of the organ of Corti. Trans Am Acad Ophthalmol Otolaryngol 57:366–383.

Schuknecht HF (1964) Further observations on the pathology of presbycusis. Arch Otolaryngol 80:369–382.

Schuknecht HF (1993) Pathology of the Ear, 2nd ed. Philadelphia: Lea and Febiger.

Schuknecht HF, Gacek MR (1993) Cochlear pathology in presbycusis. Ann Otol Rhinol Laryngol 102:1–16.

Schuknecht HF, Watanuki K, Takahashi T, Belal AA, Kimura RS, Jones DD (1974) Atrophy of the stria vascularis, a common cause for hearing loss. Laryngoscope 84:1777–1821.

Schulte BA, Schmiedt RA (1992) Lateral wall Na,K-ATPase and endocochlear potentials decline with age in quiet-reared gerbils. Hear Res 61:35–46.

Schwaller B, Meyer M, Schiffmann S (2002) 'New' functions for 'old' proteins: the role of the calcium-binding proteins calbindin D-28K, calretinin and parvalbumin, in cerebellar physiology. Studies with knockout mice. Cerebellum 1:241–258.

Seidman MD (2000) Effects of dietary restriction and antioxidants on presbyacusis. Laryngoscope 110:727–738.

Seidman MD, Ahmad N, Bai U (2002) Molecular mechanisms of age-related hearing loss. Ageing Res Rev 1:331–343.

Seldon HL, Clark GM (1991) Human cochlear nucleus: comparison of Nissl-stained neurons from deaf and hearing patients. Brain Res 551:185–194.

Shimada A, Ebisu M, Morita T, Takeuchi T, Umemura T (1998) Age-related changes in the cochlea and cochlear nuclei of dogs. J Vet Med Sci 60:41–48.

Siemens J, Lillo C, Dumont RA, Reynolds A, Williams DS, Gillespie PG, Muller U (2004) Cadherin 23 is a component of the tip link in hair-cell stereocilia. Nature 428:950–955.

Sikora MA, Morizono T, Ward WD, Paparella MM, Leslie K (1986) Diet-induced hyperlipidemia and auditory dysfunction. Acta Otolaryngol 102:372–381.

Sinclair DA (2005) Toward a unified theory of caloric restriction and longevity regulation. Mech Ageing Dev 126:987–1002.

Snell KB, Frisina DR (2000). Relations among age-related differences in gap detection and speech perception. J. Acoust Soc Am 107: 1615–1626.

Snell KB, Mapes FM, Hickman ED, Frisina DR (2002). Word recognition in competing babble and the effects of age, temporal processing, and absolute sensitivity. J Acoust Soc Am 112: 720–727.

Someya S, Yamasoba T, Weindruch R, Prolla TA, Tanokura M (2007) Caloric restriction suppresses apoptotic cell death in the mammalian cochlea and leads to prevention of presbycusis. Neurobiol Aging 28(10): 1613–1622.

Spencer JT (1973) Hyperlipoproteinemia in the etiology of inner ear disease. Laryngoscope 83:639–678.

Spicer SS, Schulte BA (2002) Spiral ligament pathology in quiet-aged gerbils. Hear Res 172:172–185.

Spicer SS, Schulte BA (2005) Pathologic changes of presbycusis begin in secondary processes and spread to primary processes of strial marginal cells. Hear Res 2005: 225–240.

Squier TC (2001) Oxidative stress and protein aggregation during biological aging. Exp Gerontol 36:1539–1550.

Squier TC, Bigelow DJ (2000) Protein oxidation and age-dependent alterations in calcium homeostasis. Front Biosci 5:d504–526.

Stamataki S, Francis HW, Lehar M, May BJ, Ryugo DK (2006) Synaptic alterations at inner hair cells precede spiral ganglion cell loss in aging C57BL/6J mice. Hear Res 221: 104–118.

Starr A, Picton TW, Kim R (2001) Pathophysiology of auditory neuropathy. In: Sininger Y, Starr A (eds) Auditory Neuropathy: A New Perspective on Hearing Disorders San Diego: Singular, pp. 67–81.

Suzuka Y, Schuknecht HF (1988) Retrograde cochlear neuronal degeneration in human subjects. Acta Otolaryngol Suppl 450:2–20.

Suzuki K, Kaneko M, Murai K (2000) Influence of serum lipids on auditory function. Laryngoscope 110:1736–1738.

Swartz KP, Walton JP, Crummer GC, Hantz EC, Frisina RD (1992) P3 event-related potentials and performance of healthy old and Alzheimer's dementia subjects for music perception tasks. Psychomusicol 11: 96–118.

Swartz KP, Walton JP, Hantz EC, Goldhammer E, Crummer GC, Frisina RD (1994) P3 event-related potentials and performance of young and old subjects for music perception tasks. Int J Neurosci 78:223–239.

Sweet RJ, Price JM, Henry KR (1988) Dietary restriction and presbyacusis: periods of restriction and auditory threshold losses in the CBA/J mouse. Audiology 27:305–312.

Tachibana M, Yamamichi I, Nakae S, Hirasugi Y, Machino M, Mizukoshi O (1984) The site of involvement of hypertension within the cochlea. Acta Otolaryngol 97:257–265.

Tadros SF, Frisina ST, Mapes F, Kim S-H, Frisina DR, Frisina RD (2005a) Loss of peripheral right ear advantage in age-related hearing loss. Audiol Neuro-Otol, 10:44–52.

Tadros SF, Frisina ST, Mapes F, Frisina DR, Frisina RD (2005b) Higher serum aldosterone correlates with lower hearing thresholds: a possible protective hormone against presbycusis. Hear Res 209:10–18.

Takumida M, Anniko A (2005) Radical scavengers: a remedy for presbycusis. A pilot study. Acta Otolaryngol 125:129–1293.

Tarnowski BI, Schmiedt RA, Hellstrom LI, Lee FS, Adams JC (1991) Age-related changes in cochleae of Mongolian gerbils. Hear Res 54:123–134.

Toescu EC (2005) Normal brain ageing: models and mechanisms. Philos Trans R Soc Lond B 360:2347–2354.

Toppila E, Pyykko I, Starck J (2001) Age and noise-related hearing loss. Scand Audiol 30:236–244.

Torre P, Cruickshanks KJ, Klein BEK, Nondahl DM (2005) The association between cardiovascular disease and cochlear function in older adults. J Speech Lang Hear Res 48:473–481.

Trune DR, Kempton JB (2002) Female MRL.MpJ-Fas[lpr] autoimmune mice have greater hearing loss than males. Hear Res 167:170–174.

Trune DR, Kempton JB, Gross ND (2006) Mineralocorticoid receptor mediates glucocorticoid treatment effects in the autoimmune mouse ear. Hear Res 212:22–32.

Uchida Y, Nakashima T, Ando F, Niino N, Shimokada H (2005) Is there a relevant effect of noise and smoking on hearing? A population-based aging study. Int J Audiol 44:86–91.

Ulchlova L (1983) Stria vascularis in acoustic trauma. Arch Otorhinolaryngol 237: 133–138.

Unal M, Tamer L, Dogruer ZN, Yildirim H, Vayisoglu Y, Camdevirin H (2005) N-acetyltransferase 2 gene polymorphism and presbycusis. Laryngoscope 115:2238–2241.

Varghese GI, Zhu XX, Frisina, RD (2005) Age-related declines in contralateral suppression of distortion product otoacoustic emissions utilizing pure tones in CBA/CaJ mice. Hear Res 209:60–67.

Vaughan N, James K, McDermott D, Griest S, Fausti S (2005) A 5-year prospective study of diabetics and hearing loss in a veteran population. Otol Neurotol 27:37–43.

von Zglinicki T, Martin-Ruiz CM (2005) Telomeres as biomarkers for aging and age–related disease. Curr Mol Med 5:197–203.

Wackym PA, Linthicum FH (1986) Diabetes mellitus and hearing loss: clinical and histopathologic relationships. Am J Otol 7:176–182.

Walton JP, Frisina RD, Ison JE, O'Neill WE (1997) Neural correlates of behavioral gap detection in the inferior colliculus of the young CBA mouse. J Comp Physiol A 181:161–176.

Wang Y, Liberman MC (2002) Restraint stress and protection from acoustic injury in mice. Hear Res 165:96–102.

Wang Y, Manis P B (2005) Synaptic transmission at the cochlear nucleus end bulb synapse during age-related hearing loss in mice. J Neurophysiol 94: 1814–1824.

Wangemann P (2002) K$^+$ recycling and the endocochlear potential. Hear Res 165:1–9.

Whitlock NA, Agarwal N, Ma JX, Crosson CE (2005) Hsp27 upregulation by HIF-1 signaling offers protection against retinal ischemia in rats. Inv Ophthalmol Visual Sci 46:1092–1098.

Willott JF (1991). Aging and the Auditory System: Anatomy, Physiology, and Psychophysics Sam Diego: Singular.

Willott JF, Bross LS (1990) Morphology of the octopus cell area of the cochlear nucleus in young and aging C57BL/6J and CBA/J mice. J Comp Neurol 300:61–81.

Willott JF, Bross LS (1996) Morphological changes in the anteroventral cochlear nucleus that accompany sensorineural hearing loss in DBA/2J and C57BL/6J mice. Dev Brain Res 91:218–226.

Willott JF, Turner JG (1999) Prolonged exposure to an augmented acoustic environment ameliorates age-related auditory changes in C57BL/6J and DBA/2J mice. Hear Res 135:78–88.

Willott JF, Jackson LM, Hunter KP (1987) Morphometric study of the anteroventral cochlear nucleus of two mouse models of presbycusis. J Comp Neurol 260:472–480.

Willott JF, Parham K, Paris Hunter K (1991) Comparison of the auditory sensitivity of neurons in the cochlear nucleus and inferior colliculus of young and aging C57BL/6J and CBA/J mice. Hear Res 53: 78–94.

Willott JF, Bross LS, McFadden SL (1992) Morphology of the dorsal cochlear nucleus in C57BL/6J and CBA/J mice across the life span. J Comp Neurol 321:666–678.

Willott JF, Milbrandt JC, Seegers Bross L, Caspary DM (1997) Glycine immunoreactivity and receptor binding in the cochlear nucleus of C57BL/6J and CBA/CaJ mice: effects of cochlear impairment and aging. J Comp Neurol 385:405–414.

Willott JF, Turner JG, Carlson S, Ding D, Bross LS, Falls WA (1998) The BALB/c mouse as an animal model for progressive sensorineural hearing loss. Hear Res 115:162–174.

Wright CG, Schuknecht HF (1972) Atrophy of the spiral ligament. Arch Otolaryngol 96:16–21.

Wu W-J, Sha S, McLaren JD, Kawamoto K, Raphael Y, Schacht J (2001) Aminoglycoside ototoxicity in adult CBA, C57BL, and BALB mice and the Sprague-Dawley rat. Hear Res 158:165–178.

Xiu A–P, Kikuchi T, Minowa O, Katori Y, Oshima T, Noda T, Ikeda K (2002) Late-onset hearing loss in a mouse model of DFN3 non-syndromic deafness: morphologic and immunohistochemical analyses. Hear Res 166:150–158.

Yamashita H, Shimogori H, Sugahara K, Takahashi M (1999) Cell proliferation in spiral ligament of mouse cochlea damaged by dihydrostreptomycin sulfate. Acta Otolaryngol 119:322–325.

Yoshida N, Kristiansen A, Liberman MC (1999) Heat stress and protection from permanent acoustic injury in mice. J Neurosci 19:10116–10124.

Zettel ML, Frisina RD, Haider SEA, O'Neill WE (1997) Age-related changes in calbindin D–28K and calretinin immunoreactivity in the inferior colliculus of CBA/CaJ and C57B1/6 mice. J Comp Neurol 386:92–110.

Zettel ML, O'Neill WE, Trang TT, Frisina RD (2001) Early bilateral deafening prevents calretinin up-regulation in the dorsal cortex of the inferior colliculus of aged CBA/CaJ mice. Hear Res 158:131–138

Zimmermann CE, Burgess BJ, Nadol JB (1995) Patterns of degeneration in the human cochlear nerve. Hear Res 90:192–201.

7
Patterns and Mechanisms of Noise-Induced Cochlear Pathology

Donald Henderson, Bohua Hu, and Eric Bielefeld

1. Introduction

Exposure to high levels of noise is the most common cause of hearing loss in the adult population. The exact number of people at risk is not known because it is not only the people in workplace settings (10 million in the United States alone), but also people who listen to loud music, hunt, or ride noisy vehicles are also at risk. The magnitude and frequency profile of the hearing loss is related to the parameters of the noise: intensity, duration, and temporal characteristics of the exposure (continuous, intermittent, or impulse noise). In addition, the effects of noise can be exacerbated by exposure to chemicals such as organic solvents or by taking certain ototoxic drugs. The cochlear pathology associated with noise-induced hearing loss (NIHL) involves all the cellular systems of the cochlea. The pattern and extent of the cochlear pathology (i.e., the number of missing cells, transient changes in VIIIth nerve synapses, etc.) are also influenced by whether the cochlea is examined immediately after the exposure or at 20–60 days later when the hearing loss and cochlear damages have stabilized.

2. Cochlear Pathology

The cochlea is a complex biological system that is highly energy consumptive. When operating normally, the coding and transduction of sounds require the endocochlear potential provided by the stria vascularis and the integration of the movement of the basilar membrane with the action of the outer hair cells (OHCs), inner hair cells (IHCs), and the neural fibers of the VIIIth nerve. As seen in Fig. 7.1, the cochlea is vulnerable to noise exposure at each of the cellular systems, i.e., supporting cells, vascular supply, sensory cells, and nerve cells.

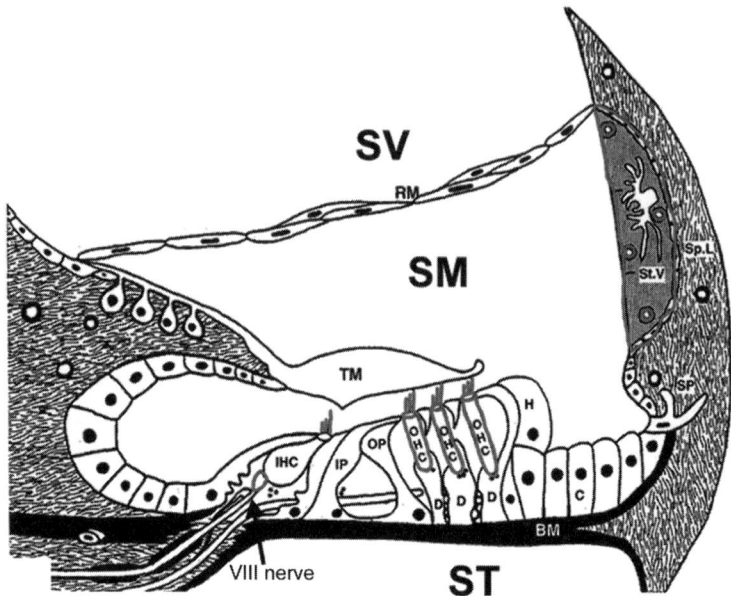

FIGURE 7.1. Cross section of a turn of the cochlea showing the major cell types. The most vulnerable cells are the outer hair cells (OHC); Deiters' cells (D); pillar cells (IP and OP); inner hair cells (IHC), and stria vascularis (St.V.).

3. Supporting Cells

The cochlear partition is 200 times stiffer at the base than at the apex (von Békésy 1952), which creates an impedance gradient from base to apex that allows the cochlea to function as a highly tuned acoustic analyzer (i.e., high frequencies create maximum vibration at the base; low frequencies at the apex). The mechanical characteristics of the cochlea are critical for proper processing of acoustic stimuli. Figure 7.2 provides examples of mechanical damage caused by noise exposure. The left side of Fig. 7.2 shows the normal orientation of outer pillar cells; the cells are anchored at the basilar membrane and angle up to link with the inner pillar cells. The right side shows noise-damaged outer pillar cells after exposure to 155-dB peak equivalent impulse noise. The interference phase-contrast photo was taken within 30 min after an exposure and the pillar cells are ripped from the basilar membrane. The pillar cells are major supporting elements to organ of Corti and their detachment will lead to an abrupt change in the basilar membrane impedance gradient, which will affect both sensitivity and mechanical tuning in the region of the pillar damage. The permanent consequences of the pillar cell pathology are not known. It may lead to cell death, or the pillar cells may reattach.

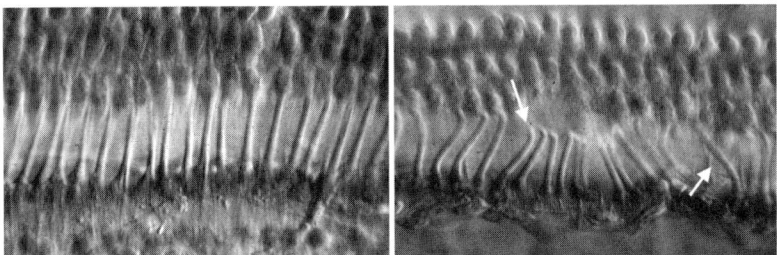

FIGURE 7.2. Phase-contrast views of outer pillar cells. Left panel shows normal anatomy. Right panel is after exposure to 50–155 dB pSPL impulses. Notice (arrows) the detachment at the level of the cuticular plate.

Noise exposure can also distort or break OHC stereocilia (Fig. 7.3). Normally the tallest stereocilia are embedded in the tectorial membrane (Fig. 7.1) and the movement of the cochlear partition causes a shearing motion between the bottom of the tectorial membrane and the top of the organ of Corti, which directly moves the stereocilia either toward the stria vascularis or back toward the modiolus (Davis 1952). The integrity of the stereocilia connection with the tectorial membrane is critical in the transduction process, because the movement of the stereocilia toward the stria vascularis opens mechanically gated transduction channels in the top of the stereocilia, permitting K^+ from the endolymph to enter the stereocilia to depolarize the sensory cells (Hudspeth and Jacobs 1979). Saunders et al. (1985), Pickles et al. (1985), and Liberman et al (1985) have reviewed stereocilia pathology and its consequences for VIIIth nerve firing patterns and hearing loss.

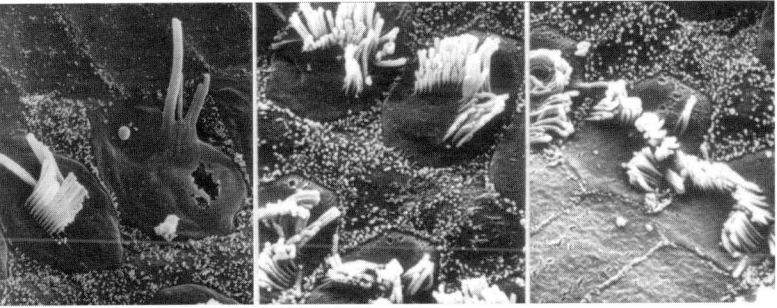

FIGURE 7.3. Examples of outer hair cell (OHC) stereocilia pathology, including fused stereocilia (left), disrupted bundles (middle), and collapsed stereocilia (right). (From Henderson et al. 2006)

4. Sensory Cells (Inner and Outer Hair Cells) and Transduction

Traditionally, the OHC are considered the most vulnerable cells in the cochlea. While the OHCs are sensory cells, they are innervated by fewer than 5% of the afferent VIII nerve fibers. The OHCs' most prominent nerve endings are the efferent fibers from the medial superior olivary nuclei. The pioneering work of Brownell (1984) showed that the OHC are motile and either elongate or contract because of the synchronized depolarization caused by the stereocilia movement. Because the OHCs are linked to the basilar membrane, the motility of the OHCs enhances both the sensitivity and tuning of the organ of Corti (Cody and Russell 1985). Figure 7.4 shows the organ of Corti with normal and damaged OHCs. In the chinchilla (*chinchilla laniger*), it is common to have a hearing loss up to 40 or 50 dB with complete losses of OHC. Also, it is common to have normal appearing VIIIth nerve fibers with scattered losses of OHCs and IHCs. Spiral ganglion fibers are found missing only when there is a region of the cochlea with complete loss of OHC.

Following their entry through transduction channels, the K^+ ions are circulated out of the bottom of the hair cells through the Henson and Claudius cells and by the fibrocytes in the spiral ligament back to the stria vascularis (Spicer and Schulte 1996). These mechanisms are discussed in detail in Wangemann, Chapter 3).Wang et al. (2002) showed that for certain noise exposures, the type II fibrocytes of the spiral ligament are among the cells most susceptible to damage and their pattern of loss most closely corresponds to the spectrum of the noise exposure (Fig. 7.5). The relationship between fibrocyte viability, hair cell loss, and hearing loss is not well understood. For example, it is not known whether fibrocytes regenerate or whether a specific loss can be compensated by functioning fibrocytes bordering on each side of the lesion.

The IHC are quite resistant to the effects of high-level noise exposure. When there is a complete loss of OHC in a region, the IHC and VIII nerve fibers

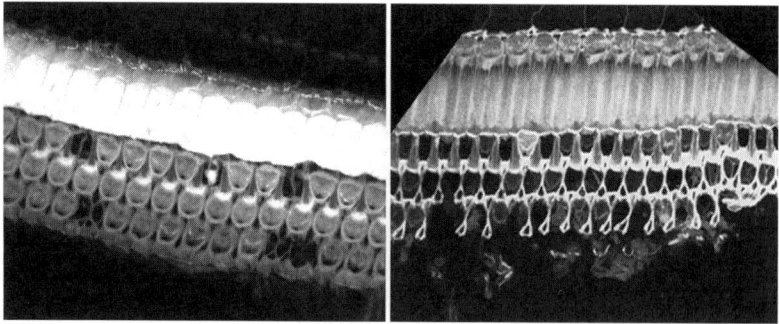

FIGURE 7.4. Surface preparation of the organ of Corti. Tissue is stained with FITC-labeled phalloidin to label actin in cells. Left: Mild cochlear damage with two missing OHC in the first row and three in the third row. Right: More severely damaged OHC. Notice that inner hair cells (arrow) appear to be intact.

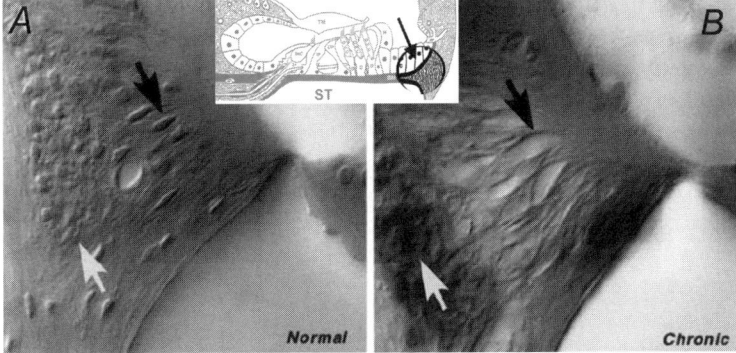

FIGURE 7.5. Fibrocytes located in the outer sulcus region (inset). Left: Distribution of normal fibrocytes. Right: Loss of fibrocytes after exposure to traumatic noise. Note that the lost cells are localized to the region of the cochlea associated with the spectrum of the exposure. (From Wang et al. 2002.)

can also be lost. In the chinchilla model of NIHL, permanent losses of 40 dB or greater are typically associated with both IHC and OHC lesions. There are, however, temporary excitotoxic effects of noise as seen at the IHC/VIIIth nerve synapse (Puel et al. 1996, 1998; Pujol et al. 1990, 1993). The VIIIth nerve synapse swells in reaction to high-level noise exposure due to the high rate of synaptic activity associated with intense noise exposures (Fig. 7.6). A buildup

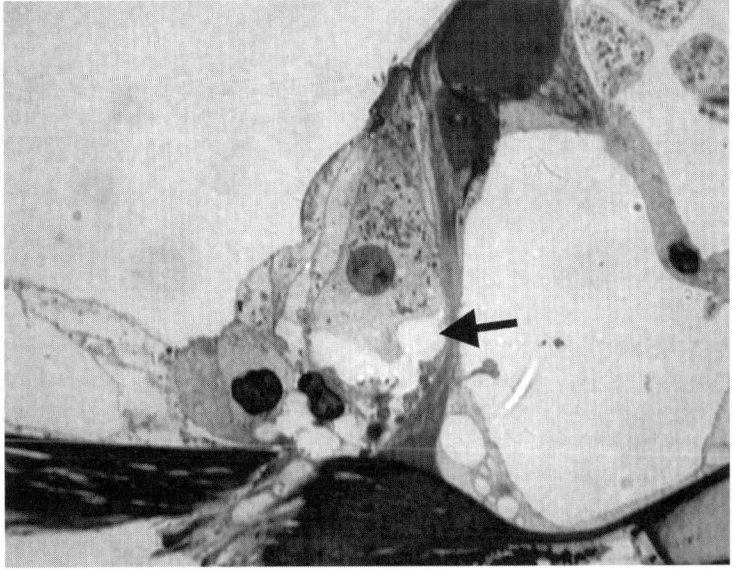

FIGURE 7.6. Inner hair cell (IHC) 30 min after exposure to traumatic noise. Notice the swollen VIIIth nerve endings and the distorted outline at the base of the IHC.

of glutamate in the VIIIth nerve dendrites occurs because of the inability of the IHC/VIIIth nerve synapse to recycle glutamate (see Wangemann, Chapter 3). Consequently, glutamate that accumulates in the dendritic terminal creates the condition of excitotoxicity, characterized by swelling of the postsynaptic cell bodies and dendrites (Kandel et al. 2000; Pujol et al. 1990). Interestingly, these excitotoxic reactions at the synapse are repairable and may contribute only to the temporary threshold shift (TTS) component of the hearing loss (Zheng et al. 1997b).

5. Cochlear Vascular System

The role of the cochlear vascular system in NIHL is complicated. Under normal homeostatic conditions, cochlear blood flow (CBF) is controlled by a combination of factors, including systemic changes (Miller and Dengerink 1988), sympathetic influence over the cochlear vasculature (Laurikainen et al. 1994, 1997), and local autoregulation (Miller et al. 1995; Konishi et al. 1998). The degree to which each of these factors affects CBF under normal conditions and under traumatic conditions is currently unclear. As detailed in a review by Miller and Dengerink (1988), CBF was once thought to be a passive response to systemic blood flow in the body. Clearly, CBF is influenced strongly by systemic changes in the body, but the cochlea has its own mechanisms of altering blood flow that enable it to modulate or fine tune the blood supply with which it is provided. . The relative contribution of the two local elements of sympathetic nervous innervation and autoregulation is not completely clear. Cochlear sympathetic influence on blood flow is mediated heavily by a bilateral innervation from the stellate ganglion (Laurikainen et al. 1993,1997), as well as a secondary influence from unilateral–ipsilateral innervation from the superior cervical ganglion (Ren et al. 1993).

Autoregulation refers to the cochlear vasculature's local intrinsic factors that can alter blood blow. The cochlear vasculature is very sensitive to the carbon dioxide content of the blood (Kallinen et al. 1991; Ugnell et al. 2000), but a number of additional factors have been implicated in local autoregulation of CBF, including: nitric oxide (Fessenden and Schacht 1998), prostaglandins (Nagahara et al. 1988), and tropomyosin (Konishi et al. 1998).

The vascular system's response to a potentially traumatic noise varies with the type of noise. For example, the trauma associated with a high-level impulse noise can be an instantaneous mechanical failure that occurs before the vascular system reacts; by contrast, long-duration noise exposure can modulate blood flow in patterns that vary with the intensity and duration of the exposure (Perlman and Kimura 1962; Thorne and Nuttall 1987; Yamane et al. 1995; Lamm and Arnold 2000; Miller et al. 2003). With a continuous noise exposure, there may be first an increase in blood and then a decrease or active blocking of cochlear vessels (Fig. 7.7). What remains unclear is the extent to which CBF

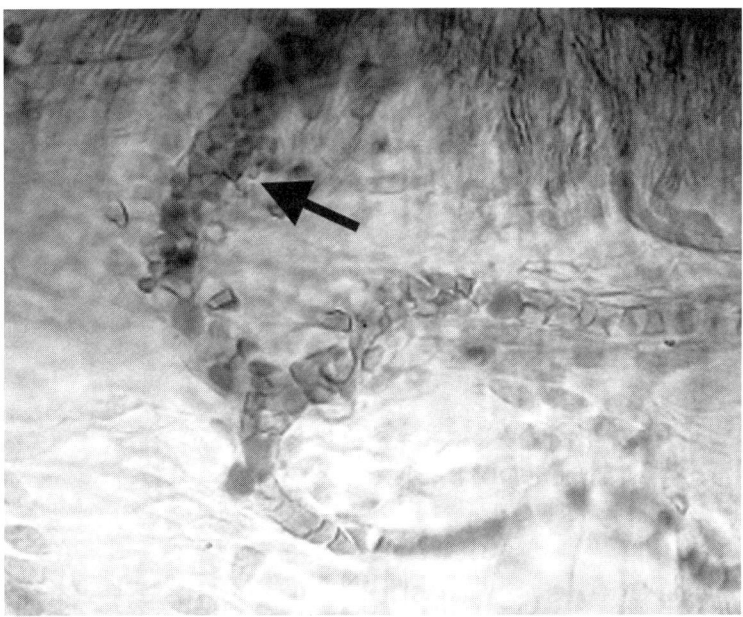

FIGURE 7.7. A capillary from spiral ligament 30 min after traumatic noise exposure. Note the accumulation of red blood cells blocking blood flow.

changes during and after noise might be influencing the cochlear pathology, or, conversely, whether the CBF changes are the result of noise-induced cochlear pathology (Miller et al. 2003). The actual influence of cochlear blood flow on the magnitude of NIHL is made more difficult to determine because of the complexity of the network of vessels supplying the cochlea. However, it is clear that reducing O_2 by breathing low levels of CO (Chen and Fechter 1999) during an exposure increases the hearing loss. In addition, interruption of the sympathetic nervous innervation of the cochlea reduces susceptibility to noise, although the effect is fairly small (Borg 1982; Hildesheimer et al. 1991, 2002). The possible influence of changes in CBF on noise-induced cochlear pathology is discussed further in Section 10.

6. Acoustic Characteristics of Noise and Patterns of HL and Cochlear Pathology

The patterns of hearing loss and cochlear pathology are related to both the response of the basilar membrane and the acoustic characteristics of the noise. In a classic study, Davis and colleagues showed that short duration (1–20 min) higher-frequency exposures (i.e., 4 kHz) produced a temporary hearing loss

(TTS) that was focused at 1/2 to 1 octave above 4 kHz (Davis et al. 1950). By contrast, low-frequency exposures (i.e., 0.5-kHz tones) produced a hearing loss that spanned the 500–8000 Hz range.

Large-scale epidemiological studies of NIHL in workplace or laboratory studies with white noise both show a hearing loss centered at 4 kHz. The 4-kHz notch is a hallmark of NIHL and is the consequence of both acoustic transformations and biochemical factors. When broad-band, high-level noise enters the pinna and external meatus, the spectrum of the noise is changed to a band-passed noise tuned to approximately 3 kHz. The transformation is the result of quarter wave resonance response of the pinna and external meatus [resonant frequency = (speed of sound)/(4 × length of the EAM)]. As the 3-kHz band of noise stimulates the cochlea, the displacement of the traveling wave frequency occurs not at the normal 3 kHz but at a place 1/2 to 1 octave above the center frequency (Sellick et al. 1982). Consequently, a flat spectrum noise primarily stresses the 4-kHz region of the cochlea. Consequently, a broad-band noise produces a high-frequency "notch" audiogram because of the mechanical transformation at the EAM and basilar membrane.

7. High-Level Transients

Exposure to impulse noise (gunfire or any other explosive event) or high-level impact noise (two hard objects hitting together forcefully) can lead to direct and pervasive mechanical failure in the cochlea (see Henderson and Hamernik 1986 for review). Figure 7.8 shows a scanning electron microscope view of a chinchilla cochlea 30 minutes after exposure to impulse noise at 155 dB pSPL. Several points are interesting. First, the organ of Corti is ripped from the basilar membrane to a relatively large extent (arrows). The sensory cells of the detached organ of Corti will not recover and the tissue will be digested by surrounding tissue. Second, at a point basal to the detached organ of Corti, there is a cleft between the first and second row of OHCs. This cleft will allow endolymph to enter the organ of Corti, creating large osmotic and ionic changes across the OHC membranes which cause additional cell death. These mechanical failures as seen in Fig. 7.8 happen because of the extreme acceleration and displacement associated with impulse noise.

Impulse and impact noise can have another, more subtle, but also traumatic, effect on the cochlea. Noise exposures that are a combination of moderate levels of impact/impulse noise and continuous noise are much more traumatic to the ear than the simple additive effects of either noise alone (see review by Henderson and Hamernik 1986). There are examples of combination exposures in working populations. For example, construction workers exposed to a combination of continuous noise and impact noise develop larger hearing losses

FIGURE 7.8. Scanning electron microscopic view of partially dissected cochlea showing OHC separated from the basilar membrane. Right view shoes the region adjacent the dissociated OHC. Note the split between the rows of OHCs, which provides an open channel for endolymph to enter the organ of Corti. Also, IHCs appear normal.

than would be expected on the basis of their A-weighted noise dose (Sweeney et al. 2005).

8. Temporary Effects of Noise

After a damaging noise is turned off, hearing recovers either to preexposure levels or to some permanent threshold shift (PTS). The pathology associated with PTS has been well studied, and key features have been described in this chapter. The pathological changes associated with noise-induced temporary threshold shift (TTS) are not as well understood. Possible pathologies underlying TTS include a detachment of the tectorial membrane from the tips of the stereocilia (Nordmann et al. 2000), excitotoxicity of the VIIIth nerve synapses at the IHC (see Section 4), and partial depolymerization of actin in the supporting cells (Fig. 7.9). The biological basis of TTS is an interesting open question. Studies that correlate TTS with PTS do not show strong correlations, suggesting that the fundamental pathologies of TTS are different from those of PTS (Ward 1966).

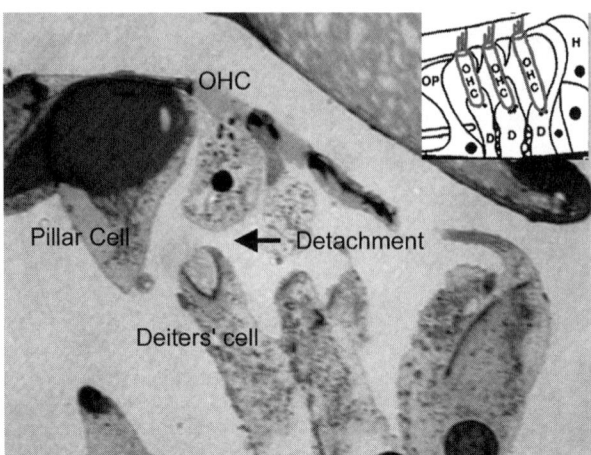

FIGURE 7.9. Section of organ of Corti showing the tops of pillar cells and an OHC separated from the Deiters' cup. Notice that the normally cylindrical OHC has shrunk and is more oval shaped. The OHC nucleus is shrunken and going through apoptosis. The animal was exposed to impulse noise at 155 dBpSPL. (From Henderson et al. 2006)

9. Dynamics of Cochlear Pathology

We have known for some time that the cochlear pathology, especially hair cell loss, continues to increase for approximately 30 days after an exposure (Hamernik et al. 1984; Bohne 1999). Recent studies by Hu et al. (2002) have shown that after a high-level, short-duration exposure, there can be a small focal lesion that primarily involves OHCs. In the next few days, the lesion continues to expand, primarily in the basal direction and with cells dying by both necrosis and apoptosis (Fig. 7.10).

The detailed anatomical studies of the growth of the cochlear pathology have used primarily high-level, short-duration exposures. In an interesting study of the relationship between long-term TTS or PTS (Bohne and Clark 1982), chinchillas were exposed to noise for 24 hours a day for up to 6 months. The animals developed a stable level of threshold shift by 24 hours and this level did not significantly change for up to 6 months; consequently the change in hearing sensitivity is referred to as asymptotic threshold shift (ATS). When the chinchillas were removed after only several days of exposure, hearing completely recovered and there was essentially no permanent hearing loss or cochlear pathology. However, if the subjects were exposed for 6 months, the recovery of hearing sensitivity was minimal and there was a large hair cell loss. Mills and colleagues (1981) systematically studied the ATS phenomenon and found that the level of ATS at any frequency was determined by the spectral level at that frequency; and for noise exposures at the threshold for creating hearing loss, the

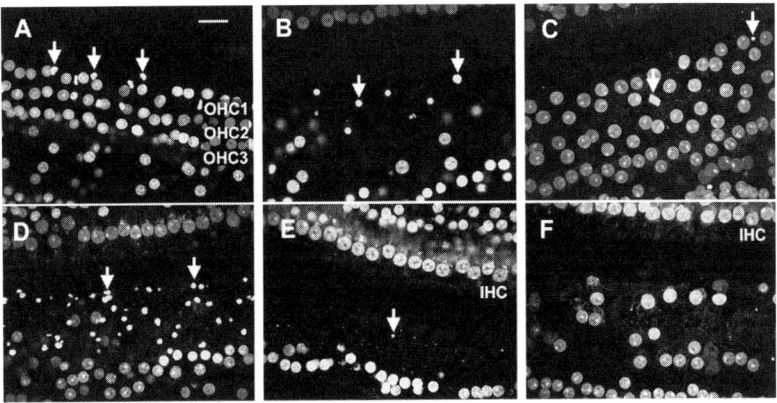

FIGURE 7.10. Top row shows the organ of Corti 30 min after a traumatic noise exposure. Bottom row is 2 days later. (**B**) is focused on the center of the lesion where there are many missing cells and a few cells with condensed nuclei that are going through apoptosis. (**A**) is at the basal margin of lesion **B** and most of the OHCs are present, but there are a few shrunken nuclei. (**C**) is at the apical part of the lesion. Most cells are present, and there are only two shrunken nuclei. Two days later (**E**), the center of the lesion shows no OHCs. The basal margin of the lesion (**D**) shows that all OHC are shrunken and will eventually die. The cells at the apical margin of the lesion (**F**) are stable.

level of ATS grew at the rate of 1.7 dB increase in hearing loss for each dB of increase in the noise level.

For higher-level impact noise as found in factories and construction sites or impulse noise as found in military settings, the relationship between the impulse/impact level and hearing loss has a growth constant of 3–5 dB. In systematic studies of impact noise, Henderson and colleagues (1991) found that, for lower level impacts (99–115 dBA), the increase in hearing loss with increase in impact level was about 1.9 dB. However, above 120 dB pSPL, the hearing loss grows at a rate of 3–5 dB for each 1-dB increase in peak level. The data suggest that when the peak level of an exposure exceeds a "critical level" the mode of cochlear damage shifts to a direct mechanical failure, particularly at the points of adhesion between cells in the organ of Corti (as seen in Fig. 7.8). The "critical level" is not fixed but varies with both the species and signature of the impact or impulse noise. Spoendlin (1985) postulated that the "critical level" for guinea pigs was approximately 120 dBA for noise bursts of 100 ms; for chinchillas exposed to impact noise (50–150 ms) the "critical level" is between 119 and 125 dB peak and for impulse noise the "critical level" is between 150 and 155 dB peaks. Given the differences in sensitivity and conductive function, the critical level for humans is probably approximately 10 dB higher (Mills and colleagues 1981).

10. Noise as a Stressor to the Cochlea

The cochlea normally operates at a high level of metabolism (Thalmann et al. 1975). The major demand for energy is at the stria vascularis, which constantly extrudes K^+ ions as it maintains the ionic balance and polarity of the endolymph (see review by Wangemann 2002 and Wangemann, Chapter 3). The high energy demands are supported by a large population of mitochondria in marginal, intermediate, and basal cells of stria vascularis.

The mitochondrial electron transport chain has long been recognized as a major source of reactive oxygen species (ROS) inside of cells. Under normal physiological conditions, 98% of molecular oxygen (O_2) consumed by the mitochondria is used to promote phosphorylation of ADP to generate ATP. The 1% to 2% of O_2 that is not consumed is converted to superoxide ($O_2 \cdot^-$) or hydrogen peroxide (H_2O_2) at mitochondrial or extramitochondrial locations (Chance et al. 1979). In a number of pathological processes or in the presence of drugs, toxins, electron chain inhibitors and uncouplers, the mitochondrial generation of ROS can increase several-fold (Turrens et al. 1982). With high-level noise exposure, there is a large increase in cochlear ROS generation because of two factors. First, high-level noise drives cochlear metabolism at a much faster and demanding rate. Therefore, the number of free radicals generated increases. Second, to exacerbate the situation, noise also influences CBF (Miller and Dengerink 1988). When the blood flow is reduced, ischemia develops in the organ of Corti and there is a shortage of O_2 for mitochondrial operation leading to an even greater rate of superoxide generation. Conversely, with reperfusion (the return of blood flow to its preischemic level) there is an increased blood flow which also increases availability of O_2 to the mitochondria, resulting in another burst of superoxide generation (Halliwell and Gutteridge 1999). Finally, if the cells of the cochlea are damaged, then cellular contents can be spilled into the extracellular matrix. Trace amounts of iron from the cell create the condition for the Fenton reaction which can produce the highly reactive and toxic hydroxyl radical ($OH^\bullet$) from hydrogen peroxide (Beauchamp and Fridovich 1970). Redox homeostasis is discussed in detail in Wangemann, Chapter 3.

Several studies have reported increased activity of reactive oxygen species (ROS; free radicals) following traumatic noise exposure in eluates from the cochlea (Ohlemiller et al. 1999) or localized to marginal cells of the stria vascularis (Yamane et al. 1995). In the organ of Corti, Nicotera et al. (1999) found increased ROS activity around the basal pole of the OHC and along the neural plexus under the IHC (Fig. 7.11) (Henderson et al. 2006). Increased ROS activity continues for several days (Fig. 7.12) after exposure to traumatic noise (Hu et al. 2002; Yamashita et al. 2004). The persistent ROS activity is interesting because hair cells continue to die for days after an exposure (Bohne 1976; Hamernik et al. 1985).

The significance of the free radical activity in the cochlea raises a fundamental question: Is the free radical activity the consequence of dying cells or is the cell death initiated by increased free radical activity? One approach to this question

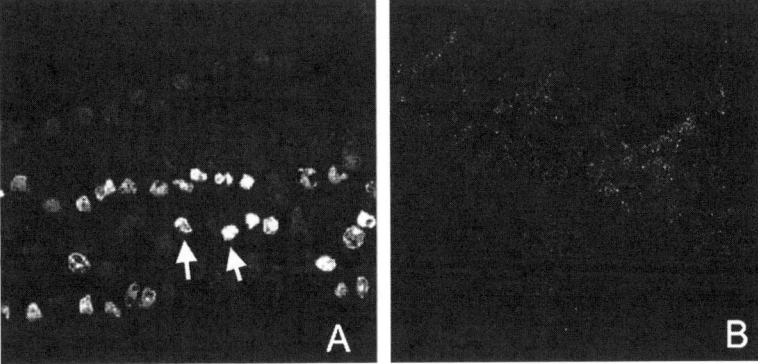

FIGURE 7.11. OHC region of the cochlea stained with dihydroethidene (DE) 15 min after traumatic exposure. DE reacts with superoxide. Notice the reaction in the OHC. The right panel shows the same of the cochlea, but at 1 h after the exposure. There is no DE reaction, which suggests that superoxide is present during and after an exposure, but that superoxide is not a factor for long after the exposure.

was to stress the cochlea with increased free radical activity by exposing the cochlea to paraquat, an herbicide that reacts with molecular O_2 to create $O_2^{\cdot-}$ radicals (Nicotera et al. 2004; Bielefeld et al. 2005). Figure 7.13 shows the comparison of cell death patterns induced by paraquat and exposure to noise. There are similarities in the pattern of the two pathologies. Both have damaged OHCs while the IHCs are relatively intact. Additional experiments by Bielefeld et al. (2005) show that paraquat creates a high-frequency based hearing loss. The significance of the paraquat experiments is that the superoxide activity and the activity of other downstream ROS are a sufficient cause to create a pattern of cochlear pathology very similar to the pathology found with noise exposure, but without the mechanical stress associated with noise.

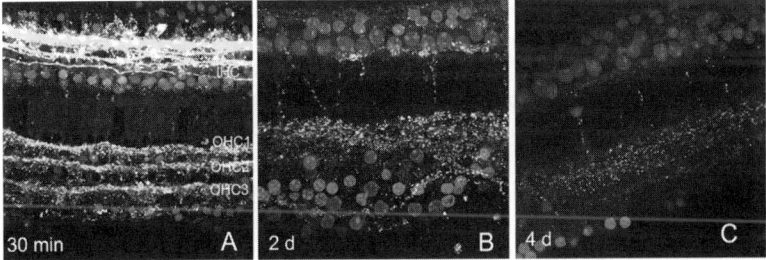

FIGURE 7.12. Organ of Corti at labeled with dichlorofluorescein at 30 min (**A**), 2 days (**B**), and 4 days (**C**) after noise exposure. Notice the bright reaction at 30 min, but the reaction persists for at least 4 days. (From Henderson et al. 2006.)

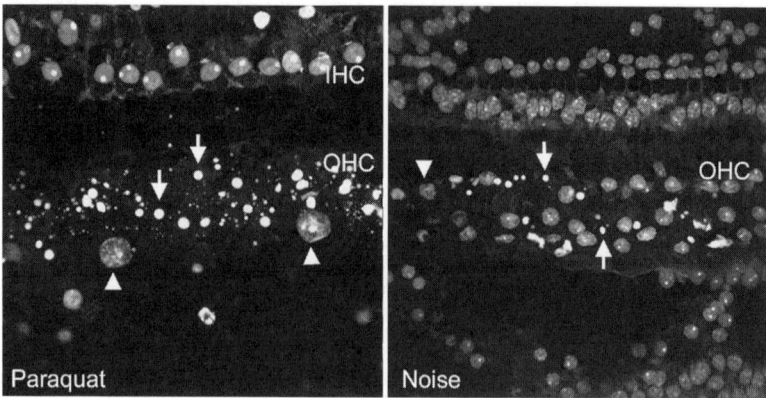

FIGURE 7.13. The left panel shows the organ of Corti stained with propidium iodide 1 h after paraquat was placed on the round window. The right panel shows the organ of Corti 18 h after exposure to traumatic noise. Notice that both conditions produce a very similar pattern of histology. The IHCs are intact after each treatment. The OHCs are in the process of dying through apoptosis (arrows) and necrosis (arrowheads).

11. Pathways of Sensory Cell Death

In the last few years, Hu et al. (2002) and Nicotera et al. (2003) have reported that noise exposures can produce both necrotic and apoptotic cell death. Both types of cell death are illustrated in Fig. 7.14. The OHC with swollen nuclei are dying by necrosis. The cell membrane has been compromised, Ca^{2+} and water has leaked into the cell and expanded the cell volume. The cell eventually ruptures and spills its contents into the surrounding area. The trace elements of the cell content will be available to react with H_2O_2 and create the very reactive and toxic $OH^{\bullet}$ (Ohlemiller et al. 1999). OHCs with condensed nuclei

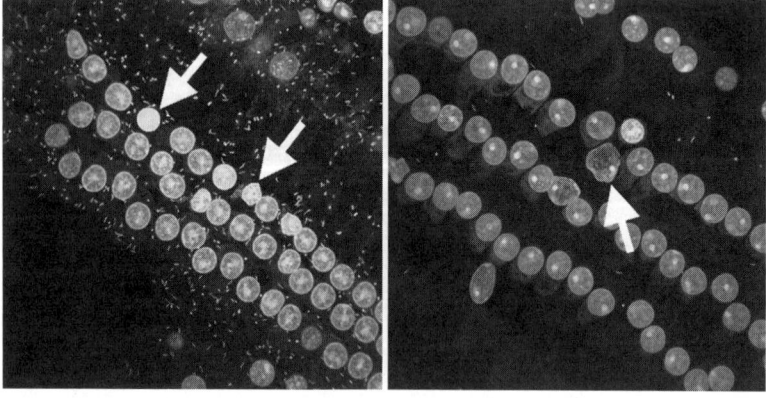

FIGURE 7.14. Examples of apoptotic cell death (left) and necrotic cell death (right).

are dying by apoptosis, in which the proteins of the cell are being disassembled. Apoptosis is an active process and both the cell membrane and mitochondria continue to function. In a normal functioning body, apoptosis is a very useful mechanism for ridding tissues of unwanted cells. For example, in the developing brain, apoptosis is used to reduce drastically the number of neurons in order to maximize the efficiency of the pathways that remain. For this streamlining to be effective, the cells must be eliminated in an organized and controlled way. The cell death pathway of apoptosis allows for controlled cell death that prevents damage to the neighboring cells that survive. In the case of noise trauma, OHC death is problematic because the cells do not regenerate. Therefore, the loss of OHC, even through the controlled death of apoptosis, leaves the cochlea in an impaired state because OHCs are essential for maximal sensitivity and tuning.

Apoptosis is regulated by a family of enzymes called caspases (see also Green, Altschuler, and Miller, Chapter 10). Apoptosis can be initiated by cell death signals from the mitochondria, nucleus, or cell membrane. Caspase-8 is an initiator related to cell death signals from the cell mechanisms; caspase-9 is generated by cell death signals at the mitochondria; caspase-3 is an effector caspase associated with the final stages of apoptosis (see Cohen 1997 for a review of caspase activity in cell death). Figure 7.15 shows a cell labeled with propidium iodide (PI), a stain that is taken up by the nucleus of a dying or fixed cell. Notice that all the darker red shrunken nuclei are also colabeled for caspase-3 (green staining), and the swollen nuclei of necrotic cells do not express caspase-3. Nicotera et al. (2003) reported that after a noise exposure both caspase-8 and -9 were expressed, which implies that apoptosis in hair cells can be triggered into cell death through multiple pathways (Fig. 7.16). There is also

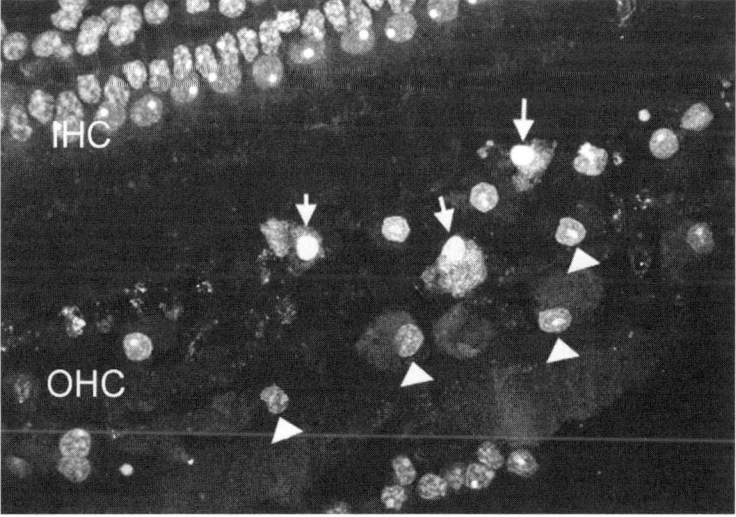

FIGURE 7.15. Apoptotic cells labeled for caspase-3, an effector caspases at the terminal end of cell death.

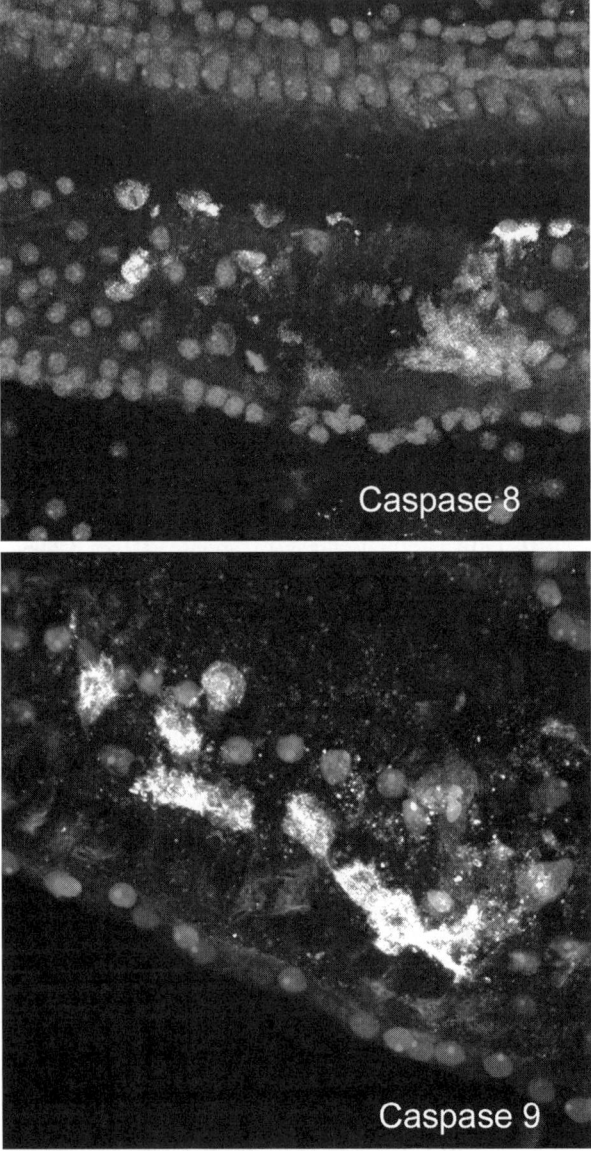

FIGURE 7.16. Organ of Corti from noise-exposed cochleae. Top is labeled for caspase-8, the bottom for caspase-9.

evidence that calcium homeostasis is central in OHC response to traumatic noise. Vicente-Torres and Schacht (2006) reported increased levels of phosphatase calcineurin and Bcl-xL/Bcl-2-associated death promoter (BAD) following noise exposure. Local application of FK506 and cyclosporin A, calcineurin inhibiting agents, provided significant protection from noise (Minami et al. 2004).

The course of apoptosis has a short latency. Fifteen minutes after a 1-h noise exposure hair cells were already missing and other cells showed both apoptosis and necrotic-like changes (Hu et al. 2002). Given the 1-h period of the exposure, it is difficult to say when apoptosis begins. To better define the latency of the cell death, the noise exposure was changed to a 1-min series of impulses at 155 dB peak SPL. Cochleae were evaluated at 5 min and 30 min after the exposure. Interestingly, at 5 min after the exposure, there was a small lesion consisting of only apoptotic cells, but at 30 minutes, the size of the lesion expanded and both apoptotic and necrotic cells were found (Fig. 7.17). The necrotic cells may have been cells that begin to die by apoptosis, but convert to necrosis because of a lack of energy to finish the active apoptosis process.

It is clear that the extent of the OHC lesion continues to expand for days after the exposure (Hu et al. 2002; Yang et al. 2004). The direction of expansion is primarily from the center of the lesion toward the basal end of the cochlea and the mechanism driving the expansion is apoptosis, likely to be driven by lipid peroxidation (Yamashita et al. 2004) because lipid peroxidation is self perpetuating (Halliwell and Gutteridge 1999).

12. Cell Death and Impulse Noise

Impulse noise from gunfire, explosions, and so forth generate peak levels of 150 dB SPL or greater. Figure 7.9 illustrates an extreme reaction to such an exposure. There can be much less dramatic examples of mechanical damage. Exposure to impulse noise produces a proliferation of ROS similar to exposure to continuous noise (i.e., concentration at the base of OHCs and neural plexus under IHCs (see Fig. 7.12). Some pathological changes are characteristic of impulse noise, such as disassociation of the OHCs from their supporting Deiters cup (Fig. 7.9). Notice several changes: the OHC has shortened in length and its diameter is larger; the nucleus has migrated from the basal pole to the middle of the cell and, most importantly, the nucleus has shrunk. This may be an example of anoikis, a form of apoptosis where the triggering signal is a loss of attachment to the extracellular matrix. Using the same exposure, the chinchilla's cochlea expresses *p53*, a tumor suppressor gene that regulates the cellular response to DNA damage by mediating cell cycle arrest, DNA repair, and cell death (Ko and Prives 1996). The mechanisms involved in *p53*-mediated cell death remain controversial, and regulation of *p53* function is complicated. However, DNA damage and cell stress events including oxidative stress (from ROS) are known to activate *p53* (Finkel and Holbrook 2000).

13. Cochlear Responses to Stress from Noise

The cochlea has several lines of defense against stresses from high level noise. At a general body systems level, the auditory system responds to high level noise by triggering both the acoustic middle ear reflex (Henderson 1993; Quaranta

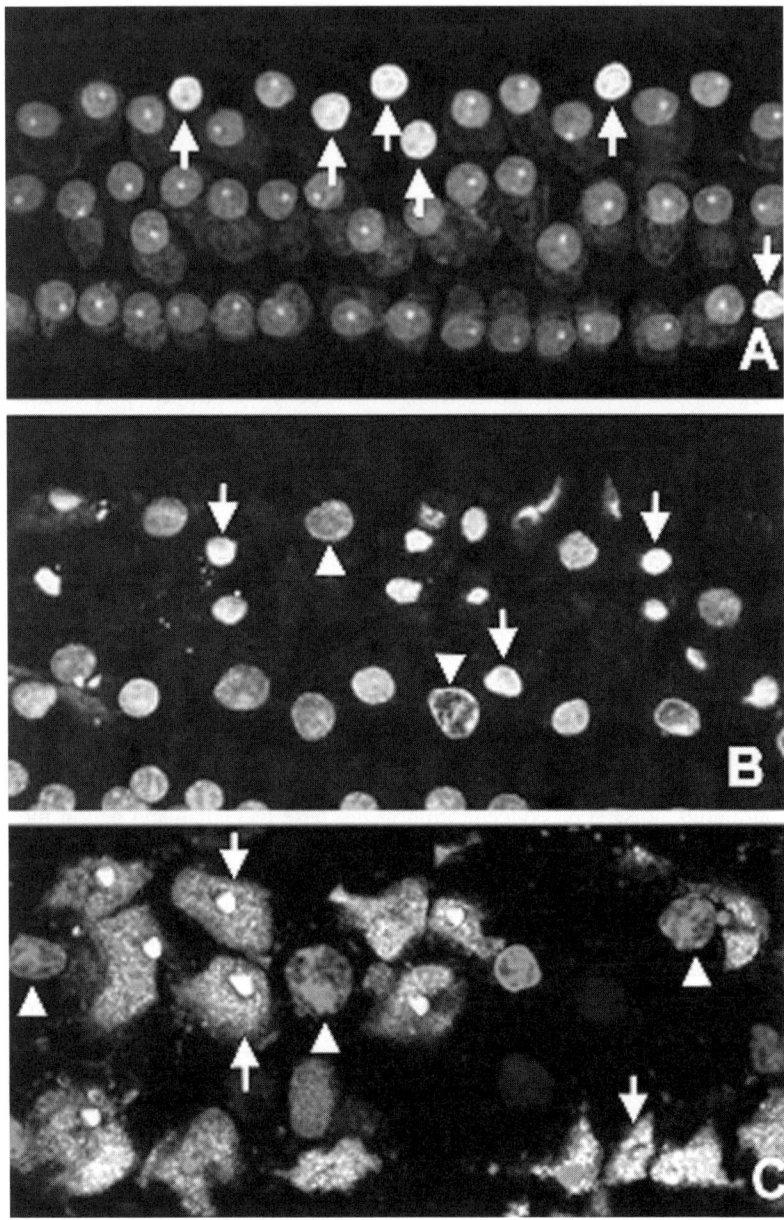

FIGURE 7.17. Top view shows propidium iodide labeled organ of Corti 5 min after noise exposure. Note the number of early apoptotic cells. Middle view shows the same area in a cochlea analyzed 30 min after exposure. Note the presence of both apoptotic and necrotic cells. Bottom view shows an organ of Corti colabeled with propidium iodide and caspase-3. Only shrunken nuclei (apoptotic cells) are labeled for caspase-3.

et al. 1998) and activating the medial olivary system (Rajan and Johnstone 1983; Reiter and Liberman 1995; Rajan 1996; Zheng et al. 1997a). At the level of the organ of Corti, the ear can respond with the expression of cell survival factors, such as heat shock proteins (Yoshida et al. 1999) or bcl-2 (Niu et al. 2003). The ear can also increase the activity of the protective antioxidant system, as demonstrated in a study by Jacono et al. (1998). In the study, chinchillas were exposed to one of three conditions: a conditioning exposure of a 500-Hz octave band of noise for 6 h per day for 10 days, the conditioning exposure plus a 2-day rest period and then a high level noise exposure of 4 h duration, or only the 4-h high level exposure. Interestingly, all three exposures led to increases in the concentration of the antioxidant enzymes catalase, glutathione reductase, and γ-glutamyl cysteine synthetase in both stria vascularis and the organ of Corti. However, the largest increase in each of the three antioxidant enzymes was in the group that had both the conditioning exposure and the traumatic exposure. It can be argued that the conditioning effects in which the prior prophylactic exposure to moderate noise levels rendered the cochlear antioxidant system more effective. The Jacono et al. (1998) study is a key link in developing pharmacological approaches to protecting the ear from noise. More on protection is found in Green, Altschuler, and Miller, Chapter 10).

14. Summary

Noise causes damage throughout the cochlea but for hearing losses up to about 50 dB the sensory targets are primarily the OHCs, especially in the basal third of the cochlea. Noise causes this damage by creating a large increase in toxic ROS, which in turn initiates cell death by both necrosis and apoptosis. The cell death process continues, primarily by apoptosis, for days after a traumatic noise exposure, albeit at a progressively decreasing rate.

A better understanding of the parameters of cell death (i.e., triggers for the initiation of apoptosis, the driving force behind prolonged cell death after an exposure, factors that influence apoptosis versus necrosis) are interesting issues from a scientific perspective and are already providing direction for the eventual development of drugs for prevention and treatment of acquired hearing loss.

References

Beauchamp C, Fridovich I (1970) A mechanism for the production of ethylene from methional. The generation of the hydroxyl radical by xanthine oxidase. J Biol Chem 245:4641–4646.

Bielefeld EC, Hu BH, Harris KC, Henderson D (2005) Damage and threshold shift resulting from cochlear exposure to Paraquat-generated superoxide. Hear Res 207: 35–42.

Bohne BA (1976) Mechanisms of noise damage in the inner ear. In: Henderson D, Hamernik RP, Dosanjh D, Mills, J (eds) Effects of Noise on Hearing. New York: Raven Press, pp. 41–68.

Bohne BA, Clark WW (1982) Growth of hearing loss and cochlear lesion with increasing duration of noise exposure. In: Hamernik RP, Henderson D, Salvi RJ (eds) New Perspectives on Noise-Induced Hearing Loss. New York: Raven Press, pp. 283–302.

Borg E (1982) Protective value of sympathectomy of the ear in noise. Acta Physiol Scand 115:281–282.

Brownell WE (1984) Microscopic observation of cochlear hair cell motility. Scan Electron Microsc(Pt 3):1401–1406.

Chance B, Sies H, Boveris A (1979) Hydroperoxide metabolism in mammalian organs. Physiol Rev 59:527–605.

Chen GD, Fechter LD (1999) Potentiation of octave-band noise induced auditory impairment by carbon monoxide. Hear Res 132:149–159.

Cody AR, Russell IJ (1985) Outer hair cells in the mammalian cochlea and noise-induced hearing loss. Nature 315:662–665.

Cohen GM (1997) Caspases: the executioners of apoptosis. Biochem J 326 (Pt 1):1–16.

Davis H (1952) Neuroanatomy and neurophysiology in the cochlea. Trans Am Acad Ophthalmol Otolaryngol 56:630–634.

Davis H, Morgan CT, Hawkins JE Jr, Galambos R, Smith FW (1950) Temporary deafness following exposure to loud tones and noise. Acta Otolaryngol Suppl 88:1–56.

Fessenden JD, Schacht J (1998) The nitric oxide/cyclic GMP pathway: a potential major regulator of cochlear physiology. Hear Res 118:168–176.

Finkel T, Holbrook, NJ (2000) Oxidants, oxidative stress and the biology of ageing. Nature 408:239–247.

Halliwell, B, Gutteridge J (1999) Free Radicals in Biology and Disease. Oxford: Oxford University Press.

Hamernik RP, Turrentine G, Roberto M, Salvi R, Henderson D (1984) Anatomical correlates of impulse noise-induced mechanical damage in the cochlea. Hear Res 13:229–247.

Hamernik RP, Turrentine G, Roberto M (1985) Mechanically induced morphological changes in organ of Corti. In Salvi RJ, Henderson D, Hamernik RP, Colletti, V (eds) Basic and Applied Aspects of Noise Induced Hearing Loss. New York: Plenum Press, pp. 69–84.

Henderson D, Hamernik RP (1986) Impulse noise: critical review. J Acoust Soc Am 80:569–584.

Henderson D, Subramaniam M, Gratton MA, Saunders SS (1991) Impact noise: the importance of level, duration, and repetition rate. J Acoust Soc Am 89:1350–1357.

Henderson D, Subramaniam M, Boettcher FA (1993) Individual susceptibility to noise-induced hearing loss: an old topic revisited. Ear Hear 14:152–168.

Henderson D, Bielefeld EC, Harris KC, Hu BH (2006) The role of oxidative stress in noise-induced hearing loss. Ear Hear 27:1–19.

Hildesheimer M, Sharon R, Muchnik C, Sahartov E, Rubinstein M (1991) The effect of bilateral sympathectomy on noise induced temporary threshold shift. Hear Res 51:49–53.

Hildesheimer M, Henkin Y, Pye A, Heled S, Sahartov E, Shabtai EL, Muchnik C (2002) Bilateral superior cervical sympathectomy and noise-induced, permanent threshold shift in guinea pigs. Hear Res 163:46–52.

Hu BH, Henderson D, Nicotera TM (2002) Involvement of apoptosis in progression of cochlear lesion following exposure to intense noise. Hear Res 166:62–71.

Hudspeth AJ, Jacobs R (1979) Stereocilia mediate transduction in vertebrate hair cells (auditory system/cilium/vestibular system). Proc Natl Acad Sci USA 76:1506–1509.

Jacono AA, Hu B, Kopke RD, Henderson D, Van De Water TR, Steinman HM (1998) Changes in cochlear antioxidant enzyme activity after sound conditioning and noise exposure in the chinchilla. Hear Res 117:31–38.

Kallinen J, Didier A, Miller JM, Nuttall A, Grenman R (1991) The effect of CO2- and O2-gas mixtures on laser Doppler measured cochlear and skin blood flow in guinea pigs. Hear Res 55:255–262.

Kandel ER, Schwartz JH, Jessell TM (2000) Principles of Neural Science, 4th ed. New York: McGraw-Hill Health Professions Division.

Ko LJ, Prives C (1996) p53: puzzle and paradigm. Genes Dev 10:1054–1072.

Konishi K, Yamane H, Iguchi H, Takayama M, Nakagawa T, Sunami K , Nakai Y (1998) Local substances regulating cochlear blood flow. Acta Otolaryngol Suppl 538:40–46.

Lamm K, Arnold W (2000) The effect of blood flow promoting drugs on cochlear blood flow, perilymphatic pO(2) and auditory function in the normal and noise-damaged hypoxic and ischemic guinea pig inner ear. Hear Res 141:199–219.

Laurikainen EA, Kim D, Didier A, Ren T, Miller JM, Quirk WS, Nuttall, AL (1993) Stellate ganglion drives sympathetic regulation of cochlear blood flow. Hear Res 64:199–204.

Laurikainen EA, Costa O, Miller JM, Nuttall AL, Ren TY, Masta R, Quirk WS, Robinson PJ (1994) Neuronal regulation of cochlear blood flow in the guinea-pig. J Physiol 480:563–573.

Laurikainen EA, Ren T, Miller JM, Nuttall AL, Quirk WS (1997) The tonic sympathetic input to the cochlear vasculature in guinea pig. Hear Res 105:141–145.

Liberman MC, Dodds LW, Learson DA (1985) Structure–function correlation in noise-damaged ears. In: Salvi RJ, Henderson D, Hamernik RP, Colletti, V (eds) Basic and Applied Aspects of Noise Induced Hearing Loss. New York: Plenum Press, pp. 163–178.

Miller JM, Dengerink H (1988) Control of inner ear blood flow. Am J Otolaryngol 9:302–316.

Miller JM, Ren TY, Nuttall AL (1995) Studies of inner ear blood flow in animals and human beings. Otolaryngol Head Neck Surg 112:101–113.

Miller JM, Brown JN, Schacht J (2003) 8-iso-prostaglandin Γ(2alpha), a product of noise exposure, reduces inner ear blood flow. Audiol Neurootol 8:207–221.

Mills JH, Adkins WY, Gilbert RM (1981) Temporary threshold shifts produced by wideband noise. J Acoust Soc Am 70:390–396.

Minami SB, Yamashita D, Schacht J, Miller JM (2004) Calcineurin activation contributes to noise-induced hearing loss. J Neurosci Res 78:383–392.

Nagahara K, Aoyama T, Fukuse S, Noi O (1988) Effects of prostaglandins on perilymphatic oxygenation. Enhancement of cochlear autoregulation by prostacyclin. Acta Otolaryngol Suppl 456:143–150.

Nicotera T, Henderson D, Zheng XY, Ding DL, McFadden SL (1999) Reactive oxygen species, apoptosis and necrosis in noise-exposed cochleas of chinchillas. Paper presented at the 22nd Annual Midwinter Meeting of the Association for Research in Otolaryngology, St. Petersburg, FL.

Nicotera TM, Hu BH, Henderson D (2003) The caspase pathway in noise-induced apoptosis of the chinchilla cochlea. J Assoc Res Otolaryngol 4:466–477.

Nicotera TM, Ding D, McFadden SL, Salvemini D, Salvi R (2004) Paraquat-induced hair cell damage and protection with the superoxide dismutase mimetic m40403. Audiol Neurootol 9:353–362.

Nordmann AS, Bohne BA, Harding GW (2000) Histopathological differences between temporary and permanent threshold shift. Hear Res 139:13–30.

Ohlemiller KK, Wright JS, Dugan LL (1999) Early elevation of cochlear reactive oxygen species following noise exposure. Audiol Neurootol 4:229–236.

Perlman H, Kimura R (1962) Cochlear blood flow in acoustic trauma. Acta Otolaryngolica 54:99–110.

Pickles JO, Comis SD, Osborne MP (1985) The morphology of stereocilia and their cross-links in relation to noise damage in the guinea pig. In: Salvi RJ, Henderson D, Hamernik RP, Colletti, V (eds) Basic and Applied Aspects of Noise Induced Hearing Loss. New York: Plenum Press, pp. 31–42.

Puel JL, d'Aldin CG, Saffiende S, Eybalin M, Pujol R (1996) Excitotoxicity and plasticity of IHC-auditory nerve contributes to both temporary and permanent threshold shift. In: Axelsson A, Borchgrevink HM, Hamernik RP, Hellström PA, Henderson D, Salvi RJ (eds) Scientific Basis of Noise-induced Hearing Loss. New York: Thieme, pp. 36–42.

Puel JL, Ruel J, Gervais d'Aldin C, Pujol R (1998) Excitotoxicity and repair of cochlear synapses after noise-trauma induced hearing loss. NeuroReport 9:2109–2114.

Pujol R, Puel JL, d'Aldin CG, Eybalin M (1990) Physiopathology of the glutaminergic synapses in the cochlea. Acta Otolaryngol Suppl 476:32–36.

Pujol R, Puel JL, Gervais d'Aldin C, Eybalin M (1993) Pathophysiology of the glutamatergic synapses in the cochlea. Acta Otolaryngol 113:330–334.

Quaranta A, Portalatini P, Henderson D (1998) Temporary and permanent threshold shift: an overview. Scand Audiol Suppl 48:75–86.

Rajan R (1996) Involvement of cochlear efferent pathways in protective effects elicited with binaural loud sound exposure in cats. J Neurophysiol 74:582–597.

Rajan R, Johnstone BM (1983) Crossed cochlear influences on monaural temporary threshold shifts. Hear Res 9:279–294.

Reiter ER, Liberman MC (1995) Efferent-mediated protection from acoustic overexposure: relation to slow effects of olivocochlear stimulation. J Neurophysiol 73:506–514.

Ren TY, Laurikainen E, Quirk WS, Miller JM, Nuttall AL (1993) Effects of electrical stimulation of the superior cervical ganglion on cochlear blood flow in guinea pig. Acta Otolaryngol 113:146–151.

Saunders JC, Canlon B, Flock A (1985) Mechanical changes in stereocilia following overstimulation. In: Salvi RJ, Henderson D, Hamernik RP, Colletti, V (eds) Basic and Applied Aspects of Noise Induced Hearing Loss. New York: Plenum Press, pp. 11–30.

Sellick PM, Patuzzi R, Johnstone BM (1982) Measurement of basilar membrane motion in the guinea pig using the Mossbauer technique. J Acoust Soc Am 72:131–141.

Spicer SS, Schulte BA (1996) The fine structure of spiral ligament cells relates to ion return to the stria and varies with place-frequency. Hear Res 100:80–100.

Spoendlin H (1985) Histopathology of noise deafness. J Otolaryngol 14:282–286.

Sweeney MH, Fosbroke D, Goldenhar LM, Jackson LL, Lushniak BD, Merry L, Schneider, S, Stephenson, M (2005) Health consequences working in construction. In: Coble R, Hinze J, Haupt T (eds) Construction Safety and Health Management. Columbus, OH: Prentice-Hall, pp: 178–196.

Thalmann R, Miyoshi T, Kusakari J, Ise I (1975) Normal and abnormal energy metabolism of the inner ear. Otolaryngol Clin North Am 8:313–333.

Thorne PR, Nuttall AL (1987) Laser Doppler measurements of cochlear blood flow during loud sound exposure in the guinea pig. Hear Res 27:1–10.

Turrens JF, Freeman BA, Levitt JG, Crapo JD (1982) The effect of hyperoxia on super-oxide production by lung submitochondrial particles. Arch Biochem Biophys 217: 401–410.

Ugnell AO, Hasegawa M, Lundquist PG, Andersson R (2000) Effect of carbon dioxide on cochlear blood flow in guinea pigs. Acta Otolaryngol 120:11–18.

Vicente-Torres MA, Schacht J (2006) A BAD link to mitochondrial cell death in the cochlea of mice with noise-induced hearing loss. J Neurosci Res 83:1564–1572.

von Békésy G (1952) Direct observation of the vibrations of the cochlear partition under a microscope. Acta Otolaryngol 42:197–201.

Wang Y, Hirose K, Liberman MC (2002) Dynamics of noise-induced cellular injury and repair in the mouse cochlea. J Assoc Res Otolaryngol 3:248–268.

Wangemann P (2002) K^+ cycling and the endocochlear potential. Hear Res 165:1–9.

Ward WD (1966) The use of TTS in derivation of damage risk criteria for noise exposure. Intern Aud 5:309–313.

Yamane H, Nakai Y, Takayama M, Konishi K, Iguchi H, Nakagawa T, Shibata S, Kato A, Sunami K, Kawakatsu C (1995) The emergence of free radicals after acoustic trauma and strial blood flow. Acta Otolaryngol Suppl 519:87–92.

Yamashita D, Jiang HY, Schacht J, Miller JM (2004) Delayed production of free radicals following noise exposure. Brain Res 1019:201–209.

Yang WP, Henderson D, Hu BH, Nicotera TM (2004) Quantitative analysis of apoptotic and necrotic outer hair cells after exposure to different levels of continuous noise. Hear Res 196:69–76.

Zheng XY, Henderson D, Hu BH, Ding DL, McFadden SL (1997a) The influence of the cochlear efferent system on chronic acoustic trauma. Hear Res 107:147–159.

Zheng XY, Henderson D, Hu BH, McFadden SL (1997b) Recovery of structure and function of inner ear afferent synapses following kainic acid excitotoxicity. Hear Res 105:65–76

8
Drug-Induced Hearing Loss

Leonard P. Rybak, Andra E. Talaska, and Jochen Schacht

1. Introduction

1.1 "Ototoxic" Drugs

"Ototoxicity," drug-induced damage to the auditory or vestibular parts of the inner ear, has probably been in existence since our ancestors began using herbs as remedies for their ailments (Schacht and Hawkins 2006). After all, some of the most powerful ototoxic drugs known today are derived from natural sources: the aminoglycosides, synthesized by soil-dwelling bacteria. Other prominent ototoxic drugs, although causing only temporary threshold shifts, are quinine and salicylate, both derived from tree bark. While the recognition of the cochlear and vestibular detriments exerted by drugs goes back centuries, the problem of ototoxicity was catapulted into the medical and public awareness in 1944 with the arrival of streptomycin, the first aminoglycoside antibiotic (Schatz et al. 1944). Hailed as the long-sought cure for tuberculosis and other gram-negative infections, streptomycin also very quickly revealed its destructive power to the vestibular system (Hinshaw and Feldman 1945).

Since then, with the growing appreciation of potential side effects to the inner ear, other drugs were found that affected hearing or balance. Antimicrobial agents such as chloramphenicol, erythromycin, polymyxin B, and vancomycin have been sporadically associated with ototoxic side effects, as have topical disinfectants such as chlorhexidine. Cisplatin brought the success of cancer chemotherapy at the price of hearing loss in many patients. Loop diuretics (ethacrynic acid, bumetanide, and furosemide) gained unfavorable prominence in part for their own reversible effects on the auditory system but mostly as potentiating agents when given together with aminoglycoside antibiotics. The combination of these two classes of drugs has devastating effects on the auditory system even if the concentration of either drug alone would prove innocuous. Also of concern as potentially ototoxic agents are organometals such as organotins and organic mercury preparations as well as the industrial solvents toluene and styrene. While the organometals can have profound toxicity by themselves, the solvents tend to interact adversely with noise exposure, jeopardizing those who work in industrial environments.

Because of the sheer number of patients affected and because of the irreversible nature of their effects on the inner ear, aminoglycoside antibiotics and cisplatin command the most attention today among potentially ototoxic medications. This chapter therefore focuses on these classes of drugs.

2. History

2.1 Cisplatin

Cisplatin was first synthesized by Peyrone in 1845, and hence is also known as Peyrone's chloride (Rosenberg 1980). Its chemical structure was determined in 1893 by Werner as an inorganic complex consisting of a central atom of platinum surrounded by chloride and amine groups in the *cis* position (Fig. 8.1a). Its biological effects, however, appear to have gone unnoticed for a century. In 1965, Rosenberg et al. discovered some unusual effects in experiments with *Escherichia coli* subjected to a current that was delivered between platinum electrodes. Individual cells that normally are rods of about 1 by 5 μm were elongated into filaments up to 300 times their original length under the influence of this current, due to a stable form of platinum released from the electrodes. The principal compound involved was Peyrone's chloride, or *cis*-dichlorodiammine platinum (II).

The introduction of cisplatin as an antineoplastic agent was primarily based on studies beginning in the 1960s showing its effectiveness in retarding the growth of sarcoma 180 in mice and increasing the survival of mice bearing the highly metastatic L1210 leukemia (Rosenberg et al. 1969). Among several platinum compounds, cisplatin had the greatest efficacy against a wide variety of animal tumors. The drug exhibited: (1) marked antitumor activity; (2) broad-spectrum activity against drug-resistant as well as drug sensitive tumors; (3) efficacy against slowly growing as well as rapidly growing tumors; (4) activity against a tumor insensitive to "S" phase inhibitors; (5) induction of regression of transplantable tumors induced by viruses as well as chemicals; (6) activity in a variety of species; (7) effectiveness against disseminated as well as solid tumors; and (8) potency in rescuing animals with advanced tumors who were near death (Rosenberg 1973). Subsequent studies revealed efficacy against

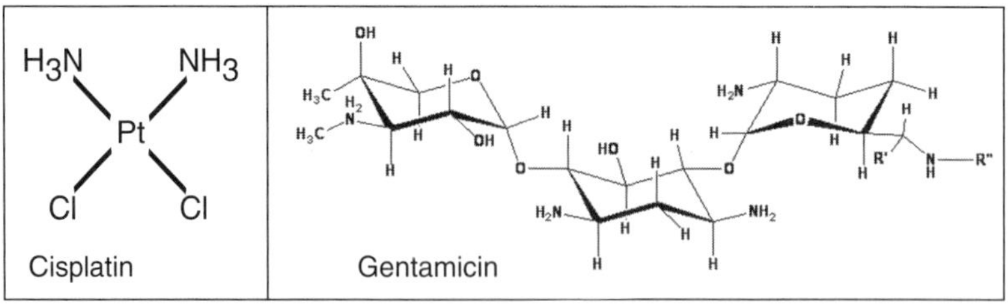

FIGURE 8.1. Structures of cisplatin and gentamicin.

testicular tumors and a variety of other solid tumors, particularly ovarian, bladder, and head and neck malignancies as well as malignant neoplasms in children (Rozencweig et al. 1977). Cisplatin also has some benefits against brain tumors (Feun et al. 1984).

The major toxicity in a phase I clinical trial was to the kidney, occurring in 25% to 61% of patients depending on the dose (Talley et al. 1972; DeConti et al. 1973). Other toxicities discovered during those trials included nausea and vomiting followed by prolonged periods of anorexia, leukopenia, and thrombocytopenia which manifested itself as late as 4 weeks after treatment, hyperuricemia, and hearing impairment primarily in the high frequencies (Kovach et al. 1973; Lippman et al. 1973).

Today, combination chemotherapy is the cornerstone of cancer treatment regimens. The goal is to achieve a synergistic effect of cisplatin with other antineoplastic agents in order to increase efficiency and lower toxicity. These combinations may also reduce the chance of resistance of tumors by inhibiting repair of platinum–DNA adducts or actually increasing the formation of platinum–DNA adducts. Such drug combinations are useful to treat cancer of the lung, ovary, testis, cervix, head and neck, gastrointestinal tract (stomach, esophagus, colon, and pancreas), and bladder, and metastatic cancers of the prostate or breast, mesothelioma, metastatic melanoma, and malignant gliomas (Boulikas and Vougiouka 2004).

2.2 Aminoglycosides

In contrast to the century-long delay between the synthesis of cisplatin and its clinical application, the therapeutic value of aminoglycoside antibiotics as a potent antibiotic against gram-negative bacteria was recognized within a year following the discovery of streptomycin (Schatz et al. 1944; Hinshaw and Feldman 1945). The first trials also discovered the adverse effects ototoxicity and nephrotoxicity. The characterization of streptomycin was followed by the isolation and semisynthetic production of a wide variety of aminoglycosides including neomycin, tobramycin, gentamicin, kanamycin, dihydrostreptomycin, and netilmicin but the ototoxic and the nephrotoxic potentials were never eliminated. Nevertheless, their broad antibacterial spectrum and efficacy against and hitherto untreatable diseases, such as tuberculosis, made the aminoglycoside antibiotics some of the most successful drugs in the second half of the 20th century.

Aminoglycoside antibiotics are low-molecular-weight compounds with similar structures consisting of several, usually three, rings. These rings are cyclitols (a saturated six-carbon ring) and five- or six-membered sugars that are linked via glycosidic bonds. The hallmark of aminoglycosides is the presence of amino groups attached to the various rings of the structure (Fig. 8.1b). These amino groups and the additional hydroxyl groups convey the major chemical properties, namely a highly polar basic character and high water solubility. Aminoglycoside antibiotics are produced by different strains of soil actinomycetes, "-mycins" by

Streptomyces, and "-micins" by *Micromonospora*. Although all aminoglycosides carry the suffix "-mycin" or "-micin," this suffix is not a chemical or therapeutic classification and does not exclusively denote an aminoglycoside. The anticancer drug bleomycin, for example, or the macrolide antibiotic erythromycin are unrelated to aminoglycosides in structure, therapeutic indications, and potential side effects.

Although new generations of antibiotics have emerged in the last decades, the aminoglycosides maintain a leading role in the treatment of enterococcal, mycobacterial, and severe gram-negative bacterial infections. Gentamicin is frequently used in newborn infants against potentially life-threatening sepsis and in adults, for urinary tract infections and the prophylactic treatment against pulmonary infections in cystic fibrosis patients are major applications (Pong and Bradley 2005). Further, aminoglycoside antibiotics are recommended by the World Health Organization as part of the combination therapy against multi-drug-resistant tuberculosis (World Health Organization 2005). A critical aspect that drives the worldwide popularity of aminoglycoside antibiotics is their use in developing countries. Aminoglycoside antibiotics are nonallergenic and can be readily used in emergency situations, and, importantly, they can be produced very inexpensively and frequently are the only affordable option in less affluent countries.

3. Mechanisms of Therapeutic Action

3.1 Cisplatin

The dichloro compound cisplatin undergoes hydration inside the cell, which is facilitated by the low intracellular concentration of chloride ions. This hydrated form is more reactive to targets within the cell, the primary target being DNA. Its configuration is reminiscent of the mustard-type bifunctional alkylating agents which can become strong electrophiles and can form covalent linkages by alkylating DNA. The 7-nitrogen atom of guanine is especially susceptible to the formation of a covalent bond with these agents so that these drugs can form inter- and intrastrand cross-links with DNA (Rozencweig et al. 1977). Platinated DNA adducts inhibit replication, inhibit transcription, arrest the cell cycle, prevent DNA repair, and induce cell death by apoptosis. The distortion of the DNA helix by platinum binding can also cause binding of several classes of proteins to the modified DNA. These include the high-mobility group (HMG) box proteins which may play a role in the antineoplastic activity of cisplatin (Wang and Lippard 2005). Platinated DNA intrastrand crosslinks can be removed by cellular repair mechanisms such as nucleotide excision repair, while interstrand crosslinks are likely eliminated by recombination repair mechanisms (McHugh et al. 2001). The extent of tolerance to DNA lesions caused by cisplatin may decide the fate of a particular cell, survival, or apoptosis (Liedert et al. 2006).

3.2 Aminoglycosides

While cisplatin can link to DNA in essentially all eukaryotic cells, amino-glycoside antibiotics target prokaryotic organisms. The antibacterial action is rather complex and is ultimately bactericidal, not merely bacteriostatic as is, for example, that of penicillin. A major contributing mechanism appears to be inhibition of protein synthesis accomplished through binding to the 30S small ribosomal subunit and specifically the A-site of the 16S rRNA. The A-site recruits the proper tRNA anti-codon and proofreads this match. Analysis of the crystal structures of the 30S bound to streptomycin, spectinomycin, paromycin, and hygromycin B have revealed that the different classes of aminoglycosides have slightly different interactions with the ribosome, bearing the efficacy of multi-antibiotic treatment for resistant bacteria. The consequence of aminogly-coside binding is an increase in the frequency of mismatches of amino acids and a decrease in tRNA dissociation, effectively halting the protein elongation process (Vakulenko and Mobashery 2003; Magnet and Blanchard 2005). Because eukaryotic ribosomes differ structurally from their bacterial counterparts, this action should be specifically directed against prokaryotes. However, mammalian mitochondrial RNA contains similar subunits and may thus represent a potential target for aminoglycoside antibiotics (see section on the 1555 mutation).

4. Adverse Side Effects

4.1 Cisplatin

Side effects of cisplatin treatment include nausea and vomiting, nephroto-xicity, neurotoxicity, and ototoxicity. The kidney actively accumulates cisplatin, probably through a carrier-mediated transport (Kawai et al. 2005), and causes decreased perfusion of the kidney, with death of cells in the proximal and distal tubules and loop of Henle. Such irreversible kidney damage may occur in up to one-third of patients (Taguchi et al. 2005).

Neurotoxicity is another troublesome side effect of cisplatin therapy. It clinically manifests as numbness and tingling of the limbs. The neuropathy may continue to progress even after cessation of chemotherapy (Grunberg et al. 1989). Clinical regimens exist to reduce both nephrotoxicity and neurotoxicity (Santoso et al. 2003; Umapathi and Chaudhry 2005).

Cisplatin-induced hearing loss is typically bilateral and begins at the higher frequencies, progressing to lower frequencies with continued treatment. The hearing loss may be gradual and cumulative or it may appear suddenly after a single treatment. Cisplatin causes ototoxicity almost exclusively in the cochlea, in contrast to aminoglycosides, all of which have cochleotoxic and vestibulotoxic potential.

4.2 Aminoglycosides

Aside from nephro- and ototoxicity, side effects of aminoglycoside treatment are rather infrequent. Nephrotoxicity affects about 20% of patients (Swan 1997) and is due to an accumulation of the drugs in the proximal tubules. As with cisplatin, necrosis of cells occurs in the proximal and distal tubules and the loop of Henle, potentially resulting in renal failure. Nephro- and ototoxicity are not necessarily expressed together in experimental animals or in patients.

Ototoxicity generally is bilateral and manifests itself only after days or weeks of systemic treatment, and the severity of hearing loss may still increase after drug administration has ended. An exception to this rule is found in patients carrying mitochondrial mutations that predispose to rapid hearing loss, oftentimes from a single dose of the drugs. Ototoxicity involves either the auditory (cochleotoxicity) or vestibular system (vestibulotoxicity), or both, and the toxic potential and organ preference varies among the different aminoglycosides. Neomycin is regarded as highly toxic; gentamicin, kanamycin, and tobramycin are of intermediate toxicity; and amikacin and netilmicin are somewhat less toxic. As to organ preferences, amikacin or neomycin may target primarily the cochlea while gentamicin is considered more vestibulotoxic in the human and therefore frequently used for vestibular ablation in Ménière's disease (Blakley 1997). Small changes in structure may greatly influence the pattern of toxicity: streptomycin mostly targets the vestibular system in the human inner ear while dihydrostreptomycin targets the cochlea. Such preferences are not predictable by any structure–activity relationship and are also not related to any site-specific uptake mechanism or drug levels in the tissues (see section on "Pharmacokinetics").

5. Incidence of Ototoxicity

5.1 Cisplatin

Because the loss of hearing begins at the high-frequency range, it may not be detected with conventional audiometry or initially perceived by the patient. Seventy-one percent of patients with cisplatin-induced hearing loss had hearing deficits at frequencies of 8000 Hz or higher (Fausti et al. 1993). While a high-frequency loss may not result in any communication problems for the patient, such an assessment illustrates the ototoxic potential. In fact, some studies peg the incidence close to 90% or 100% (Benedetti-Panici et al. 1993). A mild and early hearing loss may be partially reversible, but when the hearing loss is profound it tends to be permanent (Vermorken et al. 1983).

5.2 Aminoglycosides

The reported incidence of ototoxicity varies considerably among different clinical studies, due to varying treatment regimens and, mostly, varying assessment and definitions of "hearing loss." Nevertheless, hearing loss induced by the most commonly used aminoglycosides may occur in about 20% of patients; balance may be affected in about 15% (Fee 1980; Moore et al. 1984; Lerner et al. 1986). In cystic fibrosis patients who receive repeated courses of aminoglycoside therapy as prophylaxis against pneumonia caused primarily by *Pseudomonas aeroginosa*, the reported range of cochleotoxicity is quite variable but may approach 17% (Mulheran et al. 2001). An early prospective study of kanamycin-induced hearing loss in tuberculosis treatment found hearing loss in 80% of the patients (Brouet et al. 1959) but a lower incidence appears to be associated with current therapeutic regimens (de Jager and van Altena 2002).

Just as with cisplatin, most data must be considered conservative estimates because hearing loss by aminoglycosides also begins at the highest frequencies which are not routinely monitored. Using high-frequency testing procedures, a prospective study in 53 patients determined hearing loss in 47% of the ears studied (Fausti et al. 1992).

The problem of ototoxicity is even more prevalent in developing countries where aminoglycosides are frequently available over the counter and where safety precautions such as monitoring of serum levels or auditory function are rarely available if at all. In the absence of such auditory monitoring, no firm data on the incidence of hearing loss exist but the frequency of ototoxicity must be considerably higher than in industrialized countries, as can be extrapolated from reported cases of complete deafness by aminoglycosides. In an area of southern China, two thirds of all deaf-mutism was due to the administration of aminoglycosides to children (Lu 1987). Even more recently, aminoglycoside-induced deafness was confirmed in 15 of 77 deaf children (Zhang et al. 1997).

6. Risk Factors

6.1 Cisplatin

There is considerable individual variation in susceptibility to cisplatin and aminoglycoside ototoxicity. For cisplatin, the severity of hearing loss appears to be related to the magnitude of the cumulative dose (Bokemeyer et al. 1998) and the rate of intravenous injections (Vermorken et al. 1983). Age appears to be an important risk factor for cisplatin-induced ototoxicity, with both children younger than 5 years of age and elderly patients more susceptible to cisplatin-induced hearing loss than are younger adults (Laurell and Jungnelius 1990; Li et al. 2004). Hearing loss may increase even years after cisplatin therapy has been completed (Bertolini et al. 2004). Nutritional and metabolic status can influence the probability of hearing loss from cisplatin. Patients with hypoalbuminemia and anemia develop more severe hearing losses than do those with normal albumin and hemoglobin levels (Blakley et al. 1994). Additional factors such as renal insufficiency and preexisting hearing loss may increase the probability of cisplatin ototoxicity, as can noise, cranial irradiation, and other concomitant drugs, such as high-dose vincristine (Bokemeyer et al. 1998).

6.1.1 Genetic Predisposition

A genetic predisposition to cisplatin-induced hearing loss may be related to mitochondrial mutations. Five of 20 cancer survivors with cisplatin ototoxicity were found to cluster in a rare European J mitochondrial haplogroup that has been associated with Leber's Hereditary Optic Atrophy (Peters et al. 2003). In contrast, a variety of other mutations including those that predispose to aminoglycoside ototoxicity appear not to contribute to the susceptibility to cisplatin-induced hearing loss (Knoll et al. 2006).

6.2 Aminoglycosides

It is intriguing that studies on diverse patient groups (and hence different drug regimens) do not find a direct correlation between ototoxicity and the level of dosage and the duration of treatment. This includes hearing loss in cystic fibrosis (Mulheran et al. 2001) and tuberculosis patients (de Jager and van Altena 2002) as well as vestibular damage (Black et al. 2004). Thus, individual susceptibility may play a decisive role in addition to very few confirmed risk factors such as concurrent nephrotoxicity (Halmagyi et al. 1994). Another confounding factor, at least in animal studies, is the nutritional and physiological state of the subject. Animals stressed by infections or fed a less than optimal diet display an enhanced susceptibility to aminoglycosides (Garetz et al. 1994; Lautermann et al. 1995). Notably, drug–drug interactions pose a major threat, for example, the concomitant administration of the loop diuretic ethacrynic acid together with an aminoglycoside that can lead to a precipitous severe hearing loss (Mathog and Klein 1969). A detrimental interaction of aminoglycosides with noise exposure has also been discussed, and nontoxic doses of amikacin may indeed impair recovery from noise-induced hearing loss (Tan et al. 2001).

During specific stages in development, the ear appears to be at particular risk from aminoglycoside antibiotics. This "critical period" coincides with the development of the inner ear both in altricial and precocial mammals. For example, the ototoxic effect of kanamycin was weak in the rat before the onset of cochlear potentials (8th postnatal day) but strong thereafter (Marot et al. 1980). Because the drugs also can penetrate the placental barriers, hair cell loss can be induced in utero during and after the functional differentiation of the cochlea in the guinea pig (Raphael et al. 1983).

6.2.1 Genetic Predisposition

Mitochondrial mutations are a well-defined risk factor in aminoglycoside-induced hearing loss, and a single injection of an aminoglycoside may already lead to profound deafness in carriers. The existence of families with multiple individuals with maternally inherited susceptibility to aminoglycoside-induced deafness was noticed several decades ago in China (Hu et al. 1991) and traced to an A1555G mutation in the 12S ribosomal RNA (Prezant et al. 1993). Subsequently, the same mutation was found in families with aminoglycoside ototoxicity from all ethnic backgrounds and geographic origins (reviewed by Fischel-Ghodsian 2005). This mutation may account for deafness in about 20% of patients with aminoglycoside ototoxicity while several other mitochondrial mutations make additional but only minor contributions.

The relationship between mitochondrial mutations and aminoglycoside ototoxicity remains enigmatic, but a consideration of the pattern of pathology is informative. In the absence of the 1555 mutation, an aminoglycoside such as gentamicin can display both cochlear and vestibular toxicity with a preponderance of the latter in the human. In contrast, patients with the 1555 mutation may (in the absence of aminoglycoside treatment) develop nonsyndromic hearing loss but little or no vestibular problems. Profound hearing loss without a concomitant vestibular component is the result of aminoglycoside treatment in patients with mitochondrial mutations. It thus seems that aminoglycoside ototoxicity is not enhanced by the mutation but that aminoglycosides are a stressor that triggers the phenotypic expression of the 1555 mutation. Requiring a trigger for its pathology would not be surprising, as the 1555 mutation has variable expression and onset in the carrier population, sometimes never manifesting itself.

7. Animal Models

7.1 Cisplatin

Most of the in vivo studies of cisplatin ototoxicity have been performed in rodent models including the rat, guinea pig, gerbil, hamster, mouse, and chinchilla. Systemically applied cisplatin causes high-frequency hearing loss and loss of outer hair cells (OHCs) primarily in the basal turn of the cochlea (Fig. 8.2). Deterioration in the distortion product otoacoustic emissions and a reduction in the endocochlear potential suggest dysfunction of both the OHCs and stria vascularis (Alam et al. 2000). Unfortunately, a major difficulty with most models

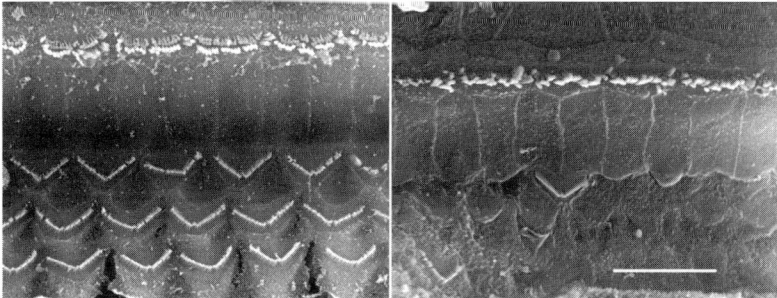

FIGURE 8.2. Cochlear damage by cisplatin. Left panel: Scanning electron micrograph of the hook region of the cochlea of a rat harvested 72 h after treatment with intraperitoneal saline. The single row of inner hair cells and the three rows of outer hair cells are well preserved. Right panel: Scanning electron micrograph of the hook region of the cochlea of a rat processed 72 h after treatment with 16 mg/kg of cisplatin by intraperitoneal infusion. Note the extensive loss of outer hair cells with preservation of the inner hair cells. Scale bar: 10 μm. Initial damage by aminoglycoside antibiotics shows a similar pattern.

is the high mortality rate of 33% to 50% (Ravi et al. 1995; Campbell et al. 1996; Reser et al. 1999; Li et al. 2002; Iraz et al. 2005).

Since clinical regimens for cisplatin consist of several courses of injections with rest intervals in between treatment days, a two-cycle model for cisplatin-induced ototoxicity in rats was developed. Each cycle consists of 4 days of cisplatin intraperitoneal injections (1 mg/kg twice a day) on days 1–4 (first cycle) and days 15–18 (second cycle). In addition, animals were injected with 10 ml of saline subcutaneously daily for hydration. These animals develop 40–50 dB elevation of auditory brainstem response (ABR) threshold at 16 and 20 kHz after the second cycle with no mortality. This model eliminates potentially confounding factors that may select the survival of a particular cohort of animals and may represent the best current model for cisplatin ototoxicity (Minami et al. 2004).

Several in vitro models to demonstrate cisplatin ototoxicity have also been used as tools to investigate mechanisms of toxicity. These include organotypic cultures of the organ of Corti of the neonatal rat (Kopke et al. 1997; Zhang et al. 2003), acutely isolated hair cells from the guinea pig cochlea (Dehne et al. 2001), hair cell lines derived from the immortomouse cochlea (Devarajan et al. 2002), mouse spiral ganglion cell cultures, and cultured type I spiral ligament fibrocytes (Liang et al. 2005). Such models can serve as useful screening techniques but because of genetic, pharmacokinetic, and metabolic differences, they may not necessarily be relevant to the processes that occur with cisplatin in the whole animal or in the human. Results obtained using these tissues would need to be confirmed in the appropriate animal model. A similar concern arises for the study of mechanisms of aminoglycoside ototoxicity in such in vitro systems.

7.2 Aminoglycosides

Inner ear pathology induced by aminoglycosides in experimental animals and humans is remarkably similar. Hair cells in the inner ears of all vertebrate classes as well as those of the lateral line organs of fish and larval amphibia are susceptible to aminoglycosides. Traditionally, the guinea pig had been the major animal model for aminoglycoside-induced hearing loss, and the cochlear pathology and pathophysiology have been well characterized. Thanks to recent developments in molecular biology, the mouse has become the favorite biomedical research subject. However, early studies in adult mice produced little if any auditory or vestibular deficits from doses of aminoglycosides that would induce severe trauma in other animal models (Henry et al. 1981).

An adult mouse model that avoids confounding aminoglycoside ototoxicity with developmental issues has only recently been developed (Wu et al. 2001). Interestingly, mice would not survive gentamicin treatment but tolerated kanamycin very well. Much higher doses of kanamycin were necessary in these mice than in previous model animals but the familiar pattern of base to apex loss of hair cells and high-frequency threshold shifts were identical to other animal models and the human as well. Although the required dosing with kanamycin was high in comparison, serum levels of the drug were comparable to levels achieved with other antibiotics in models such as the guinea pig, suggesting a higher rate of elimination rather than an intrinsic resistance to aminoglycosides was the reason behind earlier failures in establishing an adult mouse model of aminoglycoside ototoxicity.

8. Pathophysiology

8.1 Cisplatin

Although cisplatin ototoxicity is mostly documented as increased thresholds in ABR recordings, with greatest effects in the higher frequencies (Rybak et al. 2000), its actions may also include a reduction of the endocochlear potential (Ravi et al. 1995; Tsukasaki et al. 2000; Klis et al. 2000) and elevation of the thresholds for both the compound action potential and cochlear microphonic (van Ruijven et al. 2005a). Likewise, distortion product otoacoustic emissions are diminished, indicative of damage to OHCs (Alam et al. 2000). Vestibular disturbances are rare.

8.2 Aminoglycosides

Akin to the damage found with cisplatin, hearing loss is initially confined to the high frequencies and then progresses to lower frequencies (Aran and Darrouzet 1975) including frequencies in the human speech range. Histori-cally, vestibular disturbances were noted and investigated before the cochlear effects (Caussé et al. 1949) because streptomycin, the first aminoglycoside to be discovered, is primarily vestibulotoxic. Following systemic drug adminis-tration, the compound action potential threshold is elevated, and the depression of the cochlear microphonic potential and distortion product otoacoustic emissions suggest the involvement of OHCs.

In contrast to cisplatin, the endocochlear potential (EP) may be maintained close to normal during the early course of aminoglycoside treatment (Davis 1958; Komune et al. 1987). A significant decline in EP occurs only after the stria is substantially atrophied, usually weeks after the manifestations of hair cell damage.

9. Cochlear Pathology

9.1 Cisplatin

Only a few human temporal bone studies have been published for cisplatin pathology (Wright and Schaefer 1982; Strauss et al. 1983; Schuknecht et al. 1993; Hinojosa et al. 1995; Hoistad et al. 1998). Despite differences in dose and duration of cisplatin treatment, they clearly delineate OHC loss as the hallmark

TABLE 8.1. Patterns of cisplatin-induced damage to the cochlea in humans and experimental animals.

Cochlear structure	Human	Animal
Outer hair cells	Consistent damage starting from basal turn	Consistent damage starting from basal turn
Stria vascularis	Affected; variable	Variable
Supporting cells	Little change	Damage
Spiral ganglion cells	Major loss at variable locations	In vivo: Shrinkage only In vitro: Cell death

of pathology (Table 8.1). The loss begins in the basal turn and progresses to the apex of the cochlea with the innermost row of hair cells being more susceptible. Spiral ganglion cells also degenerate, again mostly in the basal turn. Inner hair cells (IHCs) are less consistently affected, and damage to stria vascularis is variable. The extent of damage appears dose-dependent and may, in extreme cases, include supporting cells of the cochlea. In contrast, the neuroepithelium of the saccule, utricle and the semicircular canals as well as the vestibular ganglion seem to be spared.

Studies in experimental animals largely reflect the pathology seen in humans with at least three major targets in the cochlea: the organ of Corti, the spiral ganglion cells and the tissues of the lateral wall, stria vascularis and spiral ligament. The loss of OHCs occurs initially in the most basal region of the cochlea (Fig. 8.2). As the ototoxic damage increases, the loss of OHCs progresses more apically (Schuknecht et al. 1993). OHC loss is most pronounced in the first row and least in the third row in the basal turn. With increasing doses or prolonged administration, the hair cell loss progresses apically and eventually involves the IHCs and supporting cells. IHCs show damage and degeneration only after all three rows of OHCs in the same region have degenerated (Marco-Algarra et al. 1985; Barron and Daigneault 1987). When hair cells are lost, protrusion of the supporting cells into the space of Nuel and into the tunnel of Corti occurs. This can eventually result in complete replacement of the sensory cells by a layer of epithelial cells (Estrem et al. 1981; Laurell and Bagger-Sjöbäck 1991).

Cisplatin ototoxicity is frequently accompanied by deleterious effects on the stria vascularis of the basal turn, such as strial edema, bulging, and depletion of organelles from the cytoplasm. Such damage and eventual atrophy occurs primarily in the marginal cells, with some mild changes in the intermediate cells (Meech et al. 1998; Campbell et al. 1999). On the other hand, some studies have failed to detect any morphological changes in the stria vascularis after cisplatin treatment (Fleischman et al. 1975; Boheim and Bichler 1985; DeGroot et al. 1997). These discrepancies could be owed to dose-intensity and timing.

The type I spiral ganglion cells can undergo detachment of their myelin sheaths. The time sequence of damage to spiral ganglion cells and the OHCs in guinea pig follow a similar time course, suggesting that injury to both areas occurs in parallel, rather than sequentially (van Ruijven et al. 2005a).

9.2 Aminoglycosides

The pathology of ototoxicity in the cochlea and the vestibular organ has long been established with the hair cells in both tissues as the primary targets of all aminoglycosides (Caussé et al. 1949; Rüedi et al. 1952; see also Hawkins 1976, for a review of early studies). In the organ of Corti, pathology is first evident as a loss of OHCs at the base of the cochlea, comparable to the action of cisplatin (Fig. 8.2) progressing toward the apex. Superimposed on this progression is a lateral gradient whereby OHCs of the first (innermost) row are affected before those of the second row and the third row (Hawkins 1976). IHCs are more resistant than OHCs and they generally disappear only after OHCs in their immediate vicinity are lost.

While damage to hair cells is the early characteristic of aminoglycoside toxicity, prolonged drug treatment can turn the entire cochlear sensory epithelium into a nonspecialized squamous epithelium (Hawkins 1976). Changes also occur in the stria vascularis (Rüedi et al. 1952), which becomes thinner and loses some marginal cells (Hawkins 1973), but strial changes do not appear to be prerequisite to hair cell damage (Forge and Fradis 1985).

Nerve fibers degenerate subsequent to hair cell loss in experimental animals and humans (Hawkins et al. 1967; Johnsson et al 1981), and pathological changes may continue long after drug treatment has been terminated (Webster and Webster 1981; Leake and Hradek 1988). In addition, examinations of human temporal bones have suggested that spiral ganglion cells may be affected in the absence of obvious morphological damage to hair cells (Hinojosa and Lerner 1987; Song et al. 1998). Although neuronal loss is a well established potential consequence of aminoglycoside treatment, its extent appears variable. For example, the density of spiral ganglion cells had remained high in the inner ears of some patients who were deafened by aminoglycoside during life (Nadol 1997). The reasons behind such variability may reflect differential neuronal survival capability in different species or simply in different individuals. Neurotrophic factors which regulate development and maintenance of neurons can enhance this capability also in experimental aminoglycoside deafness (Green et al., Chapter 10).

10. Vestibular Pathology

10.1 Cisplatin

As aggressive as cisplatin is against cochlear structures and function, it lacks any major detrimental effects on the vestibular system. Although some functional assays have detected vestibular deficits in cisplatin treated patients

(Prim et al. 2001), others have found none in animals (Myers et al. 1993); even high doses of cisplatin did not induce significant morphological damage on the vestibular neuroepithelium in the guinea pig (Schuknecht et al. 1993; Sergi et al. 2003).

10.2 Aminoglycosides

Vestibular pathology is a major side effect of aminoglycoside antibiotics. It is again the hair cells that are affected in the vestibular organs, and the initial damage occurs in the apex of the cristae and the striolar regions of the maculi (Caussé et al. 1949; Lindemann 1969). From there, hair cell loss progresses toward the periphery of the vestibular receptor organ with type I hair cells affected earlier than the type II hair cells (Wersäll et al. 1969). The otoconial membrane and otolith structures may also be affected.

Like their cochlear counterparts, afferent nerve endings and ganglion cells will eventually also deteriorate (Li et al. 1995) and may be protected by nerve growth factors (Altschuler, Chapter 10). Regeneration of vestibular hair cells has been observed in mammalian species (Forge et al. 1993, 1998) and even the cochlea may have a latent capacity to regenerate hair cells (see Raphael and Heller, Chapter 11).

11. Pharmacokinetics

11.1 Cisplatin

Following systemic injections, more than ninety percent of cisplatin is bound to serum protein, and this cisplatin–protein complex is biologically inactive (Gormley et al. 1979). About 25% of administered cisplatin is eliminated from the body during the first 24 hours, with renal clearance for more than 90%. This clearance follows triphasic pattern, with an initial plasma half-life ($t_{1/2}$) of 20–30 min, a second phase $t_{1/2}$ of 60 min, and a terminal $t_{1/2}$ of more than 24 h (Himmelstein et al. 1981). Cisplatin preferentially collects in the liver, kidneys, and large and small intestines, with little penetration of the central nervous system (Vermorken et al. 1984).

Aiding the antitumor activity of cisplatin is the fact that a large amount of cisplatin may remain in brain tumors after intra-arterial administration. Up to a 10-fold greater amount of radioactivity was detected in brain tumor tissue compared with normal brain using scintigraphic imaging following radiolabeled cisplatin intraarterial injection. After intravenous injection, however, the differential localization of label in tumors was seldom greater than twice that of normal brain (Shani et al. 1989).

The cellular uptake of cisplatin was initially thought to occur by passive diffusion (see review by Wang and Lippard 2005) but more recent studies have established a direct link between the cellular regulation of copper and platinum concentrations. Cisplatin resistance in tumor cells has been associated with mutations or deletions in copper transporter genes controlling drug uptake (*Ctr 1*) and drug efflux (*ATP7B* and *ABCC2*). These observations are supported by the fact that both copper and cisplatin can prevent the uptake of each other (Ishida et al. 2002). Consistent with its targets in the cochlea, platinated DNA has been localized to the nuclei of OHCs, marginal cells of the stria vascularis, and the cells in the spiral ligament (van Ruijven et al. 2005b). No mention was made regarding uptake into vestibular structures.

11.2 Aminoglycosides

Aminoglycosides exhibit negligible binding to serum proteins. After systemic application, they reach peak plasma levels by 30–90 min and their plasma half-life ranges from 2 to 6 h. The drugs are excreted essentially unaltered. Aminoglycoside antibiotics enter the inner ear via the bloodstream within minutes after an injection and may reach a plateau as early as after 0.5–3 h (Tran Ba Huy et al. 1986). Drug concentrations in the inner ear typically remain at one-tenth of peak serum levels (Henley and Schacht 1988) but are less efficiently cleared from the inner ear than from serum. Clearance is biphasic and the half-life of the second phase may exceed 30 days (Tran Ba Huy et al. 1986), and even 11 months after cessation of treatment gentamicin could be found in hair cells (Dulon et al. 1993). This difference in half-lives in serum and cochlear tissues incorrectly gave rise to the idea of an "accumulation" of aminoglycosides in the inner ear and was held responsible for the organ specific toxicity of these drugs. However, the concentrations reached in the inner ear by different aminoglycosides correlate neither with the magnitude of their ototoxic potential (Ohtsuki et al. 1982) nor with their preferential vestibular or cochlear toxicity (Dulon et al. 1986).

The precise mechanisms of aminoglycoside uptake into hair cells remain enigmatic. Even localization studies in the inner ear are contradictory and differences have been reported whether immunocytochemistry or radioactive or fluorescently tagged drugs were employed. While a preferential uptake into hair cells can be seen in some studies, others find a more widespread distribution in the cochlea (Imamura and Adams 2003; see also the review by Steyger 2005).

Potential transport mechanisms include a polyamine-like transport consistent with the polyamine-like nature of the drugs (Williams et al. 1987) and a vesicular transport at the base of hair cells (Lim 1986). The possibility of endocytotic uptake at the apex of the hair cell was suggested by early observations that lysosomes appear in the subcuticular portion of hair cells soon after systemic treatment of guinea pigs with kanamycin (Darrouzet and Guilhaume 1974). An uptake mechanism at the apical region of hair cells is also supported by the apparent involvement of myosin VII-A (Richardson et al. 1997). The myosin VII mutation affects the turnover of the apical plasma membrane of hair cells but not their basolateral membrane and explants from the inner ear of mutant mice do not take up aminoglycosides. Another candidate transporter, the glycoprotein megalin, has been suggested by evidence from the proximal tubules of the kidney

(Moestrup et al. 1995). Megalin is present in the inner ear, but it is more widely distributed than the established pattern of aminoglycoside uptake and ototoxicity would suggest and it may even be absent from OHCs (Ylikoski et al. 1997; Mizuta et al. 1999).

Fluorescently tagged aminoglycosides (Dulon et al. 1989; Arbuzova et al. 2000) provide a convenient means, together with the advanced imaging techniques, to follow time course and localization of the drugs. Recent studies using these techniques confirmed a relatively rapid uptake into cochlea and primarily so into the sensory cells (Dai et al. 2006). There was, however, some diffuse fluorescence in the inner and outer pillar cells and in part of the pharyngeal processes of the Dieters cells. A base to apex gradient of uptake, interestingly, could only be detected at early time points following the injection that was obliterated later. Lateral wall tissues likewise transiently took up labeled aminoglycosides. Another potential route is an entry through the mechanoelectrical transducer channel (Marcotti et al. 2005). Although several studies have suggested a correlation between the development of hair cell sensitivity to aminoglycosides and onset of mechanotransduction, these events have been dissociated in the zebrafish lateral line (Santoso et al. 2006). Several other ion channels of the TRP (transient receptor potential) class are aminoglycoside permissive (trpp1, trpa1, trpv4) and may contribute to the overall pattern of cellular distribution of these drugs (Steyger 2005).

From the sum of the evidence, it seems reasonable to conclude that more than one mechanism of uptake for aminoglycosides operates in the inner ear. Because there is apparently also no exclusive uptake into hair cells, the reason for the differential sensitivity of inner ear sensory cells (and OHCs in particular) must be more complex. It may include the extreme persistence of the drugs or an intrinsic susceptibility to their actions, notably to reactive oxygen species as described later (Sha et al. 2001).

12. Biochemical Actions

12.1 Cisplatin

In addition to binding to nuclear DNA, cisplatin also binds to a variety of other molecules (Bose et al. 2002) including mitochondrial DNA and membrane phospholipids, and can alter microtubule formation and disrupt the cytoskeleton (Gonzalez et al. 2001; Fuertes et al. 2003). Binding to glutathione and other sulfhydryl-containing molecules, such as metallothionins can lead to lipid peroxidation. In fact, a major hypothesis of cisplatin cytotoxicity centers on the formation of reactive oxygen species, which is considered in detail in the text that follows.

12.2 Aminoglycosides

Early studies were replete with actions of aminoglycoside antibiotics on metabolic pathways including effects on DNA, RNA, enzymes and other proteins, lipids, and metabolic intermediates. Today most of these are considered secondary events and not causally related to the mechanisms that trigger aminoglycoside toxicity. One of the earliest adverse reactions of aminoglycoside antibiotics was an antagonism of these drugs with calcium, originally discovered as a neuromuscular blocking action (Vital-Brazil 1957). The underlying mechanism was later confirmed as a block of calcium channels, and aminoglycoside antibiotics were widely used as experimental tools to elucidate calcium channel function (Corrado et al. 1989). For example, aminoglycosides may block N-type and P/Q-type channels in neurons (Pichler et al. 1996) as well as prevent calcium-entry into hair cells (Dulon et al. 1989). They also block transduction channels at the tips of stereocilia (Kroese et al. 1989). This action does not directly lead to hair cell death (Kossl et al. 1990), and such acute actions may not be causally related to the chronic ototoxicity of the drugs.

Aminoglycoside–calcium interactions are yet another example of the complexity of the actions of these drugs. On the one hand, aminoglycoside antibiotics are confirmed calcium channel blockers; on the other hand, gentamicin was able to increase intracellular calcium in explants of the chick sensory epithelium (Hirose et al. 1999). Aminoglycoside antibiotics frequently exert apparently contradictory actions, often dose- or tissue-dependently, for example stimulating or inhibiting free radical formation (Priuska and Schacht 1995) or stimulation or inhibiting lipid metabolism (McDonald and Mamrack 1995). Aminoglycoside antibiotics are also agonists at calcium sensing receptors that respond to these drugs by a mobilization of intracellular calcium (Ward et al. 2005). Such an activation of calcium-sensing receptors and elevation of intracellular calcium has been suggested to contribute to the renal toxicity of aminoglycoside antibiotics by disrupting homeostasis and initiating of calcium-dependent cell death pathways.

13. Oxidative Stress

The overproduction of reactive oxygen species (ROS; free radicals) and the resulting redox imbalance in the cell now appears to be a common mechanism by which many forms of stress cause damage to the inner ear, including age, noise, and ototoxic drugs. Mitochondrial respiration and oxidative enzymatic processes produce reactive oxygen species in all cells under normal conditions. While some ROS are simply byproducts of metabolism (such as mitochondrial "leakage" of superoxide radicals during respiration), others are essential metabolites or physiological mediators and second messengers (such as nitric oxide). Detrimental consequences can arise from an overproduction of ROS whereby, for example, an increased level of superoxide radicals can lead to the formation of hydrogen peroxide. Hydrogen peroxide can be catalyzed in a Fenton-type reaction by iron to form the hydroxyl radical, which is highly reactive and can cause peroxidation products including the highly toxic aldehyde, 4-hydroxynonenal.

Another source of radicals, reactive nitrogen species, can arise from the activation or induction of the enzyme nitric oxide synthase. The different forms of this enzyme serve a variety of physiological processes in normal tissue physiology and the product of the enzymatic reaction, nitric oxide, is an important second messenger molecule. When produced in excess, nitric oxide is not only potentially damaging as a free radical, but it can also combine with superoxide to produce the highly reactive and destructive peroxynitrite, which can react with proteins to form nitrotyrosine.

Balancing the adverse potential of ROS and maintaining redox homeostasis is the cellular antioxidant system. Perturbations of this physiological balance may result from a direct interaction of ROS with antioxidants such as glutathione (scavenging) or, at the level of antioxidant enzyme activity by (1) direct binding of a toxic drug to essential sulfhydryl groups within the enzymes; (2) depletion of copper and selenium, which are essential for superoxide dismutase and glutathione peroxidase activities; (3) increased ROS and organic peroxides which inactivate antioxidant enzymes; and (4) depletion of glutathione and the cofactor NADPH, which are required for detoxifying glutathione peroxidase and glutathione reductase activities.

The increased production of ROS and the resulting depletion of antioxidant capacity can initiate cell death pathways through the activation of redox-sensitive transcription factors or through calcium influx within cochlear cells, leading to pathological changes resulting in apoptosis. Chapter 3 by Wangemann gives a detailed account of homeostasis and homeostatic perturbations.

13.1 Cisplatin

Several lines of evidence indicate the formation of reactive oxygen species and a resulting redox imbalance in cisplatin-treated tissues. Cisplatin generates reactive oxygen species in cochlear tissue explants (Clerici et al. 1996; Kopke et al. 1997) including superoxide anion (Dehne et al. 2001) which may also be involved in nephrotoxicity (McGinness et al. 1978). The origin of the cochlear superoxide is speculative but an isoform of NADPH oxidase, NOX 3, could be a major source of its generation and constitute part of the pathway leading to cisplatin-mediated hair cell damage (Mukherjea et al. 2006). NOX 3 was upregulated following systemic cisplatin administration in the rat cochlea and after in vitro cisplatin application to hair cell lines derived from the immortomouse. Another source for superoxide anion may be xanthine oxidase although it has not been directly demonstrated that its activity increases in the cochlea after cisplatin exposure. Rather, this hypothesis is based on the partial protection against cisplatin afforded by allopurinol, an inhibitor of this enzyme (Lynch et al. 2005a, b). Nevertheless, in agreement with the emergence of reactive oxygen species, 4-hydroxynonenal has been detected immunohistochemically in the cochlea after cisplatin treatment (Lee et al. 2004a) as has malondialdehyde, an indicator of lipid peroxidation (Rybak et al. 2000).

In addition to superoxide, reactive nitrogen species, such as nitric oxide, may play a role in cisplatin ototoxicity, at least in the lateral wall of the cochlea and in the spiral ganglion cells. Increased nitric oxide levels have been found in cochlear extracts of rat cochleae after treatment with cisplatin (Kelly et al. 2003) and in the stria vascularis the onset of apoptosis was correlated with increased immuno-labeling of 4-hydroxynonenal, nitrotyrosine, and iNOS (Lee et al. 2004a).

As a consequence of the increased formation of reactive oxygen and nitrogen species, cochlear tissues are depleted of glutathione and antioxidant enzymes (superoxide dismutase, catalase, glutathione peroxidase, and glutathione reductase) (Ravi et al. 1995; Rybak et al. 2000). The resulting redox imbalance will then trigger pathways of cell death.

13.2 Aminoglycosides

Oxidative stress is also part of the toxic action of aminoglycosides although the details of ROS formation differ. In retrospect, experimental studies in the 1950s and 1960s can be interpreted as indicating the participation of ROS in aminoglycoside ototoxicity. Substances like 2,3-dimercaptopropanol were found to protect against the side effects of streptomycin but positive results were also met with failures laying such attempts to rest (see Federspil 1979 for a review of the early literature). The speculations were revived by the ability of a radical scavenger to protect from the auditory side effects of kanamycin (Pierson and Moller 1981) only to be challenged by a lack of protection by another radical scavenger (Bock et al. 1983). It took another decade to establish that antioxidants could limit ototoxicity and that free radicals were generated in tissues exposed to aminoglycosides (Lautermann et al. 1995; Clerici et al. 1996; Hirose et al. 1997). Clear evidence now exists for the potential of aminoglycoside antibiotics to catalyze ROS formation both nonenzymatically and by stimulating enzymatic reactions.

A step toward the understanding one of the mechanisms of ROS formation was the observation that gentamicin was able to accelerate iron-mediated formation of free radicals (Priuska and Schacht 1995). Since iron-catalyzed oxidations can be greatly accelerated by chelators, it was postulated that gentamicin and iron may form redox-active complexes. These complexes reduce molecular oxygen to superoxide radicals at the expense of electrons provided by polyunsaturated fatty acid with arachidonic acid being a particularly suitable coreactant (Sha and Schacht 1999a). The availability of free arachidonic acid is low in an intracellular environment where most is esterified to phospholipids. Polyphosphoinositides generally contain arachidonate as one of their fatty acid esters, and it had long been known that aminoglycosides strongly bind to phosphatidyl inositol 4,5-bisphosphate (Schacht 1979), suggesting the possibility that the lipid itself could provide reactive electrons through its arachidonic acid content. Redox-active ternary complexes between $Fe^{2+/3+}$, gentamicin, and arachidonic acid, capable of producing superoxide radicals, indeed exist (Lesniak et al. 2005).

Enzymatic mechanisms may be an additional source of ROS in aminoglycoside toxicity. One of the initial reactions of aminoglycoside antibiotics in vivo is an activation of redox-dependent molecular signaling pathways linked to Rho-GTPases (Jiang et al. 2006b). The activity of Rac-1, a member of the family of Rho-GTPases, is enhanced by aminoglycosides, thereby leading to activation of the NADPH oxidase complex which enzymatically promotes the formation of the superoxide radicals. Such an action would be akin to the stimulation of the NOX 3 isoform of NADPH oxidase by cisplatin.

Although acute exposure of tissues to aminoglycosides in vitro or by local application may generate nitric oxide, for example in vestibular structures (Takumida et al. 2000), evidence for such a reaction is notably absent from the chronically exposed mouse cochlea (Jiang et al. 2005). This resembles the situation with cisplatin, which also has the capability of enhancing NO formation in the stria vascularis and spiral ganglion cells (Liu et al. 2006) but apparently does not do so in hair cells.

The question whether the formation of ROS causally relates to cell death or simply represents an epiphenomenon is best answered by the ability of iron chelators and antioxidants to protect against aminoglycoside ototoxicity (see below and Green, Altschuler, and Miller, Chapter 10).

14. Biochemical Basis of Genetic Susceptibility

Considering the enhanced aminoglycoside ototoxicity due to the mitochondrial mutations, it is interesting to notice that these mutations increase the structural similarity of the mitochondrial RNA in this region to the bacterial ribosomal RNA (Prezant et al. 1993), which is a target site of the antimicrobial actions of the drugs. Binding experiments have proven that aminoglycoside binding to the mitochondrial 12S ribosomal RNA is enhanced (Hamasaki and Rando 1997), potentially resulting in altered protein synthesis in the mitochondria. However, it has not yet been established whether protein synthesis is indeed affected in the cochlea in vivo in response to aminoglycosides. Furthermore, the vestibular system is not involved in the enhanced response to aminoglycosides (Tono et al. 2001), leading to more unresolved questions on how the mutation interacts with proposed mechanisms of toxicity.

15. Pathways of Cell Death

On oxidative insult, a plethora of molecular pathways can be activated or attenuated, leading to cell death or survival. These pathways of cell death and survival are a complex network of interfacing signaling systems involving the activation of a variety of transcription factors and proteases, and the participation of intracellular organelles such as mitochondria or lysosomes (Leist and Jäättelä 2001). Of particular interest in the context of drug-induced hearing loss are caspases, a family of proteases that are inactive in the basal state and promote apoptotic cell death upon activation. Also frequently investigated is the Bcl-2 family of intracellular proteins consisting of both pro-apoptotic and anti-apoptotic members. The anti-apoptotic members include Bcl-2 and Bcl-X_L while the pro-apoptotic proteins include two classes, the Bax and Bak subfamily and the BH-3 proteins such as Bid, Bad, Bim/Bod, and PUMA (Huang and Strasser 2000). Apoptosis is controlled within the cell by a balance between pro- and anti-apoptotic Bcl-2 family proteins (Cheng et al. 2005). When apoptosis is triggered, Bax can translocate from the cytoplasm to the mitochondria and increase the permeability of the mitochondrial membrane. This can lead to loss of mitochondrial membrane potential, generation of ROS and leakage of cytochrome c into the cytoplasm (Cheng et al. 2005). The p53 tumor suppressor gene, induced by DNA damage, is another important mediator of cell death (Oren 1999). When the DNA repair mechanisms fail, p53 can be activated and upregulate Bax. Alternatively, p53 can translocate to the mitochondria and damage them directly leading to loss of membrane potential and induction of apoptosis (Cheng et al. 2005). Also of relevance, the c-jun NH2-terminal kinases (JNKs) are a group of the mitogen-activated protein (MAP) kinases generally involved in apoptotic events. JNKs are activated following a variety of cell insults, such as irradiation, excitotoxic damage and inflammatory cytokines. Chapter 10 by Green, Altschuler, and Miller provides a comprehensive account of cell death pathways.

15.1 Cisplatin

Morphological and histochemical evidence points to apoptosis as the major form of cisplatin-induced cell death in the inner ear. Apoptotic markers such as terminal transferase dUTP nick end labeling (TUNEL)-staining label primarily the OHCs in the organ of Corti, the stria vascularis, spiral ligament and the spiral ganglion cells (Alam et al. 2000; Watanabe et al. 2003; Liang et al. 2005). Several pathways, individually or on concert, may contribute to this pattern of demise. Caspases appear pivotal, and members of the intrinsic apoptosis caspase cascade, the Bcl-2 family proteins and caspase-9, are activated after cisplatin treatment but c-Jun N-terminal kinases may play a lesser role.

15.1.1 Caspases

General inhibitors of caspases prevent hair cell death after cisplatin exposure (Liu et al. 1998) and, specifically, cochlear hair cells were preserved from cell death and hearing loss was prevented in guinea pigs treated with cisplatin by concomitant perilymphatic perfusion of inhibitors of caspase-3 and caspase-9 (Wang et al. 2004). Likewise, in cochlear and utricular organotypic cultures explanted from postnatal day 3–4 rats cisplatin caused a dose-dependent loss of hair cells and increased immunolabeling for caspase-1 and caspase-3.

Caspase-3 (as well as other caspases) can be activated by caspase-8 which is closely linked to the death domain-containing receptors in the cell membrane

(Cheng et al. 2005). Although an in vitro study of immortalized mouse hair cells (HEI-OC1 cells) reported an early but transient increase in caspase-8 activity after application of cisplatin (Deravajan et al. 2002), immunohistochemical studies revealed no significant caspase-8 activation in guinea pigs after systemic cisplatin administration, and a caspase-8 inhibitor was unable to protect guinea pig OHCs from apoptosis (Wang et al. 2004). Therefore, caspase-3 might be activated by caspase-9, an upstream caspase which, in turn, is activated by cytochrome c released from permeabilized mitochondria.

Cell death mechanisms in the stria vascularis appear to be similar to those in hair cells. Caspase-3 and caspase-activated deoxyribonuclease were detected by immunohistochemistry in the stria vascularis and spiral ligament (Watanabe et al. 2003). Local application of cisplatin in adult mouse resulted in apoptotic cell death of marginal cells 3 days after treatment and immunostaining was positive for caspases-3 and -9, but not for caspase-8. The finding that cytochrome c was redistributed in affected marginal cells suggested a caspase-dependent, mitochondrion-mediated pathway in marginal cells.

15.1.2 Bcl-2 Family Proteins

The expression of Bcl-2 is reduced and that of Bax is increased in auditory hair cells from gerbils treated with cisplatin (Alam et al. 2000). Bax was also translocated from the cytosol to the mitochondria in OHCs (Wang et al. 2004). Consistent with the activation of mitochondrial death pathways was the release of cytochrome c into the cytosol of OHCs and supporting cells in the cochleae of guinea pigs treated with cisplatin (Wang et al. 2004). Similar observations have been made in vitro in hair cell lines treated with cisplatin (Deravajan et al. 2002).

15.1.3 Other Mechanisms

Exposure of mouse hair cell lines and organotypic cultures of the organ of Corti to cisplatin in vitro increased $p53$ expression (Deravajan et al. 2002; Zhang et al. 2003). Conversely, deletion of the $p53$ gene prevents cisplatin-induced caspase-3 activation, cytochrome c translocation, and hair cell death (Cheng et al. 2005).

Cisplatin can also induce apoptosis in the fibrocytes of the lateral wall by activation of potassium channels, leading to potassium efflux, reducing intracellular ionic and osmotic strength, which in turn can trigger apoptosis by activating pro-apoptotic enzymes, such as caspases and pro-apoptotic nucleases. These cellular losses can then affect ion transport and endocochlear potential generation in the stria vascularis, changes that in turn may affect other cells within the cochlea (Liang et al. 2005).

Studies using JNK inhibitors suggest that the JNK pathway is not involved in the hair cell death induced by cisplatin, but rather that it may play a role in the repair of DNA and in the maintenance of cisplatin-damaged hair cells (Wang et al. 2004).

15.2 Aminoglycosides

Cell death by aminoglycosides includes both apoptosis and necrosis of hair cells in cochlear and vestibular organs (Nakagawa et al. 1998; Ylikoski et al. 2002; see also Forge and Schacht 2000). Both caspase-dependent and caspase-independent pathways appear to contribute to hair cell pathology and there seems to be some variance depending on the experimental model studied, i.e., chronic vs. acute treatment, or in vitro vs. in vivo (Table 8.2).

15.2.1 Caspase Activation

A hallmark of canonical apoptosis is the activation of caspase-3, and indeed, such activation is sometimes used as a criterion to assess in vitro aminoglycoside sensitivity. Studies implicating caspase-3 as the ring leader of hair cell death have been performed primarily in vitro with explant cultures and isolated cells and/or in the vestibular system (e.g., Matsui et al. 2004). Few exceptions include a couple in vivo studies using a single, severely high dose of gentamicin in the chick basilar papilla and the guinea pig vestibular system (Shimizu et al. 2003; Mangiardi et al. 2004). In addition, one chronic in vivo model in the rat found caspase-3 to be activated in only a few cochlear OHCs (Ladrech et al. 2004). Other caspases, usually involved in the same mitochondrial mediated cell death cascade as caspase-3, have also been implicated in aminoglycoside induced ototoxicity. Specifically, caspase-3 activation is dependent upon the initiator caspase, caspase-9 which forms the apoptosome complex with Apaf1 and cytochrome c released from the mitochondria. Cytochrome c release can be observed in vivo in an acute single dose model and in vitro using fluorescent labeling techniques following aminoglycoside exposure (Mangiardi et al. 2004; Matsui et al. 2004). Cyclosporin A, an inhibitor of mitochondrial PT pores and, therefore, release of cytochrome c, alleviates aminoglycoside cochleotoxicity by way of blocking this caspase cascade in vitro (Dehne et al. 2002). Cunningham and colleagues found similar results in the vestibular system whereby inhibition of Bcl-2 leads to a block of release of cytochrome c, and therefore activation of caspase-9, offering some protection of hair cell viability (Cunningham et al. 2004). The caspase pathways seem to be associated with acute models: direct application of aminoglycosides to cultures and other in vitro systems or extremely acute doses in vivo.

15.2.2 JNK Pathways

In addition to caspase-mediated apoptosis, the JNK apoptotic pathway is often implicated in aminoglycoside ototoxicity. Members of the JNK/MAPK pathway are stress response kinases that comprise a phosphorylation cascade that can lead to activation of transcription factors such as c-jun and is linked to cytochrome c activation. A number of groups have found that inhibition of the JNK/MAPK pathway with pharmacological inhibitors (e.g., CEP-1347) can offer some protection in vitro from aminoglycosides (Pirvola et al. 2000;

TABLE 8.2. Examples of aminoglycoside-induced signaling pathways observed in different model systems.

	In vivo		In vitro	
Cochlear	Chronic	Acute, developmental	Explants	Cell lines
MAPK/JNK	Ylikoski et al. 2002 Wang et al. 2003	Kalinec et al. 2005	Albinger-Hegyi et al. 2006 Wei et al. 2005 Battaglia et al. 2003 Cheng et al. 2003 Wang et al. 2003 Bodmer et al. 2002a Ylikoski et al. 2002 Bodmer et al. 2002b Pirvola et al. 2000	Kalinec et al. 2005 Wrzesniok et al. 2005 Osborn 1996
Caspases, mitochondrial	Okuda et al. 2005	Ladrech et al. 2004 Mangiardi et al. 2004	Wei et al. 2005 Dehne et al. 2002	Kalinec et al. 2003
Cathepsins, Calpains	Jiang et al. 2006a	Ladrech et al. 2004	Ding et al. 2003	
Other	Jiang et al. 2006b (Rac/Rho) Jiang et al. 2005 (NF-κB)	Ladrech et al. 2004 (PKC)		Jiang et al. 2006 (Rac/Rho)
	In vivo		**In vitro**	
Vestibular	Chronic	Acute, developmental	Explants	Cell lines
MAPK/JNK	Matsui et al. 2003	Shimizu et al. 2003	Matsui et al. 2004	
Caspases, mitochondrial			Cunningham et al. 2004 Matsui et al. 2004 Cunningham et al. 2002 Matsui et al. 2002 Forge and Li 2000	
Cathepsins, calpains		Shimizu et al. 2003	Ding et al. 2003	

Bodmer et al. 2002a; Wang et al. 2003). In addition, inhibition of JNK in the guinea pig inner ear found modest protection from ototoxicity (Ylikoski et al. 2002). In vitro, activation of c-jun is found following amino-glycoside exposure, and is linked to activation of the JNK pathway which is upstream of the transcription factor (Maroney et al. 1998; Ylikoski et al. 2002). Possible upstream activators of the JNK pathway are the small G-proteins (guanine-nucleotide binding proteins), Rho and Ras small GTPases. Inhibitors of Ras have been found to protect explant cultures from gentamicin exposure and reduce activation of c-jun (Battaglia et al. 2003), as did *Clostridium difficile* toxin B, an inhibitor of Rho, Rac, and Cdc42 (Bodmer et al. 2002b). Like the caspase pathway, which is likely downstream of JNK in aminoglycoside ototox-icity, the involvement of the JNK pathway is reported almost exclusively in cultured explants and the in vivo data are limited.

15.2.3 Caspase-Independent Pathways

Caspase-independent apoptosis and necrosis may be important in aminogly-coside ototoxicity. A dominance of caspase-independent cell death emerges in an in vivo treatment model where the onset of cochlear deficits is delayed more than one week after treatment begins and continues to develop after the cessation of treatment, as is seen in the clinical situation. In this model, release of cathepsins inferring lysosomal rupture occurs, presumably preceding cell death. Lysosomal destabilization may be the result of an observed μ-calpain activation or of sequestering of an overload of iron-bound molecules in the lysosome (Jiang et al. 2006a). Later, cathepsin-dependent cleavage of PARP1 results in fragments associated with necrotic cell death. Calpain involvement is also supported by the protective capability of leupeptin, a calpain inhibitor, to reduce hair cell death in utricle explants (Ding et al. 2003).

16. Protection

A detailed account of protecting the inner ear is provided by Green, Altschuler, and Miller (Chapter 10). Therefore only the basic concepts are briefly reviewed here.

16.1 Cisplatin

Interventions in both the early events (ROS formation) and the ensuing death pathways have been tested for cisplatin ototoxicity. As already mentioned as studies in the context of establishing a mechanism of cell death, protection against cisplatin ototoxicity has been observed with the intracochlear perfusion of inhibitors of caspase-3 and caspase-9. These agents dramatically reduced the extent of hearing loss and apoptosis of hair cells (Wang et al. 2004). The application of the p53 inhibitor, pifithrin-alpha, to organotypic organ of Corti cultures also protected the hair cells from damage. The protection was correlated with a reduction in p53 expression and caspase activation (Zhang et al. 2003).

More numerous, however, have been attempts at upstream protection of the cochlea with a variety of antioxidant compounds before death pathways have been initiated. Among them, several antioxidants containing thiol groups attenuate cisplatin ototoxicity including sodium thiosulfate, diethyldithiocarbamate, D- or L-methionine, methylthiobenzoic acid, lipoic acid, N-acetylcysteine, tiopronin, glutathione ester, and amifostine (Rybak and Whitworth 2005). Other agents that may function as free radical scavengers also protect the cochlea from cisplatin damage and hearing loss in experimental animals: α-tocopherol (alone or in combination with tiopronin), aminoguanidine, D- and L-methionine, sodium salicylate, ebselen (alone or in combination with allopurinol, an inhibitor of xanthine oxidase) (Lynch et al. 2005a, b; Rybak and Whitworth 2005).

A variety of strategies appear to be effective in animal models and need to be tested for safety and efficacy in humans. It will be especially important to ascertain whether the antitumor effect of cisplatin is compromised by the protective agent as may be the case for some thiol compounds. In such a case, a local application may resolve the problem. Furthermore, it remains to be determined whether extrapolations from in vitro or animal models can be made to the human. For example, amifostine protects against peripheral ototoxicity in the hamster (Church et al. 2004) but in clinical trials it was not effective in protecting against cisplatin-induced hearing loss. In adult patients with metastatic melanoma, ototoxicity was found to be unacceptable despite amifostine administration prior to cisplatin infusion (Ekborn et al. 2004). Also, no protection was found in children with germ cell tumors treated with amifostine in combination with cisplatin, etoposide and bleomycin (Marina et al. 2005; Sastry and Kellie 2005).

16.2 Aminoglycosides

Attempts to protect patients from the unwanted side effects of aminoglycosides date back to the early years of treatment with these drugs, for example the with the aforementioned use of 2,3-dimercaptopropanol (Federspil 1979). Other agents claimed to attenuate ototoxicity in animals include vitamins (A, C, K, and various components of the B complex), diverse amino acids, hormones (corticoids), antibiotics (fosfomycin), sulfhydryl compounds, herbal concoctions and more. None of these claims, however, ever advanced into a clinical treatment.

Rational and successful protective therapies have now been developed on the basis of the knowledge of the mechanisms of ototoxicity. Just as for cisplatin, two major lines of treatment have emerged: the restoration of the redox homeostasis and the manipulation of cell death pathways. Inhibitors of one of the many steps in the apoptotic cascade protect hair cells in cell or organ culture (Pirvola et al. 2000; Bodmer et al. 2002a). A clinically applicable therapy may be difficult to accomplish with this approach since systemic interventions in important cell signaling pathways potentially have far ranging physiological consequences, particularly in a long-term treatment as in tuberculosis. Local application to the round window membrane appears more feasible (Wang et al. 2003). Local gene therapy, virally introducing a gene coding for a neurotrophin or an antioxidant enzyme into cochlear tissues, has also succeeded in alleviating ototoxicity and may eventually become clinically promising (Kawamoto et al. 2004).

Antioxidant therapy is an established clinical approach to many of the pathologies that involve free radicals (Hershko 1992; Tanswell and Freeman 1995) and may be the currently most practicable method of protection against aminoglycoside ototoxicity. Antioxidants would scavenge the ROS and suppress toxic mechanisms at the very onset before apoptotic or necrotic mechanisms are being triggered. Indeed, both the vestibular and cochlear side effects of aminoglycosides can be attenuated by concomitant administration of various antioxidants (Song and Schacht 1996; Song et al. 1998), regardless of which individual aminoglycoside is the causative agent. Among the efficacious protective compounds was salicylate (Sha and Schacht 1999b), the active principle of aspirin (acetyl salicylate). The clinical effectiveness of aspirin was subsequently tested in a randomized double-blind placebo-controlled trial in patients receiving gentamicin for acute infections (Sha et al. 2006). Fourteen of 106 patients (13%) met the criterion of hearing loss in the placebo group while only 3/89 (3%) were affected in the aspirin group yielding a 75% reduction in risk. Aspirin did not influence gentamicin serum levels or the course of therapy. These results indicate that therapeutic protection from aminoglycoside ototoxicity is feasible and that findings from animal models may be extrapolated to the clinic. Furthermore, the fact that medications as common as aspirin can significantly attenuate the risk of gentamicin-induced hearing loss may provide a major benefit in developing countries.

17. Conclusion

Cisplatin and the aminoglycoside antibiotics, two classes of drugs with vast therapeutic potentials, are also associated with the greatest incidence of ototoxicity. Both damage the OHCs preferentially in the basal turn of the cochlea, thereby causing hearing loss in the high frequencies in patients and in experimental animals. Cisplatin, however, seems to have a low probability for vestibulotoxicity, whereas aminoglycosides, such as streptomycin and gentamicin, preferentially damage the vestibular system. Although both classes of drugs share oxidant stress as a main stimulus for cell damage and toxicity, the pathways through which cell death proceeds are particular to each. While cisplatin primarily seems to cause apoptotic cell death in the cochlea, the aminoglycosides appear to invoke multiple apoptotic and necrotic cascades. Finally, a solution to the long-standing problem of drug-induced hearing loss seems to be in sight since antioxidants protect against ototoxicity caused by both groups of therapeutic agents.

References

Alam SA, Ikeda K, Oshima T, Suzuki M, Kawase T, Kikuchi T, Takasaka T (2000) Cisplatin-induced apoptotic cell death in Mongolian gerbil cochlea. Hear Res 141:28–38.

Albinger-Hegyi A, Hegyi I, Nagy I, Bodmer M, Schmid S, Bodmer D (2006) Alteration of activator protein 1 DNA binding activity in gentamicin-induced hair cell degeneration. Neuroscience137:971–980.

Aran JM, Darrouzet J (1975) Observation of click-evoked compound VIII nerve responses before, during and over seven months after kanamycin treatment in the guinea pig. Acta Otolaryngol 79:24–32.

Arbuzova A, Martushova K, Hangyas-Mihalyne G, Morris AJ, Ozaki S, Prestwich GD, McLaughlin S (2000) Fluorescently labeled neomycin as a probe of phosphatidylinositol-4,5-bisphosphate in membranes. Biochimica Biophys Acta 1464:35–48.

Barron SE, Daigneault EA (1987) Effect of cisplatin on hair cell morphology and lateral wall Na, K-ATPase activity. Hear Res 26:131–137.

Battaglia A, Pak K, Brors D, Bodmer D, Frangos JA, Ryan AF (2003) Involvement of ras activation in toxic hair cell damage of the mammalian cochlea. Neuroscience 122:1025–1035.

Bertolini P, Lassalle M, Mercier G, Raquin MA, Izzi G, Corradini N, Hartmann O (2004) Platinum compound-related ototoxicity in children. Long-term follow-up reveals continuous worsening of hearing loss. J Pediatr Hematol Oncol 26:649–655.

Black FO, Pesznecker S, Stallings V (2004) Permanent gentamicin vestibulotoxicity. Otol Neurotol 25:559–596.

Blakley BW (1997) Clinical forum: a review of intratympanic therapy. Am J Otol 18:520–526.

Blakley BW, Gupta AK, Myers SF, Schwan S (1994) Risk factors for ototoxicity due to cisplatin. Arch Otolaryngol Head Neck Surg 120:541–546.

Benedetti Panici P, Greggi S, Scambia G, Baiocchi G, Lomonaco M, Conti G, Mancuso S (1993) Efficacy and toxicity of very high-dose cisplatin in advanced ovarian carcinoma: 4-year survival analysis and neurological follow-up. Int J Gynecol Cancer 3:44–53.

Bock GR, Yates GK, Miller JJ, Moorjani P (1983) Effects of N-acetylcysteine on kanamycin ototoxicity in the guinea pig. Hear Res 9:255–262.

Bodmer D, Brors D, Bodmer M, Ryan AF (2002a) Rescue of auditory hair cells from ototoxicity by CEP-11 004, an inhibitor of the JNK signaling pathway. Laryngorhinootologie 81:853–856.

Bodmer D, Brors D, Pak K, Gloddek B, Ryan A (2002b) Rescue of auditory hair cells from aminoglycoside toxicity by Clostridium difficile toxin B, an inhibitor of the small GTPases Rho/Rac/Cdc42. Hear Res 172:81–86.

Boheim K, Bichler E (1985) Cisplatin-induced ototoxicity: audiometric findings and experimental cochlear pathology. Arch Otorhinolaryngol 242:1–6.

Bokemeyer C, Berger CC, Hartmann JT, Kollmannsberger C, Schmoll HJ, Kuczyk MA, Kanz L (1998) Analysis of risk factors for cisplatin-induced ototoxicity in patients with testicular cancer. Br J Cancer 77:1355–1362.

Bose RN (2002) Biomolecular targets for platinum antitumor drugs. Mini Rev Med Chem 2:103–111.

Boulikas T, Vougiouka M (2004) Recent clinical trials using cisplatin, carboplatin and their combination chemotherapy drugs (review). Oncol Rep 11:559–565.

Brouet G, Marche J, Chevallier J, Liot F, LE Meur G, Bergogne (1959) Étude expéri-mentale et clinique de la kanmycine dans l'infection tuberculeuse. Re Tub Pneum 23:949–988.

Campbell KCM, Rybak LP, Meech RP, Hughes L (1996) D-Methionine provides excellent protection from cisplatin ototoxicity in the rat. Hear Res 102:90–98.

Campbell KC, Meech RP, Rybak LP (1999) D-Methionine protects against cisplatin damage to the stria vascularis. Hear Res 138:13–28.

Caussé R, Gondet I, Vallancien B (1949) Action de la streptomycine sur les cellules ciliées des organes vestibulaires de la souris. Compt Rend Soc Biol 143: 619–620.

Cheng AG, Cunningham LL, Rubel EW (2003) Hair cell death in the avian basilar papilla: characterization of the in vitro model and caspase activation. J Assoc Res Otolaryngol 4:91–105.

Cheng AG, Cunningham LL, Rubel EW (2005) Mechanisms of hair cell death and protection. Curr Opin Otolaryngol Head Neck Surg 13:343–348.

Choe WT, Chinosornvatana N, Chang KW (2004) Prevention of cisplatin ototoxicity using transtympanic N-acetylcysteine and lactate. Otol Neurotol 25:910–915.

Church MW, Blakley BW, Burgio DL, Gupta AK (2004) WR-2721 (amifostine) amelio-rates cisplatin-induced hearing loss but causes neurotoxicity in hamsters. J Assoc Res Otolaryngol 5:227–237.

Clerici WJ, Hensley K, DiMartino DL, Butterfield DA (1996) Direct detection of ototoxicant-induced reactive oxygen species generation in cochlear explants. Hear Res 98:116–124.

Corrado AP, de Morais IP, Prado WA (1989) Aminoglycoside antibiotics as a tool for the study of the biological role of calcium ions. Historical overview. Acta Physiol Pharmacol Latinoamer 39:419–430.

Cunningham LL, Matsui JI, Warchol ME, Rubel EW (2004) Overexpression of Bcl-2 prevents neomycin-induced hair cell death and caspase-9 activation in the adult mouse utricle in vitro. J Neurobiol 60:89–100.

Dai CF, Mangiardi D, Cotanche DA, Steyger PS (2006) Uptake of fluorescent gentamicin by vertebrate sensory cells in vivo. Hear Res 213:64–78.

Darrouzet J, Guilhaume A (1974) Ototoxicité de la kanamycine au jour le jour. Étude expérimentale en microscopie électronique. Rev Laryng (Bordeaux) 95:601–621.

Davis H, Deatherage BH, Rosenblut B, Fernandez C, Kimura R, Smith CA (1958) Modifi-cation of cochlear potentials produced by streptomycin poisoning and by extensive venous obstruction. Laryngoscope 68:596–627.

de Jager P, van Altena R (2002) Hearing loss and nephrotoxicity in long-term aminogly-coside treatment in patients with tuberculosis. Int J Tuberc Lung Dis 6:622–627.

DeConti RC, Toftness BR, Lange RC, Creasey A (1973) Clinical and pharmacological studies with cis-diamminedichloroplatinum (II). Cancer Res 33:1310–1315.

DeGroot JC, Hamers FP, Gispen WH, Smoorenberg GF (1997) Co-administration of the neurotropic ACTH (4-9) analogue, ORG 2766, may reduce the cochleotoxic effects of cisplatin. Hear Res 106:9–19.

Dehne N, Lautermann J, Petrat F, Rauen U, de Groot H (2001) Cisplatin ototoxicity: involvement of iron and enhanced formation of superoxide anion radicals. Toxicol Appl Pharmacol 174:27–34.

Dehne N, Rauen U, de Groot H, Lautermann J (2002). Involvement of the mitochondrial permeability transition in gentamicin ototoxicity. Hear Res 169:47–55.

Devarajan P, Savoca M, Castaneda MP, Park MS, Esteban-Cruciani N, Kalinec G, Kalinec F (2002) Cisplatin-induced apoptosis in auditory cells: role of death receptor and mitochondrial pathways. Hear Res 174:45–54.

Ding D, Stracher A, Salvi RJ (2002) Leupeptin protects cochlear and vestibular hair cells from gentamicin ototoxicity. Hear Res. 2002 164:115–26. Erratum in: Hear Res. 2003 180:129.

Dulon D, Aran JM, Zajic G, Schacht J (1986) Comparative pharmacokinetics of gentamicin, netilmicin and amikacin in the cochlea and the vestibule of the guinea pig. Antimicrob Agents Chemother 30:96–100.

Dulon D, Zajic G, Aran JM, Schacht J (1989) Aminoglycoside antibiotics impair calcium-entry but not viability and motility of cochlear outer hair cells. J Neurosci Res 24:338–346.

Dulon D, Hiel H, Aurousseau C, Erre JP, Aran JM (1993) Pharmacokinetics of gentamicin in the sensory hair cells of the organ of Corti: rapid uptake and long term persistence. C.R. Acad Sci III 316:682–687.

Ekborn A, Hansson J, Ehrsson H, Eksborg S, Wallin I, Wagenius G, Laurell G. (2004) High dose cisplatin with amifostine: ototoxicity and pharmacokinetics. Laryngoscope 114:1660–1667.

Estrem SA, Babin RW, Ryu JH, Moore KC (1981) Cis-diamminedichloroplatinum (II) ototoxicity in the guinea pig. Otolaryngol Head Neck Surg 89:638–645.

Fausti SA, Henry JA, Schaffer HI, Olson DJ, Frey RH, McDonald WJ (1992) High-frequency audiometric monitoring for early detection of aminoglycoside ototoxicity. J Infect Dis165:1026–1032.

Fausti SA, Frey RH, Henry JA, Olson DJ, Schaffer HI (1993) High-frequency monitoring for early detection of cisplatin ototoxicity. Arch Otolaryngol Head Neck Surg 119:661–666.

Federspil P (1979) Antibiotikaschäden des Ohres. Barth: Leipzig.

Fee WE (1980) Aminoglycoside ototoxicity in the human. Laryngoscope 40:1–19.

Feun LG, Wallace S, Stewart DJ, Chuang VP, Yung WK, Leavens ME, Burgess MA, Savaraj N, et al. (1984) Intracarotid infusion of cis-dichlorodiammineplatinum (II) in the treatment of recurrent malignant brain tumors. Cancer 54:794–799.

Fischel-Ghodsian N (2005) Genetic factors in aminoglycoside toxicity. Pharmacogenomics 6:27–36.

Fleischman RW, Stadnicki SW, Ethier MF, Schaeppi U (1975) Ototoxicity of cis-dichlorodiammine platinum (II) in the guinea pig. Toxicol Appl Pharmacol 33:320–332.

Forge A, Li L (2000) Apoptotic death of hair cells in mammalian vestibular sensory epithelia. Hear Res 139:97–115.

Forge A, Schacht J (2000) Aminoglycoside Antibiotics. Audiol Neurootol 5:3–22.

Forge A, Fradis M (1985) Structural abnormalities in the stria vascularis following chronic gentamicin treatment. Hear Res 20:233–244.

Forge A, Wright A, Davies SJ (1987) Analysis of structural changes in the stria vascularis following chronic gentamicin treatment. Hear Res 31:253–266.

Forge A, Li L, Corwin JT, Nevill G (1993) Ultrastructural evidence for hair cell regeneration in the mammalian inner ear. Science 259:1616–1619.

Forge A, Li L, Nevill G (1998) Hair cell recovery in the vestibular sensory epithelia of mature guinea pigs. J Comp Neurol 397:69–88.

Fuertes MA, Castilla J, Alonso C, Perez JM (2003) Cisplatin biochemical mechanism of action: from cytotoxicity to induction of cell death through interconnections between apoptotic and necrotic pathways. Curr Med Chem 10:257–266.

Garetz SL, Altschuler RA, Schacht J (1994a) Attenuation of gentamicin ototoxicity by glutathione in the guinea pig in vivo. Hear Res 77:81–87.

Gonzalez VM, Fuertes MA, Alonso C, Perez JM (2001) Is cisplatin-induced cell death always produced by apoptosis? Mol Pharmacol 59:657–663.

Gormley PE, Bull JM, LeRoy AF, Cysyk R (1979) Kinetics of *cis*-dichlorodiammine-platinum. Clin Pharmacol Ther 25:351–357.

Grunberg SM, Sonka S, Stevenson LL, Muggia FM (1989) Progressive paresthesias after cessation of therapy with very high-dose cisplatin. Cancer Chemother Pharmacol 2:62–64.

Halmagyi GM, Fattore CM, Curthoys IS, Wade S (1994) Gentamicin vestibulotoxicity. Otolaryngol Head Neck Surg 111:571–574.

Hamasaki K, Rando RR (1997) Specific binding of aminoglycosides to a human rRNA construct based on a DNA polymorphism which causes aminoglycoside-induced deafness. Biochem 36:12323–12328.

Hawkins JE (1973) Ototoxic mechanisms: a working hypothesis. Audiology 12:383–393.

Hawkins JE (1976) Drug ototoxicity. In: Keidel WD, Neff WD (eds) Handbook of Sensory Physiology. Berlin: Springer Verlag, 5(3) pp. 707–748.

Hawkins JE, Beger V, Aran JM (1967) Antibiotic insults to Corti's organ. In: Graham AB (ed) Sensorineural Hearing Processes and Disorders. Boston: Little Brown Co, pp. 411–425.

Henley CM, Schacht J (1988) Pharmacokinetics of aminoglycoside antibiotics in blood, inner ear fluids and tissues and their relationship to ototoxicity. Audiology 27:137–146.

Henry KR, Chole RA, McGinn MD, Frush DP (1981) Increased ototoxicity in both young and old mice. Arch Otolaryngol 107:92–95.

Hershko, C (1992) Iron chelators in medicine. Molec Asp Med 13:113–165.

Himmelstein KJ, Patton TF, Belt RJ, Taylor S, Repta AJ, Sternson LA (1981) Clinical kinetics on intact cisplatin and some related species. Clin Pharmacol Ther 29:658–664.

Hinojosa R, Lerner SA (1987) Cochlear neural degeneration without hair cell loss in two patients with aminoglycoside ototoxicity. J Infect Dis 156:449–555.

Hinojosa R, Riggs LC, Strauss M, Matz GJ (1995) Temporal bone histopathology of cisplatin ototoxicity Am J Otol 16:731–740.

Hinshaw HC, Feldman WH (1945) Streptomycin in treatment of clinical tuberculosis: a preliminary report. Proc Mayo Clinic 20:313–318.

Hirose K, Hockenberry DN, Rubel EW (1997) Reactive oxygen species in chick hair cells after gentamicin exposure in vitro. Hear Res 104:1–14.

Hirose K, Westrum LE, Stone JS, Zirpel L, Rubel EW (1999) Dynamic studies of ototoxicity in mature avian auditory epithelium. Ann NY Acad Sci 884:389–409.

Hoistad DL, Ondrey FG, Mutlu C, Schachern PA, Paparella MM, Adams GL (1998) Histopathology of human temporal bone after cis-platinum, radiation, or both. Otolaryngol Head Neck Surg 118:825–832.

Hu DN, Qui WQ, Wu BT, Fang LZ, Zhou F, Gu YP, Zhang QH, Yan JH, et al. (1991) Genetic aspects of antibiotic induced deafness: mitochondrial inheritance. J Med Genet 28:79–83.

Huang DC, Strasser A (2000) BH-3-Only proteins—essential initiators of apoptotic cell death. Cell 103:839–842.

Imamura S, Adams JC (2003) Distribution of gentamicin in the guinea pig inner ear after local or systemic application. JARO 4:176–195.

Iraz M, Kalcioglu MT, Kizilay A, Karatas E (2005) Aminoguanidine prevents ototoxicity induced by cisplatin in rats. Ann Lab Clin Sci 35:329–335.

Ishida S, Lee J, Thiele DJ, Herskowita I (2002) Uptake of the anticancer drug cisplatin mediated by the copper transporter Ctr1 in yeast and mammals. Proc Natl Acad Sci USA 99:14298–14302.

Jiang H, Sha SH, Schacht J (2005) The NF-κB pathway protects cochlear hair cells from aminoglycoside-induced ototoxicity. J Neurosci Res 79:644–651.

Jiang H, Sha SH, Forge A, Schacht J (2006a) Caspase-independent pathways of hair cell death induced by kanamycin in vivo. Cell Death Diff 13:20–30.

Jiang H, Sha SH, Schacht J (2006b) Rac/Rho pathway regulates actin depolymerization induced by aminoglycoside antibiotics. J Neurosci Res 83:1544–1551.

Johnsson LG, Hawkins JE Jr, Kingsley TC, Black FO, Matz GJ (1981) Aminoglycoside-induced cochlear pathology in man. Acta Otolaryngol S383:1–19.

Kalinec GM, Webster P, Lim DJ, Kalinec F (2003) A cochlear cell line as an in vitro system for drug ototoxicity screening. Audiol Neurootol 8:177–189.

Kalinec GM, Fernandez-Zapico ME, Urrutia R, Esteban-Cruciani N, Chen S, Kalinec F (2005) Pivotal role of Harakiri in the induction and prevention of gentamicin-induced hearing loss. Proc Natl Acad Sci USA 102:16019–16024.

Kawai Y, Kohda Y, Kodawara T, Gemba M (2005) Protective effect of a protein kinase inhibitor on cellular injury induced by cephaloridine in the porcine kidney cell line LLC-PK(1). J Toxicol Sci 30:157–122.

Kawamoto K, Sha SH, Minoda R, Izumikawa M, Kuriyama H, Schacht J, Raphael Y (2004) Antioxidant gene therapy can protect hearing and hair cells from ototoxicity. Mol Ther 9:173–181.

Kelly TC, Whitworth CA, Husain K, Rybak LP (2003) Aminoguanidine reduces cisplatin ototoxicity. Hear Res 186:10–16

Klis SF, O'Leary SJ, Hamers FP, De Groot JC, Smoorenburg GF (2000) Reversible cisplatin ototoxicity in the albino guinea pig. Neuro Report 11:623–626.

Knoll C, Smith RJH, Shores C, Blatt J (2006) Hearing genes and cisplatin deafness: a pilot study. Laryngoscope 116:72–74.

Komune S, Ide M, Nakano T, Morimitsu T (1987) Effects of kanamycin sulfate on cochlear potentials and potassium ion permeability through the cochlear partitions. ORL 49:9–16.

Kopke RD, Liu W, Gabaizadeh R, Jacono A, Feghali J, Spray D, Garcia P, Steinman H, et al. (1997) The use of organotypic cultures of Corti's organ to study the protective effects of antioxidant molecules on cisplatin induced damage of auditory hair cells. Am J Otol 18:559–571.

Kossl M, Richardson GP, Russell IJ (1990) Stereocilia bundle stiffness: effects of neomycin, A23187 and concanavalin A. Hear Res 44:217–230.

Kovach JS, Moertel CG, Schutt AJ, Reitemeier RG, Hahn RG (1973) Phase II study of cis-diamminedichloroplatinum (NSC-119875) in advanced carcinoma of the large bowel. Cancer Chemother Rep 57:357–359.

Kroese ABA, van den Bercken J (1982) Effects of ototoxic antibiotices on sensory hair cell functioning. Hear Res 6:183–197.

Kroese ABA, Das A, Hudspeth AJ (1989) Blockage of the transduction channels of hair cells in the bull frog' sacculus by aminoglycoside antibiotics. Hear Res 37:203–218.

Ladrech S, Guitton M, Saido T, Lenoir M (2004) Calpain activity in the amikacin-damaged rat cochlea. J Comp Neurol 477:149–60.

Laurell G, Bagger-Sjoback D (1991) Dose-dependent inner ear changes after IV administration of cisplatin. J Otolaryngol 20:158–167.

Laurell G, Jungnelius U (1990) High-dose cisplatin treatment: hearing loss and plasma concentrations. Laryngoscope 100:724–734.

Lautermann J, McLaren J, Schacht J (1995) Glutathione protection against gentamicin ototoxicity depends on nutritional status. Hear Res 86:15–24.

Leake PA, Hradek GT (1988) Cochlear pathology of long term neomycin induced deafness in cats. Hear Res 33:11–33.

Lee JE, Nakagawa T, Kim TS, Endo T, Shiga A, Iguchi F, Lee SH, Ito J (2004a) Role of reactive radicals in degeneration of the auditory system of mice following cisplatin treatment. Acta Otolaryngol 124:1131–1135.

Lee JE, Nakagawa T, Kim TS, Endo T, Shiga A, Iguchi F, Lee SH, Ito J (2004b) Mechanisms of apoptosis induced by cisplatin in marginal cells in mouse stria vascularis. ORL 66, 111–118.

Leist M, Jäättelä M (2001) Four deaths and a funeral: from caspases to alternative mechanisms. Nat Revs 2:1–10.

Lerner SA, Schmitt BA, Seligsoh, Matz R (1986) Comparative-study of ototoxicity and nephrotoxicity in patients randomly assigned to treatment with amikacin or gentamicin. Am J Med 80:98–104.

Lesniak W, Pecoraro VL , Schacht J (2005)Ternary complexes of gentamicin with iron and lipid catalyze formation of reactive oxygen species. Chem Res Toxicol 18:357–364.

Li G, Sha SH, Zotova E, Arezzo J, Van de Water T, Schacht J (2002) Salicylate protects hearing and kidney function from cisplatin toxicity without compromising its oncolytic action. Lab Invest 82:585–596.

Li L, Nevill G, Forge A (1995) Two modes of hair cell loss from the vestibular sensory epithelia of the guinea pig. J Comp Neurol 355:405–417.

Li Y, Womer RB, Silber JH (2004) Predicting cisplatin ototoxicity in children: the influence of age and cumulative dose. Eur J Cancer 40:2445–2451.

Liang F, Schulte BA, Qu C, Hu W. (2005) Inhibition of the calcium- and voltage-dependent big conductance potassium channel ameliorates cisplatin-induced apoptosis in spiral ligament fibrocytes of the cochlea. Neuroscience 135:263–271.

Liedert B, Pluim D, Schellens J, Thomale J (2006) Adduct-specific monoclonal antibodies for the measurement of cisplatin-induced DNA lesions in individual cell nuclei. Nucleic Acids Res 34:[epub March 29, 2006].

Lim D (1986) Effects of noise and ototoxic drugs at the cellular level in the cochlea: a review. Am J Otolaryngol 7:73–99.

Lindemann HH (1969) Regional differences in sensitivity of the vestibular sensory epithelia to ototoxic antibiotics. Acat Otolaryngol 67:177–189.

Lippman AJ, Helson C, Helson L, Krakoff IH (1973) Clinical trials of cis-diamminedichloroplatinum (NSC-119875). Cancer Chemother Rep 57:191–200.

Liu W, Staecker H, Stupak H, Malgrange B, Lefebvre P, Van De Water TR. (1998) Caspase inhibitors prevent cisplatin-induced apoptosis of auditory sensory cells. Neuro Report 9:2609–2614.

Lu YF (1987) Cause of 611 deaf mutes in schools for deaf children in Shanghai. Shanghai Med J 10:159.

Lynch ED, Gu R, Pierce C, Kil J (2005a) Reduction of acute cisplatin ototoxicity in rats by oral administration of allopurinol and ebselen. Hear Res 201: 81–89.

Lynch ED, Gu R, Pierce C, Kil J (2005b) Combined oral delivery of ebselen and allopurinol reduces multiple cisplatin toxicities in rat breast and ovarian cancer models while enhancing anti-tumor activity. Anticancer Drugs 16: 569–579.

Magnet S, Blanchard JS (2005) Molecular insights into aminoglycoside action and resistance. Chem Rev 105:477–497.

Mangiardi DA, McLaughlin-Williamson K, May KE, Messana EP, Mountain DC, Cotanche DA (2004) Progression of hair cell ejection and molecular markers of apoptosis in the avian cochlea following gentamicin treatment. J Comp Neurol 475:1–18.

Marco-Algarra J, Basterra J, Marco J (1985) Cis-diaminedichloroplatinum ototoxicity. An experimental study. Acta Otolaryngol 99:343–347.

Marcotti W, van Netten S, Kros CJ (2005) The aminoglycoside antibiotic dihydrostreptomycin rapidly enters hairs through the mechano-electrical transducer channels. J Physiol 576:505–521.

Marina N, Chang KW, Malogolowkin M, London WB, Frazier AL, Womer RB, Rescorla F, Billmire DF, et al. (2005) Amifostine does not protect against the ototoxicity of high-dose cisplatin combined with etoposide and bleomycin in pediatric germ-cell tumors: a Children's Oncology Group study. Cancer 104:841–847.

Maroney AC, Glicksman MA, Basma AN, Walton KM, Knight E Jr, Murphy CA, Bartlett BA, Finn JP, et al. (1998) Motoneuron apoptosis is blocked by CEP-1347 (KT 7515): a novel inhibitor of the JNK signaling pathway. J Neurosci 18:104–111.

Marot M, Uziel A, Romand R (1980) Ototoxicity of kanamycin in developing rats: relationship with the onset of the auditory function. Hear Res 2:111–113.

Mathog RH, Klein Jr WJ (1969) Ototoxicity of ethacrynic acid and aminoglycoside antibiotics in uremia. N Engl J Med 280:1223–1224.

Matsui JI, Gale JE, Warhol ME (2004) Critical signaling events during the aminoglycoside-induced death of sensory hair cells in vitro. J Neurobiol 61: 250 266.

McDonald LJ, Mamrack MD (1995) Phosphoinositide hydrolysis by phospholipase C modulated by multivalent cations La(3+), Al(3+), neomycin, polyamines, and melttin. J Lipid Mediat Cell Signal 11:81–91.

McHugh PJ, Spanswick VJ, Hartley JA (2001) Repair of DNA interstrand crosslinks: molecular mechanisms and clinical relevance. Lancet Oncol 2:483–490.

Meech RP, Campbell KC, Hughes LF, Rybak LP (1998) A semiquantitative analysis of the effects of cisplatin on the rat stria vascularis. Hear Res 124:44–59.

Minami SB, Sha SH, Schacht J (2004) Antioxidant protection in a new animal model of cisplatin-induced ototoxicity. Hear Res 198:137–143.

Mizuta K, Saito A, Watanabe T, Nagura M, Arakawa M, Shimizu F, Hoshino T (1999) Ultrastructural localization of megalin in the cochlear duct. Hear Res 129:83–91.

Moestrup SK, Cui S, Vorum H, Bregengård C, Bjrn SE, Norris K, Gliemann J, Christensen EI (1995) Evidence that epithelial glycoprotein 330/megalin mediates uptake of polybasic drugs. J Clin Invest 96:1404–1413.

Moore RD, Smith CR, Lietman PS (1984) Risk factors for the development of auditory toxicity in patients receiving aminoglycosides. J Infect Dis 149:23–30.

Mukherjea D, Whitworth CA, Nandish S, Dunaway GA, Rybak LP, Ramkumar V (2006) Expression of the kidney injury molecule (KIM) 1 in the rat cochlea and induction by cisplatin. Neuroscience 139:733–740.

Mulheran M, Degg C, Burr S, Morgan DW, Stableforth DE (2001) Occurrence and risk of cochleotoxicity in cystic fibrosis patients receiving repeated high-dose aminoglycoside therapy. Antimicrob Agents Chemother 45:2502–2509.

Myers SF, Blakley BW, Schwan S (1993) Is cis-platinum vestibulotoxic? Otolaryngol Head Neck Surg 108:322–328.

Nadol JB Jr (1997) Patterns of neural degeneration in the human cochlea and auditory nerve: implications for cochlear implantation. Otolaryngol Head Neck Surg 117:220–228.

Nakagawa T, Yamane H, Takayama M, Sunami K, Nakai Y (1998) Apoptosis of guinea pig cochlear hair cells following aminoglycoside treatment. Eur Arch Otorhinolaryngol 255:127–131.

Ohtsuki K, Ohtani I, Aikawa T, Sato Y (1982) The ototoxicity and the accumulation in the inner ear fluids of the various aminoglycoside antibiotics. Ear Res Jap 13:85–87.

Okuda T, Sugahara K, Takemoto T, Shimogori H, Yamashita H. (2005) Inhibition of caspases alleviates gentamicin-induced cochlear damage in guinea pigs. Auris Nasus Larynx 32:33–37.

Oren M (1999) Regulation of the p53 tumor suppressor protein. J Biol Chem 274: 36031–36034.

Osborn MT, Chambers TC (1996) Role of the stress-activated/c-Jun NH2-terminal protein kinase pathway in the cellular response to adriamycin and other chemotherapeutic drugs. J Biol Chem 271:30950–30955.

Peters U, Preisler-Adams S, Lanvers-Kaminsky C, Jürgens H, Lamprecht-Dinnesen A (2003) Sequence variations of mitochondrial DNA and individual sensitivity to the ototoxic effect of cisplatin. Anticancer Res 23:1249–1255.

Pichler M, Wang Z, Grabner-Weiss C, Reimer D, Hering S, Grabner M, Glossmann H, Striessnig J (1996) Plock of P/Q-type calcium channels by therapeutic concentrations of aminoglycoside antibiotics. Biochemistry 35:14659–14664.

Pierson MG, Moller AR (1981) Prophylaxis of kanamycin-induced ototoxicity by a radioprotectant. Hear Res 4:79–87.

Pirvola U, Xing-Qun L, Virkkala J, Saarma M, Murakata C, Camoratto AM, Walton KM, Ylikoski J (2000) Rescue of hearing, auditory hair cells, and neurons by CEP-1347/KT-7515, an inhibitor of c-Jun N-terminal kinase activation. J Neurosci 20:43–50.

Pong AL, Bradley JS (2005) Guidelines for the selection of antibacterial therapy in children. Pediatr Clin North Am 52:869–894.

Prezant TR, Agapian JV, Bohlman MC, Bu X Oztas S, Qiu WQ, Arnos KS, Cortopassi GA, et al. (1993) Mitochondrial ribosomal RNA mutation associated with both antibiotic induced and non-syndromic deafness. Nat Gen 4:289–294.

Prim MP, de Diego JI, de Sarria MJ, Gavilan J (2001) Vestibular and oculomotor changes in subjects treated with cisplatin. Acta Otorrinolaringol Esp 52:367–370.

Priuska EM, Schacht J (1995) Formation of free radicals by gentamicin and iron and evidence for an iron/gentamicin complex. Biochem Pharmacol 50:1749–1752.

Raphael Y, Fein A, Nebel L (1983) Transplacental kanamycin ototoxicity in the guinea pig. Arch Oto-Rhino-Laryngol 238:45–51.

Ravi R, Somani SM, Rybak LP (1995) Mechanism of cisplatin ototoxicity: antioxidant defense system. Pharmacol Toxicol 76:386–394.

Reser D, Rho M, Dewan D, Herbst L Li G, Stupak H, Zur K, Romaine J, et al. (1999) L- and D-methioniine provide long-term protection against CDDP-induced ototoxicity in vivo, with partial in vivo and in vitro retention of antineoplastic activity. Neurotoxicology 20:731–748.

Richardson GP, Forge A, Kros CJ, Fleming J, Brown SD, Steel KP (1997) Myosin VIIA is required for aminoglycoside accumulation in cochlear hair cells. J Neurosci 17:9506–9519.

Rosenberg B (1973) Platinum coordination complexes in cancer chemotherapy. Naturwissenschaften 60:399–406.

Rosenberg B (1980) Cisplatin: its history and possible mechanisms of action. In: Prestayko, AW, Crooke ST, Carter SK (eds) Cisplatin: Current Status and New Developments. New York: Academic Press, pp. 9–20.

Rosenberg B, Van Camp L, Krigas T (1965) Inhibition of cell division in Escherichia coli by electrolysis products from a platinum electrode. Nature 205:698–699.

Rosenberg B, Van Camp L, Trosko JE, Mansour VH (1969) Platinum compounds: a new class of potent antitumor agents. Nature 222:385–386.

Rozencweig M, Von Hoff DD, Slavik M, Muggia FM (1977) Cis-ciamminechloroplatinum (II): a new anticancer drug. Ann Intern Med 86:803–812.

Ruedi L, Furrer W, Graf K, Luthy F, Nager G, Tschirren B (1951) Nouvelles constatations sur la toxicité de la streptomycine et de la quinine à l'égard de l'oreille du cobaye. Rev Laryngol (Bordeaux) 72:238–264.

Ruedi L, Furrer W, Graf K, Nager G, Tschirren B, Luthy F (1952) Further observations concerning the toxic effects of streptomycin and quinine on the auditory organ of guinea pigs. Laryngoscope 62:333–357.

Rybak LP, Whitworth CA (2005) Ototoxicity: therapeutic opportunities. Drug Discovery Today, 10:1313–1321.

Rybak LP, Husain K, Morris C, Whitworth C, Somani S (2000) Effect of protective agents against cisplatin ototoxicity. Am J Otol 21:513–520.

Santoso JT, Lucci JA 3rd, Coleman RL, Schafer I, Hannigan EV (2003) Saline, mannitol, and furosemide hydration in acute cisplatin nephrotoxicity: a randomized trial. Cancer Chemother Pharmacol 52:13–18.

Sastry J, Kellie SJ (2005) Severe neurotoxicity, ototoxicity and nephrotoxicity following high-dose cisplatin and amifostine. Pediatr Hematol Oncol 22:441–445.

Schacht J (1979) Isolation of an aminoglycoside receptor from guinea pig inner ear tissues and kidney. Arch Otorhinolaryngol 224:129–134

Schacht J, Hawkins JE (2006) Sketches of Otohistory. Part 11: Ototoxicity: Drug-induced hearing loss. Audiol Neurotol 11:1–6.

Schatz A, Bugie E, Waksman SA (1944) Streptomycin, a substance exhibiting antibiotic activity against gram-positive and gram-negative bacteria. Proc Soc Exp Biol Med 55:66–69.

Schuknecht HF (1993) Pathology of the Ear. Philadelphia: Lea & Febiger, pp. 273–274.

Schweitzer VG (1993) Ototoxicity of chemotherapeutic agents. Otolaryngol Clin North Am 26: 759–785.

Sergi B, Ferraresi A, Troiani D, Paludetti G, Fetoni AR (2003) Cisplatin ototoxicity in the guinea pig: vestibular and cochlear damage. Hear Res 182:56–64.

Sha SH, Schacht J (1999a) Formation of reactive oxygen species following bioactivation of gentamicin. Free Radic Biol Med 26:341–347.

Sha SH, Schacht J (1999b) Salicylate attenuates gentamicin-induced ototoxicity. Lab Invest 79:807–813.

Sha SH, Qiu JH, Schacht J (2006) Aspirin attenuates gentamicin-induced hearing loss. New Engl J Med 354:1856–1857.

Sha SH, Zajic G, Epstein CJ, Schacht J (2001) Overexpression of SOD protects from kanamycin-induced hearing loss. Audiol Neuro-Otol 6:117–123.

Shani J, Bertram J, Russell C, Dahalan R, Chen DC, Parti R, Ahmadi J, Kempf RA, et al. (1989) Noninvasive monitoring of drug distribution and metabolism: studies with intraarterial Pt-195m-cisplatin in humans. Cancer Res 49:1877–1881.

Shimizu A, Takumida M, Anniko M, Suzuki M (2003) Calpain and caspase inhibitors protect vestibular sensory cells from gentamicin ototoxicity. Acta Otolaryngol 123:459–465.

Sone M, Schachern PA, Paparella MM (1998) Loss of spiral ganglion cells as primary manifestation of aminoglycoside ototoxicity. Hear Res 115:217–223.

Song BB, Schacht J (1996) Variable efficacy of radical scavengers and iron chelators to attenuate gentamicin ototoxicity in guinea pig in vivo. Hear Res 94:87–93.

Song BB, Sha SH, Schacht J (1998) Iron chelators protect from aminoglycoside-induced cochleo- and vestibulotoxicity in guinea pig. Free Radical Biol Med 25:189–195.

Steyger PS (2005) Cellular uptake of aminoglycosides. Volta Rev 105:299–324.

Strauss M, Towfighi J, Lord S, Lipton A, Harvey HA, Brown B (1983) Cis-platinum ototoxicity: clinical experience and temporal bone histopathology. Laryngoscope 93:1554–1559.

Swan SK (1997) Aminoglycoside nephrotoxicity. Semin Nephrol 17:27–33.

Taguchi T, Nazneen A, Abid MR, Razzaque MS (2005) Cisplatin-associated nephrotoxicity and pathological events. Contrib Nephrol 148:107–121.

Talley RW, O'Bryan RM, Gutterman JU, Brownlee RW, McCredie KB (1973) Phase I clinical trial of platinum diamminedichloro-cis (NSC-119875). Cancer Chemother Rep 57:465–471.

Tan CT, Hsu CJ, Lee SY, Liu SH, Lin-Shiau SY (2001) Potentiation of noise-induced hearing loss by amikacin in guinea pigs. Hear Res 161:72–80.

Tanswell AK, Freeman BA (1995) Antioxidant therapy in critical care medicine. New Horizons 3:330–341.

Tono T, Kiyomizu K, Matsuda K, Komune S, Usami S, Abe S, Shinkawa H. (2001) Different clinical characteristics of aminoglycoside-induced profound deafness with and without the 1555 A–>G mitochondrial mutation. ORL 63:25–30.

Tran Ba Huy P, Bernard P, Schacht J (1986) Kinetics of gentamicin uptake and release in the rat: comparison of inner ear tissues and fluids with other organs. J Clin Invest 77:1492–1500.

Tsukasaki N, Whitworth CA, Rybak LP (2000) Acute changes in cochlear potentials due to cisplatin. Hear Res 149:189–198.

Umapathi T, Chaudhry V (2005) Toxic neuropathy. Curr Opin Neurol 18:574–580.

Vakulenko SB, Mobashery S (2003) Versatility of aminoglycosides and prospects for their future. Clin Microbiol Rev 16:430–450

van Ruijven MWM (2004) The cochlear targets of cisplatin. Thesis, University of Utrecht.

van Ruijven MWM, De Groot JCMJ, Klis SFL, Smoorenburg G (2005a) Cochlear targets of cisplatin: an electrophysiological and morphological time-sequence study. Hear Res 205: 241–248.

van Ruijven MWM, De Groot JCMJ, Smoorenburg GF (2005b) Time sequence of degeneration pattern in the guinea pig cochlea during cisplatin administration. A quantitative histological study. Hear Res 197:44–54.

Vermorken JB, Kapteijn TS, Hart AA, Pinedo HM (1983) Ototoxicity of cis-diamminedichloroplatinum (II): influence of dose, schedule and mode of administration. Eur J Cancer Clin Oncol 19:53–58.

Vermorken JB, van der Vijgh WJ, Klein I, Hart AA, Gall HE, Pinedo HM (1984) Pharmacokinetics of free and total platinum species after short-term infusion of cisplatin. Can Treat Rep 68:505–513.

Vital-Brazil O and Corrado AP (1957) The curariform action of streptomycin. J Pharm Exp Ther 120:452–459.

Wang D, Lippard SJ (2005) Cellular processing of platinum anticancer drugs. Nat Rev Drug Discov 4:307–320.

Wang J, Lloyd Faulconbridge RV, Fetoni A, Guitton MJ, Pujol R, Puel JL (2003) Local application of sodium thiosulfate prevents cisplatin-induced hearing loss in the guinea pig. Neuropharmacol 45:380–393.

Wang J, Ladrech S, Pujol R, Brabet P, Van De Water TR, Puel JL (2004) Caspase inhibitors, but not c-Jun NH2-terminal kinase inhibitor treatment prevents cisplatin-induced hearing loss. Cancer Res 64: 9217–9224.

Ward DT, Maldonado-Perez D, Hollins L, Riccardi D (2005) Aminoglycosides induce acute cell signaling and chronic cell death in renal cells that express the calcium-sensing receptor. J Am Soc Nephrol 16:1236–1244.

Watanabe K, Inai S, Jinnouchi K, Baba S, Yagi T (2003) Expression of caspase-activated deoxyribonuclease (CAD) and caspase-3 (CPP32) in the cochlea of cisplatin (CDDP)-treated guinea pigs. Auris Nasis Larynx 30: 219–225.

Webster M, Webster DB (1981) Spiral ganglion neuron loss following organ of Corti loss: A quantitative study. Brain Res 212:17–30.

Wei X, Zhao L, Liu J, Dodel RC, Farlow MR, Du Y (2005) Minocycline prevents gentamicin-induced ototoxicity by inhibiting p38 MAP kinase phosphorylation and caspase 3 activation. Neuroscience 131:513–521.

Wersäll J, Lundquist PG, Björkroth B (1969) Ototoxicity of gentamicin. J Infect Dis 119:410–416.

Wersäll J, Björkroth B, Flock A, Lundquist PG (1973) Experiments on ototoxic effects of antibiotics. Adv Oto-Rhino-Laryng 20:14–41.

Williams SE, Smith DE, Schacht J (1987) Characteristics of gentamicin uptake in the isolated crista ampullaris of the inner ear of the guinea pig. Biochem Pharmacol 36:89–95.

World Health Organization (2005) WHO Report 2005. Global Tuberculosis Control: surveillance, planning, financing. Geneva: World Health Organization.

Wright CG, Schaefer SD (1982) Inner ear histopathology in patients treated with cis-platinum. Laryngoscope 92:1408–1413.

Wrzesniok D, Buszman E, Karna E, Palka J (2005) Melanin potentiates kanamycin-induced inhibition of collagen biosynthesis in human skin fibroblasts. Pharmazie 60:439–443.

Wu WJ, Sha SH, McLaren JD, Kawamoto K, Raphael Y, Schacht J (2001) Aminoglycoside ototoxicity in adult CBA, C57BL and BALB mice and the Sprague-Dawley rat. Hear Res 158:165–178.

Ylikoski, J, Pirvola, U, Suoqiang, Z, Eriksson, U, Solin, M-L, Miettinen, A, (1997) Aminoglycoside ototoxicity: high affinity receptors are expressed in secretory epithelia. Aud Neurosci 3:415–424.

Ylikoski J, Xing-Qun L, Virkkala J, Pirvola U (2002) Blockade of c-Jun N-terminal kinase pathway attenuates gentamicin-induced cochlear and vestibular hair cell death. Hear Res 163:71–61.

Zappia JJ, Altschuler RA (1989) Evaluation of the effect of ototopical neomycin on spiral ganglion cell density in the guinea pig. Hear Res 40:29–38.

Zhang M, Liu W, Ding D, Salvi R (2003) Pifithrin-alpha suppresses p53 and protects cochlear and vestibular hair cells from cisplatin-induced apoptosis. Neuroscience 120:191–205.

Zhang S, Zhao C, Yu L (1997) [Analysis of sensorineural hearing loss in 77 children]. Lin Chuang Erh Pi Yen Hou Ko Tsa Chih Journal of Clinical Otorhinolaryngology 11:252–254.

9
Central Consequences of Cochlear Trauma

D. Kent Morest and Steven J. Potashner

1. Sensorineural Hearing Loss

Sensorineural hearing loss, which involves cochlear and cochlear nerve damage, is usually accompanied by additional pathological symptoms that further degrade the quality of life. These include tinnitus (Bauer and Brozoski, Chapter 4), loudness misperceptions, and difficulties isolating important sounds in a background of noise. Sensorineural hearing loss in the adult mammal can result from mechanical, acoustic (Henderson et al., Chapter 7), or various other pathologic insults that involve the cochlear hair cells and/or the cochlear nerve. Some of the pathological insults include drugs, such as the aminoglycosides and ethacrynic acid, cisplatin, and other platinum-containing anti-cancer agents (Rybak et al. Chapter 8), and kainic acid or other excitotoxic agents, including glutamate. Toxic effects may result from oxygen deprivation, carbon monoxide exposure, metabolic disturbances, and exposure to heavy metals, e.g., lead. There are the diseases of the middle ear and the inner ear, e.g., homeostatic disorders and Ménière's disease (Wangemann, Chapter 3); diseases affecting the nervous system, which result in oxygen or metabolic deprivation, e.g., due to arterial occlusion, hemorrhage, hypoglycemia, jaundice, uremia; diseases of demyelination and other autoimmune phenomena (Gopen and Harris, Chapter 5); and genetic disorders (Shalit and Avraham, Chapter 2). Aging can affect the auditory system both centrally and peripherally (Ohlemiller and Frisina, Chapter 6).

In many of the aforementioned conditions, it is difficult to analyze the mechanisms causing hearing impairment, because it is not clear whether the causative agents damage only the cochlea or the central pathways or both. For example, it is generally assumed that ototoxic drugs, such as aminoglycosides, do not enter cells in the central nervous system, but this assumption may deserve further study, especially under the pathological conditions following cochlear nerve injury. This chapter focuses on mechanical and acoustic agents that do not directly damage the adult brain. However, even with these agents, we have discovered that, in the central pathways, there are secondary pathological processes, arising from the initial insult to the cochlea, which complicate our understanding of the underlying mechanisms.

Hearing impairment is often accompanied by additional symptoms, such as tinnitus (Bauer and Brozoski, Chapter 4), loudness misperceptions, and difficulties discerning important sounds in a noisy environment (Olson et al. 1975; Jastreboff 1990; Axelsson and Barrenas 1992; Salvi et al. 2000).The present thesis is that synaptic changes in the central auditory pathways may underlie these symptoms. Relevant reviews on the subject of plasticity and tinnitus in the central auditory system have been published in recent years (Syka 2002; Jastreboff and Jastreboff 2002; Kaltenbach et al. 2005). The focus in this chapter is on the cochlear nucleus (CN).

2. Effects of Mechanical Damage

2.1 Models

Previous approaches to sensorineural hearing loss in the adult have been shaped by studies on the responses of the cochlea and cochlear nerve to mechanical ablations, e.g., surgical or accidental wounds. These have usually damaged the spiral ganglion as well as the organ of Corti (e.g., Rasmussen et al. 1960; Osen 1970). Less frequent are reports of mechanical ablations limited to the organ of Corti. Such ablations damage the sensory epithelium and the peripheral ends of the cochlear nerve fibers innervating the hair cells (e.g., Webster and Webster 1978; Morest et al. 1997). Acoustic trauma is similar, in that it can physically disrupt the sensory epithelium and nerve endings, while sparing the ganglion cell bodies. However, it also produces overstimulation of the auditory system, which differs in some ways from mechanical deafferentation (Wailed and Lu 1982; Botcher and Salvi 1993; Kim et al. 1997; Salvi et al. 2000; Chang et al. 2002).

2.2 Plasticity in the Central Auditory System

The patterns of terminal axonal degeneration in the CN after mechanical destruction of the cochlea and spiral ganglion are known in a number of mammals. Projections have been traced from the cochlea and cochlear nerve to more or less all subdivisions of the CN (Rasmussen et al. 1960; Cohen et al. 1972; Rasmussen 1990; Morest et al. 1997). After mechanical ablations of the organ of Corti, the results are similar (Morest et al. 1997). There is a topographical mapping to the ventral CN of the inner hair cells and of the myelinated fibers innervating them. Regional ablations of the cochlea produce cochleotopic bands of terminal axonal degeneration in the CN with very little, if any, terminal degeneration between the bands (Fig. 9.1). Transsynaptic degeneration occurs in the superior olivary complex and in the central nucleus of the inferior colliculus, also in bands, which correspond to the cochleotopic map. Signs of terminal degeneration in the CN are detected within 2–4 days after cochlear ablations; they reach a peak by 2 weeks and are practically gone by

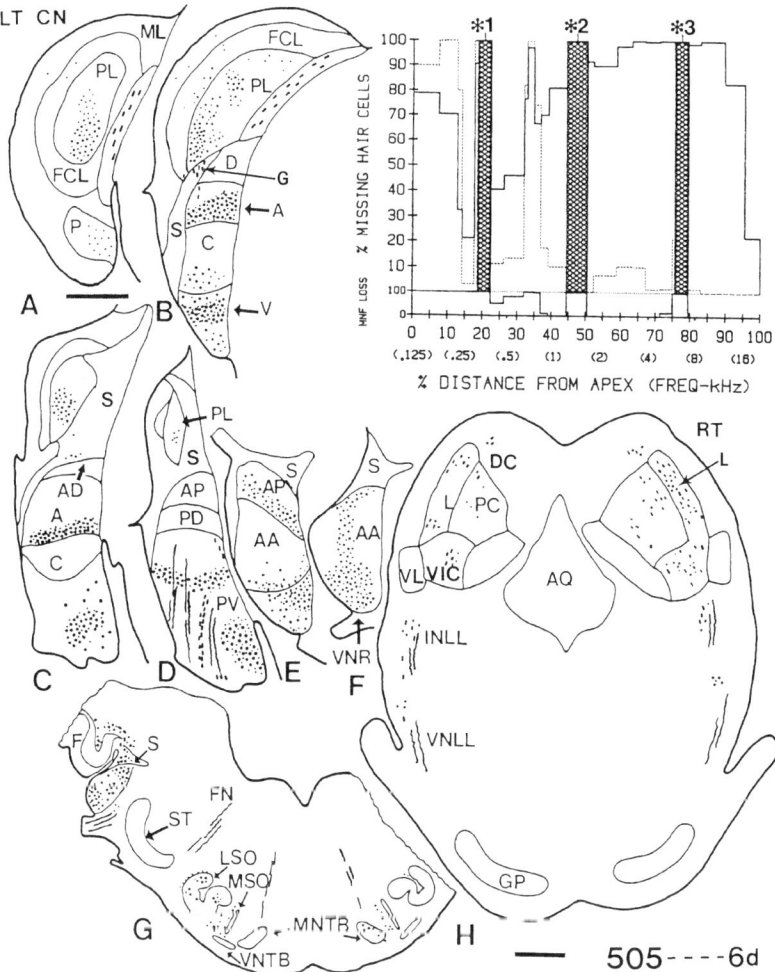

FIGURE 9.1. Mechanical lesions in a chinchilla after a 6-day survival. A cytocochleogram (upper right) illustrates the extent of damage at three different sites (*1, *2, *3). The proportion of missing inner hair cells (dashed line), outer hair cells (solid line), and peripheral processes of spiral ganglion cells, i.e., myelinated nerve fiber loss (MNF LOSS), is plotted as a function of percentage distance from the apex of the cochlea. The apical turn corresponds approximately to 0–21%, the middle turn to 21–47%, the basal turn to 47–79%, and the hook region to 79–100%. A frequency-place map is also scaled on the x-axis (Eldredge et al. 1981). Cross-hatched bars show regions in which all sensory and support cells have degenerated. Axonal degeneration is plotted on drawings of eight representative transverse sections prepared by the Nauta-Rasmussen method. Large dots indicate a terminal pattern of thick (coarse) fiber degeneration (terminal degeneration); medium-sized dots, terminal degeneration of medium-sized fibers; small dots, terminal degeneration of thin (fine) fibers. The relative amounts of degeneration correspond to the concentration of dots. Degenerated fibers of passage are shown by dashes; wiggly lines indicate normal fibers. Few degenerated fibers of passage occurred in C, E, F,

1 month. When only the organ of Corti is damaged, the time course is slower than when the ganglion is also ablated. However, in electron micrographs, early signs of abnormal endings in the CN may appear within hours (Morest, unpublished data).

When cochlear function is intact, the normal physiology of the central auditory pathways includes relatively subtle but measurable, plastic alterations (e.g., Zwislocki et al. 1958; Florentine 1976; Prosen et al. 1990; Moore 1993; Recanzone et al. 1993; Weinberger et al. 1993; Kudoh and Shibuki 1996). Hearing deficits in adults, however, can induce more striking changes. For example, following the primary lesion and cochlear nerve loss in experimental animals, axonal sprouting, the growth of new synapses, synaptic pruning, and synaptic rearrangements were observed in the CN and superior olive (Fig. 9.2A) (Benson et al. 1997; Bilak et al. 1997; Kim et al. 1997; Muly et al. 2002; Kim et al. 2004a,b,c). The patterns of altered transmitter and receptor biochemistry suggested a loss of synaptic inhibition in the CN (Komiya and Eggermont 2000; Syka 2002; Imig and Durham 2005) and an upregulation of glutamatergic excitation in the ventral CN, superior olive, and inferior colliculus (Bledsoe et al. 1995; Potashner et al. 1997; Suneja et al. 1998, 2000, 2004; Milbrandt et al. 2000). Altered processing of acoustic information was found in the auditory cortex, inferior colliculus, and CN (Robertson and Irvine 1989; Salvi et al. 1992, 2000; Botcher and Salvi 1993; Rajan et al. 1993; Bledsoe et al. 1995; Kimura and Eggermont 1999), and elevated excitability or spontaneous activity of neurons was noted in the CN, superior olive, and inferior colliculus (Gerken 1979; Wailed and Lu 1982; Gerken et al. 1984, 1985; Salvi et al. 1992; Francis and Manis 2000; Kaltenbach and Afman 2000; Brozoski et al. 2002). Correlations between these

---

FIGURE 9.1. and **G**. (From Morest et al. 1997). Abbreviations:: A, anterior part of PVCN; AA, anterior part of AVCN; AD, anterodorsal part of PVCN; AP, anteroposterior part of AVCN; AVCN, anteroventral cochlear nucleus; AQ, cerebral aqueduct; C, central part (octopus cell area) of PVCN; CN, cochlear nucleus; D, dorsal part of PVCN; DAS, dorsal acoustic stria; DC, dorsal cortex of IC; DCN, dorsal cochlear nucleus; DNLL, dorsal nucleus of lateral lemniscus; DPO, dorsal periolivary nucleus; FCL, fusiform cell layer of DCN; FL, flocculus; FN, facial nerve root; G, internal granular layer of S; GP, griseum pontis; IAS, intermediate acoustic stria; IC, inferior colliculus; INLL, intermediate nucleus of lateral lemniscus; IV, fourth ventricle; L, lateral part of central nucleus of IC; LSO, lateral superior olivary nucleus; ML, molecular layer of DCN; MNTB, medial nucleus of trapezoid body; MSO, medial superior olivary nucleus; P, posterior part of PVCN; PC, central part of central nucleus of IC; PD, posterodorsal part of AVCN; PL, deep (polymorphic layer) of DCN; PM, medial part of central nucleus of IC; PV, posteroventral part of AVCN; PVCN, posteroventral cochlear nucleus; S, small-cell shell of CN; ST, spinal trigeminal tract; V, ventral part of PVCN; VAP, ventral sector of AP; VIC, ventral part of central nucleus of IC; VL, ventrolateral nucleus of IC; VN, vestibular nerve root; VNLL, ventral nucleus of lateral lemniscus; VNTB, ventral nucleus of trapezoid body.

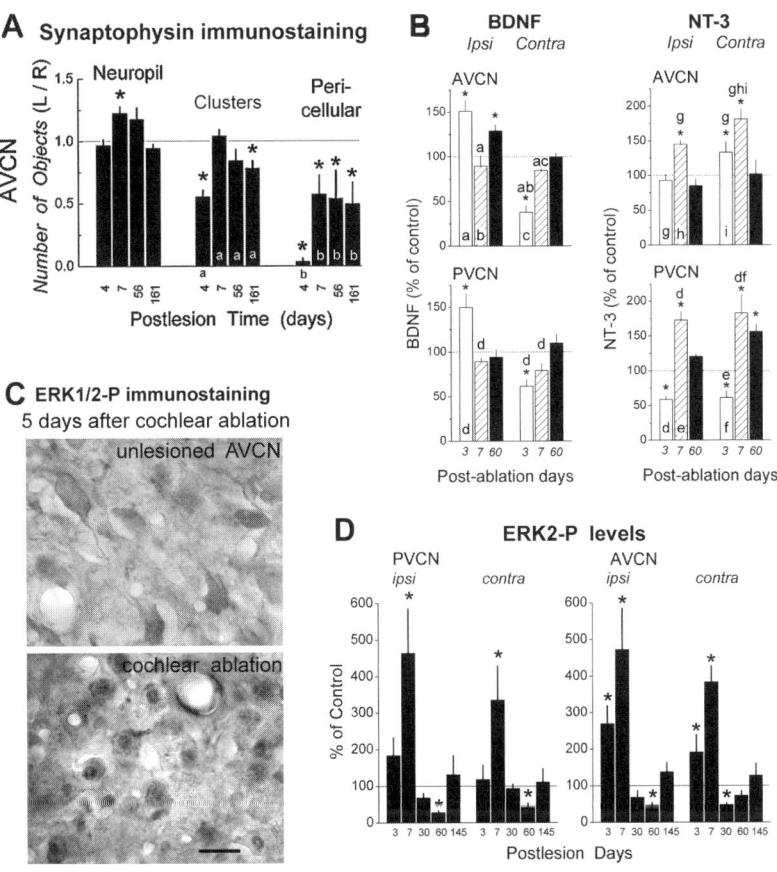

FIGURE 9.2. Guinea pig ventral CN after unilateral cochlear ablation. (**A**) In the AVCN on the ablated side, synaptophysin-containing clusters in the neuropil (Clusters) and those surrounding cell bodies (Pericellular) declined by 4 days after ablation, consistent with the loss of cochlear nerve endings. Resurgence of these profiles at 7 days is consistent with a growth of new synaptic endings. Asterisks denote a difference from the unlesioned control; bars containing lower case letters differ from those with the same letter below (paired *t*-test). Data from Benson et al. (1997). (**B**) BDNF was elevated first, at 3 days, in the ventral CN on the ablated side (IPsi), while NT3 became elevated later, at 7 days. These changes may have contributed to the survival of deafferented CN neurons and to the synaptogenesis illustrated in **A**. On the intact side (Contra), BDNF deficiencies appeared at 3 days and recovered by 7 days, while NT3 levels became elevated by 7 days. (**C**) At 5 days after ablation, during the period of synaptogenesis (**A**), and when neurotrophins were elevated (**B**), ERK1/2-P levels increased in AVCN neurons on the ablated side, and ERK1/2-P was transported mainly into the cell nucleus. Scale bar = 25 μm. (**D**) Quantification of bands in Western blots confirmed the elevation of ERK2-P levels at 3 and 7 days after ablation. Asterisks denote a difference from the unlesioned control (Mann–Whitney test). These findings (in **C** and **D**) may reflect a relationship between increased neurotrophic support and altered gene expression, via increased ERK cell signaling activity, which contributed to the growth of new synaptic endings in the AVCN during the first week after ablation. (**C** and **D** from Suneja and Potashner 2003.)

changes and some of the pathological signs (Jastreboff 1990; Kaltenbach and Afman 2000; Salvi et al. 2000; Brozoski et al. 2002) are consistent with the view that such alterations generate or contribute significantly to the symptoms in sensorineural hearing loss.

The aforementioned examples of plasticity can be viewed as altered cellular phenotypic behaviors that probably originate from molecular signals that appear when the cochlear nerve is damaged or lost. The earliest changes might be manifested in the CN and could include the loss of transmission and trophic support provided by the cochlear nerve endings. Also, signals might include biochemical factors released from the degenerating cochlear nerve and its endings or induced in other cells by cochlear nerve loss. The signals presumably generate successive cascades of cellular responses (via signal transduction pathways) and subsequent biochemical signals, followed by further cellular responses and signals. The cascades would emanate from the CN, travel through the auditory pathway to other auditory nuclei, and shape the pathobiology of the hearing loss. The development and course of CN alterations may initiate and sustain plastic changes in higher auditory nuclei. Similarly, plastic changes in the brain stem nuclei might contribute to the development of changes in the thalamus and cortex, all of which might feed back to the CN. Kaltenbach et al. (2005) suggest that the dorsal CN is the primary site for the origin of central hyperactivity after noise exposure. This does not necessarily exclude a contribution from other regions, especially the ventral CN (see also Bauer and Brozoski, Chapter 4). For it is hard to believe that the extensive remodeling described in the ventral CN has no consequences for hearing.

Our concept of the critical events for sensorineural hearing loss is supported by the previous findings, some of which indicate that neurotrophins, upregulated after sensorineural hearing loss, probably function as relatively early and long-term signals (Suneja et al. 2004). Preliminary data suggest that additional signals may consist of other cytokines (Suneja and Potashner, unpublished). However, the complement of plasticity-inducing molecules and the cells that produce them remain to be identified. Despite this gap in our knowledge, evidence suggests that such signals do act to change the behavior of central auditory cells and the function of central auditory pathways. Signals, such as cytokine and neurotrophin proteins, typically bind to and activate cell surface receptors, which in turn activate one or more signal transduction pathways. In effect, the signal transduction pathways transduce the initial set of signals to produce altered cell morphology and behavior by modifying a variety of cellular proteins, eliciting regulatory changes in cellular pathways and altering gene expression.

The activity of such a process in the central auditory nuclei is evident after sensorineural hearing loss. For example, cochlear ablation altered activity in the ERK (extracellular regulated kinase) signal transduction pathway (Fig. 9.2C,D) (Suneja and Potashner 2003), suggesting that such pathways respond to signals that appear after the lesion. In addition, cochlear ablation altered the activation of CREB (Mo et al. 2006), a protein that controls gene expression, and changed the activation of factors that control protein synthesis (Mo et al. 2006). Since these changes were roughly coincident with the post-ablation growth of new ectopic synapses mentioned previously (Benson et al. 1997; Muly et al. 2004), they suggest that signal transduction activity was responsible for these morphological rearrangements.

Cochlear ablation also brought about altered transmitter release and synaptic receptor activities in central auditory nuclei that were consistent with a loss of inhibitory glycinergic synaptic activity and an increase in excitatory glutamatergic activity (Potashner et al. 1997, 2000; Suneja et al. 1998a,b, 2000; Suneja and Potashner 2003). Moreover, many of these plasticities depended on protein kinases (Zhang et al. 2002, 2003a,b, 2004; Yan et al. 2006) that were themselves sensitive to changes in signal transduction activity. It is likely, therefore, that the regulation of synaptic transmission in central auditory nuclei is governed by signal transduction mechanisms.

3. Effects of Acoustic Overstimulation

Structural and functional reorganization may occur in the adult central nervous system in response to direct or indirect perturbations, including sensory deprivation and overstimulation (Johnson 1975; Merzenich et al. 1983; Erb and Povlishock 1991; Diamond et al. 1993; Rajan et al. 1993). At the level of the spinal cord, damage of peripheral sensory nerves in adult animals may elicit structural reorganization in the gray matter, with the sprouting of new axon terminations (Goldberger and Murray 1985; LaMotte and Kapadia 1993). In the auditory system, exposure of adult animals to intense sound can produce cochlear lesions and hearing impairment (Bohne 1976). The location of the damage in the cochlea reflects the spectral composition of the traumatizing sound (Eldredge et al. 1981). The noise-induced hearing loss may be associated with abnormal auditory functions, such as loudness recruitment or tinnitus (Axelsson and Barrenas 1992; Jastreboff and Jastreboff 2002), and with degenerative changes in the central auditory pathways (Saunders et al. 1991). Animal models have implicated prolonged hyperactivity and decrease of inhibitory transmitters in tinnitus (e.g., Kaltenbach and McCaslin 1996; Milbrandt et al. 2000; Salvi et al. 1992; Chang et al. 2002).

3.1 Fiber Degeneration and Evidence of a New Growth of Axons

Following acoustic trauma, the patterns of terminal degeneration in the CN are similar to those after ablation, but they have several differences (Kim et al. 1997). Heavy terminal degeneration still collects in bands that correspond to the tonotopic locations of the inner hair cell and myelinated fiber lesions (Fig. 9.3). However, unlike ablation, the regions between bands also contain degenerated axons, thinner and in lower numbers. Terminal degeneration can be

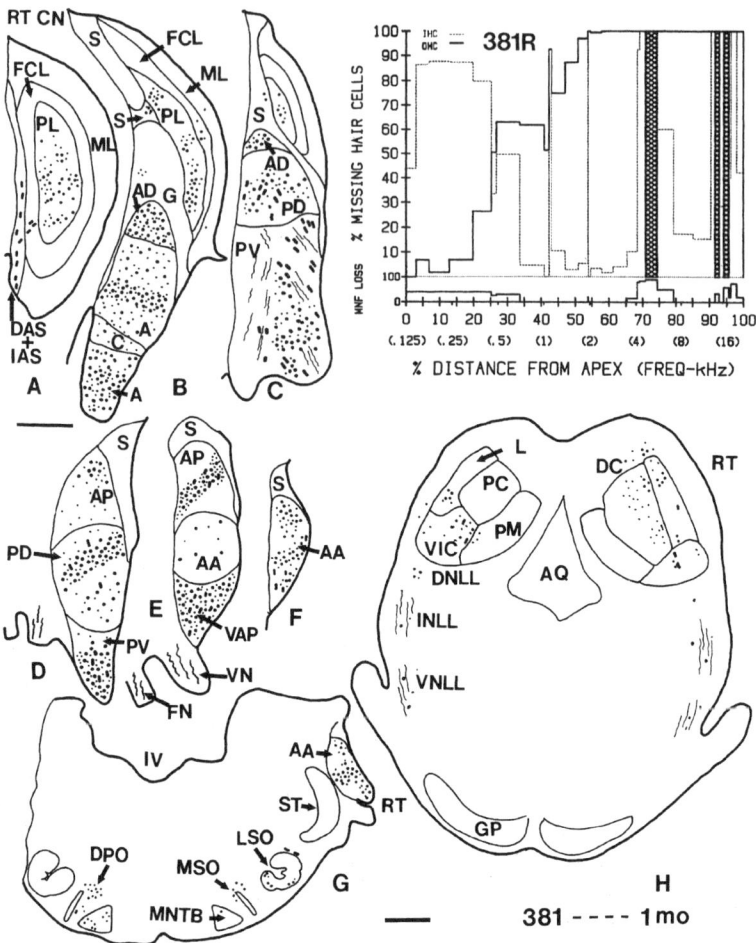

FIGURE 9.3. Chinchilla. 32-day survival, unilateral exposure. The cytocochleogram (right cochlea) plots myelinated fiber loss in three regions at 0–33%, 65–79%, and 92–100%; the basal two regions lack support cells and hair cells (hatched). The apical 35% has unusually high inner hair cell loss. Axonal degeneration is mapped in six transverse sections from caudal (**A, B**), middle (**C, D**), and rostral levels of the right CN, SOC (**G**) and IC (**H**). Large dots indicate terminal degeneration of coarse fibers; medium-sized dots, of medium-sized fibers; small dots, fine fibers. The relative amounts of degeneration correspond to the concentration of dots. Dashes, degenerated fibers of passage; wiggly lines, normal fibers. Scale for **A–F** (upper left) = 0.5 mm, for **G, H** (lower center) = 1.0 mm. (From Kim et al. 1997.) For abbreviations, see the legend to Fig. 9.1.

detected within days following a single noise exposure, but it takes at least a month to peak in chinchillas and 2 months in cats; it persists for up to 8 months in chinchillas and for more than a year in cats. Transsynaptic degeneration occurs, but not in the expected cochleotopic pattern, and unexpected zones of terminal degeneration appear.

In the early studies, cats survived for up to 3 years after acoustic damage to the middle turn of the cochlea (Morest et al. 1979, 1987; Morest 1982; Morest et al. unpublished findings). The large cochlear nerve endings, including the end-bulbs of Held in the ventral CN, disappeared gradually over a period of months. After 3 years, the coarser fiber plexus associated with cochlear nerve endings was gone. Instead there was a marked infiltration of a plexus of very fine axons, clearly visible in silver impregnations, throughout the ventral CN. This appeared to be a new growth of axons. In electron microscopic reconstructions of bushy cell bodies, following mechanical ablation of the cochlea, the largest synaptic endings with the primary-like cytology of cochlear nerve endings disappeared. A similar result followed acoustic trauma, except that there was also a large shift in the ratio of axosomatic endings to the smaller terminals, mostly without primary-like cytology.

In summary, the evidence suggests that acoustic trauma can initiate a structural reorganization of synaptic connections in the cochlear nucleus and central auditory pathways. Such a reorganization might entail axonal pruning, dendritic remodeling, and the growth of new axons and synapses. A disturbance in the ratio of excitatory to inhibitory endings provides a potential basis for explaining hyperactivity and tinnitus. Terminal degeneration continued for at least 8 months, suggesting that this may be a progressive disease.

3.2 A New Growth of Axons and Synapses

The anterior part of the posteroventral CN, dorsal subdivision (PVCNA), is a good place to begin a detailed study of the noise-induced changes, because it contains all of the major neuronal types found in the ventral CN, thus providing a veritable microcosm of the auditory brain stem. After acoustic trauma in the chinchilla, PVCNA consistently contained heavy terminal degeneration, followed by the reappearance of many normal axons, as first shown by Bilak et al. (1997). Comparisons of these axon concentrations to those in equivalent areas of unexposed and protected controls indicates that this deafferented zone initially lost half of its axons, followed by a partial recovery of small diameter axons (Kim et al. 2004a,b,c) (Fig. 9.4). During this recovery, the number of axonal endings on neuronal cell bodies increased. These findings imply that changes in the PVCNA initiated by acoustic trauma include initial deafferentation, followed by the sprouting of new, small diameter axons and the formation of new synaptic contacts. There was a marked increase in the ratio of excitatory to inhibitory synapses. Terminal degeneration and new growth of axons continued for at least 8 months, suggesting that this may be a progressive disease. The result is a reorganization which may well contribute to hyperexcitability and increased spontaneous activity and lead to tinnitus and loudness misperceptions.

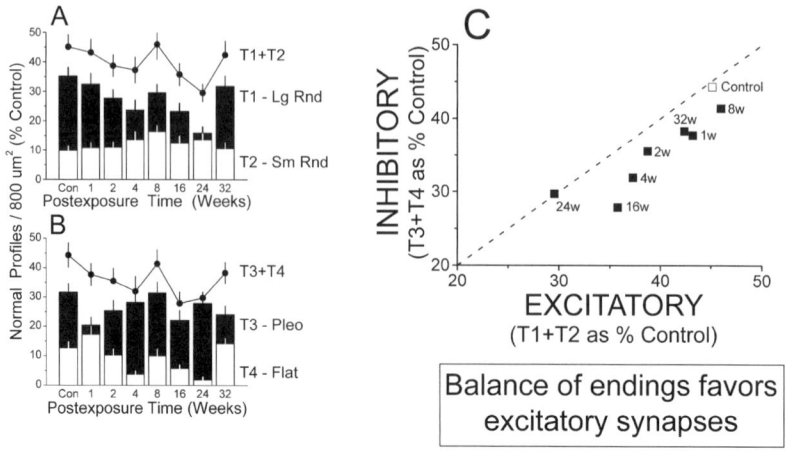

FIGURE 9.4. Normal endings after noise exposure in chinchillas. (After Kim et al. 2004c). (A) Excitatory (large+small round vesicles) compared to (B) inhibitory (pleomorphic + flat vesicles) for unexposed controls (Con) and exposed. Bars = mean ± SEM as % of total number in controls. (C) Comparison of combined sums from A and B: Kolmogorov-Smirnov, $p < 0.05$.

4. Key Molecules Underlying the Synaptic Reorganization Following Noise

There is little information available on the molecular basis for the pathological changes in the brain due to noise damage. Only a few reports address the changes following cochlear ablation or aging with respect to neurotransmitters and growth factors.

Studies of the transmitter-related receptor molecules have been reviewed by Sato et al. (2002). In the brain, the chief inhibitory transmitters are γ-aminobutyric acid (GABA) and glycine, whereas glutamate is generally excitatory. The effect of a transmitter depends on activation of its specific receptor. In the case of glutamate there are many different kinds of receptors with different properties, including ionotropic(α-amino-3-hydroxy-5-methylisoxazole-4-propionic acid [AMPA] and N-methyl-D-aspartate [NMDA]) and metabotropic (not directly linked to ion channels). In brief, the literature suggests that there may be significant changes in expression of GABA, glycine, AMPA, and NMDA receptors. In their own experiments following cochlear ablation, those authors report decreases of these receptor mRNAs in certain cells in the CN of rats by using in situ hybridization after 5 days, but there was a complete restoration by 20 days. In comparison, Potashner et al. (1997, 2000) and Suneja et al. (1998a,b), using biochemical assays in vitro, have documented an increase in glutamate release as well as a decline in glycine release and glycine receptor binding shortly after cochlear ablation in guinea pigs. Other receptors and channels are of interest, including metabotropic receptors, sodium, potassium, and calcium channels.

After noise exposure in chinchillas, Muly et al. (2004) reported an early increase in glutamatergic release and decreased uptake in the first few days in the CN. By 14 days, there was a decrease in glutamatergic uptake and release. Finally, after 90 days, there was a recovery of release and an increase in AMPA receptor binding. These findings are consistent with an early excitotoxic disturbance, followed by a dystrophic disorder. The subsequent recovery of glutamate activity and overexpression of the receptor are consistent with a dystrophic effect. Later increases in glutamatergic release and AMPA receptor activity are consistent with the new growth of axons and their synapses.

Growth factors are specialized proteins known to promote development of neurons and their connections. For example, fibroblast growth factors (FGF) have a widespread distribution in the nervous system and act over broad time periods in development. Neurotrophic factors, such as brain-derived neurotrophic factor (BDNF) and neurotrophic factor 3 (NT3) play specific roles in the differentiation of neurons and their synapses. Specific receptors for these factors have been identified. These same factors and receptors may become active in the pathological responses of the nervous system.

Some studies address the role of growth factors involved in damage to the cochlea, but few look at the effects in the CN. Studies of the role of FGFs in the development of the auditory system (e.g., Hossain et al. 1995, 1997; Zhou et al. 1996) suggest that FGF1 and FGF2 and their receptors are critical in switching spiral ganglion cells and their target cells in the CN from a proliferation mode to one favoring migration and process outgrowth (Brumwell et al. 2000; Hossain and Morest 2000; Hossain et al. 2002; Bilak et al. 2003). FGF2 can act by upregulating the TrkB receptor, which is the high-affinity receptor for BDNF, whereupon process outgrowth and targeting of sensory and central neurons are accelerated. Finally NT3, together with FGF2 and BDNF, upregulate the TrkC receptor, which is the high-affinity receptor for NT3. At that stage, axonal maturation and target selection, together with synapse formation, appear. The details differ in the chicken and mouse (Brumwell et al. 2005, Hossain et al. 2006). In the adult mouse, expression of these molecules may be upregulated in response to noise damage (Smith et al. 2002). The rationale is that, after the excitotoxic and dystrophic events which result from overstimulation, the genes for these molecules are upregulated in response to the renewed opportunity for axonal growth and new synapse formation provided by the loss of synaptic endings from the central neurons. Some support for this rationale comes from studies of cochlear ablation, where the first week after ablation was marked by elevations of BDNF in the ventral CN, followed by increases of NT3 (Fig. 9.2B) and a growth of new synapses (Fig. 9.2A) (Suneja et al. 2005).

5. Summary and Conclusions

Cochlear damage in adult animals induces changes in central auditory neurons and glia, such as degeneration of axons and dendrites, followed by growth of new axonal endings, synaptic reorganization, and altered transmitter biochemistry and physiology. These plasticities are thought to underlie the pathology and symptoms. For example, a disturbance in the ratio of excitatory to inhibitory endings in the cochlear nucleus provides a potential basis for explaining central hyperactivity and tinnitus. Also the long-term continuation of synaptic degeneration in the brain after noise damage suggests that noise-induced hearing loss may be a progressive disease.

In the research on sensorineural hearing loss in animal models, three main concepts have emerged. First, there is strong evidence that noise-induced hearing loss resulted in initial synaptic depletion, followed by synaptogenesis and synaptic rearrangements in the cochlear nucleus. This creates a net loss of inhibitory synaptic contacts and the establishment of ectopic terminals, many of which exhibit the morphology of excitatory synapses. These events provide a structural basis for the well known hyperexcitability of the auditory pathway and its deficient inhibition after sensorineural hearing loss. The findings may also indicate a structural basis for the signs and symptoms that accompany sensorineural hearing loss, such as tinnitus and loudness misperceptions.

Second, activity in the signal transduction pathways may be altered after unilateral cochlear ablation, implying that signals and signal transduction may play a role in altering gene expression, endogenous regulation, and phenotypic cell behavior in auditory nuclei with sensorineural hearing loss. Temporal correlations between the expression of signaling molecules and/or signal transduction events, on the one hand, and synaptogenesis and/or changes in synaptic biochemical activities, on the other hand, suggest that altered signal transduction activity might control synaptic reorganization. This implies that after sensorineural hearing loss, synaptic plasticity, and thus pathological changes, might be alleviated by manipulations of signal molecules and signal transduction activity.

Third, neurotrophins and cytokines have emerged as candidate signal molecules that might generate the central, plastic changes in signal transduction activity after sensorineural hearing loss. These plasticities may be correlated with altered expression of neurotrophins, growth factors, neurotransmitters and their receptors, and altered activity in signaling molecules, such as protein kinases and the cyclic-AMP response element binding protein. These discoveries allow us, for the first time in the auditory field, to study the cellular and molecular mechanisms of sensorineural hearing loss in the brain as well as the cochlea.

Acknowledgments. The authors' research is supported by NIH grants DC000127 (D.K.M.) and DC000199 (S.J.P.).

References

Axelsson A, Barrenas ML (1992) Tinnitus in noise-induced hearing loss. In: Dancer AL, Henderson D, Salvi RJ, Hamernik RP (eds) Noise-Induced Hearing LossBoston: Mosby Year Book, pp. 269–276.

Benson CG, Gross JS, Suneja SK, Potashner SJ (1997) Synaptophysin immunoreactivity in the cochlear nucleus after unilateral cochlear or ossicular removal. Synapse 25: 243–257.

Bilak M, Kim J, Potashner SJ, Bohne BA, Morest DK (1997) New growth of axons in the cochlear nucleus of adult chinchillas after acoustic trauma. Exp Neurol 147:256–268.

Bilak MM, Hossain WA, Morest DK (2003) Intracellular fibroblast growth factor (FGF-2) produces different effects than extracellular application on development of cochleo-vestibular ganglion cells in vitro. J Neurosci Res 71:629–647.

Bledsoe SC Jr, Nagase S, Miller JM, Altschuler RA (1995) Deafness-induced plasticity in the mature central auditory system. NeuroReport 7:225–229.

Bohne B (1976). Mechanisms of noise damage in the inner ear. In: Henderson D, Hamernik R, Dosanjh DS, Mills JH (eds) Effects of Noise on Hearing. New York: Raven Press, pp. 41–67.

Botcher FA, Salvi RJ. 1993. Functional changes in the ventral cochlear nucleus following acute acoustic overstimulation. J Acoust Soc Am 94: 2123–2134.

Brozoski TJ, Bauer CA, Caspary DM (2002) Elevated fusiform cell activity in the dorsal cochlear nucleus of chinchillas with psychophysical evidence of tinnitus. J Neurosci 22: 2383–2390.

Brumwell C, Hossain WA, Morest DK, Bernd P (2000) Role for basic fibroblast growth factor (FGF-2) in tyrosine kinase (Trib.) expression in the early development and innervation of the auditory receptor: in vitro and in situ studies. Exp Neurol 162:121–145.

Brumwell C, Hossain WA, Morest DK, Wolf B (2005) Biotinidase reveals the morpho-genetic sequence in cochlea and cochlear nucleus of mice. Hear Res 209:104–121.

Chang H, Chen K, Kaltenbach JA, Zhang J, Godfrey DA (2002) Effects of acoustic trauma on dorsal cochlear nucleus neuron activity in slices. Hear Res 164:59–68.

Cohen ES, Brawer JR, Morest DK (1972) Projections of the cochlea to the dorsal cochlear nucleus in the cat. Exp Neurol 35:470–479.

Diamond ME, Armstrong-James M, Ebner FF (1993) Experience-dependent plasticity in adult rat barrel cortex. PNAS USA 90:2082–2086.

Eldredge DH, Miller JD, Bohne BA (1981) A frequency-position map for the chinchilla cochlea. J Acoust Soc Amer 69:1091–1095.

Erb DE, Povlishock JT (1991) Neuroplasticity following traumatic brain injury: a study of GABAergic terminal loss, and recovery in the cat dorsal lateral vestibular nucleus. Exp Brain Res 83:253–267.

Florentine M (1976) Relation between lateralization and loudness in asymmetrical hearing loss. J Am Audiol Soc 1:243–251.

Francis HW, Manis PB (2000) Effects of deafferentation on the electrophysiology of ventral cochlear nucleus neurons. Hear Res 149:91–105.

Gerken, GM (1979) Central denervation hypersensitivity in the auditory system of the cat. J Acoust Soc Am 66:721–727.

Gerken GM, Saunders SS, Paul RE (1984) Hypersensitivity to electrical stimulation of auditory nuclei follows hearing loss in cats. Hear Res 13:249–259.

Gerken GM, Saunders SS, Simhadri-Sumithra R, Bhat KH (1985) Behavioral thresholds for electrical stimulation applied to auditory brainstem nuclei in cat are altered by injurious and noninjurious sound. Hear Res 20:221–231.

Goldberger ME, Murray M (1985) Recovery of function, and anatomical plasticity after damage to the adult, and neonatal spinal cord. In: Cotman CW (ed) Synaptic Plasticity New York: Guilford Press, pp. 77–110.

Hossain WA, Morest DK (2000) Fibroblast growth factors (FGF-1, FGF-2) promote migration and neurite growth of mouse cochlear ganglion cells in vitro: immunohisto-chemistry and antibody perturbation. J Neurosci Res 62:40–55.

Hossain WA, Zhou X, Rutledge A, Baier C, Morest DK (1995) Basic fibroblast growth factor affects neuronal migration and differentiation in normotypic cell cultures from the cochleovestibular ganglion of the chick embryo. Exp Neurol 138:121–143.

Hossain WA, Rutledge A, Morest DK (1997) Critical periods of basic fibroblast growth factor and brain-derived neurotophic factor in the development of the chicken cochleovestibular ganglion in vitro. Exp Neurol 147:437–451.

Hossain WA, Brumwell CL, Morest DK (2002) Sequential interactions of FGF-2, BNF, NT-3 and their receptors define critical periods in the development of cochlear ganglion cells. Exp Neurol 175:138–151.

Hossain WA, D'Sa C, Morest DK (2006) Site-specific interactions of neurotrophin-3 and fibroblast growth factor (FGF2) in the embryonic development of the mouse cochlear nucleus. J Neurobiol 66:897–915.

Imig TJ, Durham D (2005) Effect of unilateral noise exposure on the tonotopic distribution of spontaneous activity in the cochlear nucleus and inferior colliculus in the cortically intact and decorticate rat. J Comp Neurol 490:391–413.

Jastreboff PJ (1990) Phantom auditory perception (tinnitus): mechanisms of generation and perception. Neurosci Res 8:221–254.

Jastreboff PJ, Jastreboff MM (2002) Tinnitus and hyperacusis. In: Ballenger JJ, Snow JB Jr (eds) Ballenger's Otorhinolaryngology, Head and Neck Surgery, 16th ed. San Diego: Singular, pp. 456–471.

Johnson J (1975) A fine structural study of degenerative-regenerative pathology in the surgically deafferented vestibular nucleus of the rat. Acta Neuropath (Berlin) 33: 227–243.

Kaltenbach JA, Afman CE (2000) Hyperactivity in the dorsal cochlear nucleus after intense sound exposure and its resemblance to tone-evoked activity: a physiological model for tinnitus. Hear Res 140:165–172.

Kaltenbach JA, McCaslin DL (1996) Increases in spontaneous activity in the dorsal cochlear nucleus following exposure to high intensity sound: a possible neural correlate of tinnitus. Auditory Neurosci 3:57–78.

Kaltenbach JA, Zhang J, Finlayson P (2005) Tinnitus as a plastic phenomenon and its possible neural underpinnings in the dorsal cochlear nucleus. Hear Res 206:200–226.

Kim J, Morest DK, Bohne B (1997) Degeneration of axons in the brain stem of the chinchilla after auditory overstimulation. Hear Res 103:169–191.

Kim JJ, J Gross, SJ Potashner, DK Morest (2004a) Fine structure of degeneration in the cochlear nucleus of the chinchilla following acoustic overstimulation. J Neurosci Res 77:798–816.

Kim JJ, Gross J, Potashner SJ, Morest DK (2004b) Fine structure of long-term changes in the cochlear nucleus following acoustic overstimulation: chronic degeneration and new growth of synaptic endings. J Neurosci Res 77:817–828.

Kim JJ, J Gross J, DK Morest, SJ Potashner (2004c) A quantitative study of degeneration and new growth of axons and synaptic endings in the chinchilla cochlear nucleus following acoustic overstimulation. J Neurosci Res 77:829–842.

Kimura M, Eggermont JJ (1999) Effect of acute pure tone induced hearing loss on response properties in three auditory cortical fields in cat. Hear Res 135:146–162.

Komiya H, Eggermont JJ (2000) Spontaneous firing activity of cortical neurons in adult cats with reorganized tonotopic map following pure-tone trauma. Acta Oto-Laryngologica 6:750–756.

Kudoh M, Shibuki K (1996) Long-term potentiation of supragranular pyramidal outputs in the rat auditory cortex. Exp Brain Res 110:21–27.

LaMotte CC, Kapadia SE (1993) Deafferentation-induced terminal field expansion of myelinated saphenous afferents in the adult rat dorsal horn, and the nucleus gracilis following pronase injection of the sciatic nerve. J Comp Neurol 330:83–94.

Merzenich MM, Kaas JH, Wall JT, Sur M, Nelson RJ, Felleman DJ (1983) Progression of change following median nerve section in the cortical representation of the hand in areas 3b, and 1 in adult owl, and squirrel monkeys. Neuroscience 10: 639–665.

Milbrandt JD, Holder TM, Wilson MC, Salvi RJ, Caspary DM (2000) GAD levels and muscimol binding in rat inferior colliculus following acoustic trauma. Hear Res 147:251–260.

Mo Z, Suneja SK, Potashner SJ (2006) Phosphorylated cAMP response element-binding protein levels in guinea pig brainstem auditory nuclei after unilateral cochlear ablation. J Neurosci Res 83:1323–1330.

Moore DR (1993) Plasticity of binaural hearing and some possible mechanisms following late-onset deprivation. J Am Acad Audiol 4:277–283.

Morest DK (1982) Degeneration in the brain following exposure to noise. In: Hamernik RP, Henderson D, Salvi R (eds) New Perspectives in Noise Induced Hearing Loss. New York: Raven Press, pp. 87–93.

Morest DK (1997) Structural basis for signal processing in the mammalian cochlear nuclei. Challenge of the synaptic nests. In: Syka J (ed) The Mammalian Cochlear Nuclei: Organization and Function. New York: Plenum Press, pp. 19–32.

Morest DK, Ard MD, Yurgelun-Todd D (1979) Central auditory pathways sensitive to acoustic over-stimulation in the cat. Assoc Res Otolaryngol Abstr 2:28–29.

Morest DK. Jones DR, Kwok S, Ard MD, Bohne B, Yurgelun-Todd D (1987) Response of the brain to acoustic damage of the cochlea. Assoc Res Otolaryngol Abstr 10:3.

Morest DK, Kim J, Bohne B (1997) Neuronal and transneuronal degeneration of auditory axons in the brain stem after cochlear lesions in the chinchilla: cochleotopic and non-cochleotopic patterns. Hear Res 103:151–168

Morest DK, J Kim, SJ Potashner, BA Bohne (1998) Long-term degeneration in the cochlear nerve and cochlear nucleus of the adult chinchilla. Special Issue on Plasticity in the Central Auditory System Microsc Res Tech 41:205–216.

Muly SM, Gross JS, Morest DK, Potashner SJ (2002) Synaptophysin in the cochlear nucleus following acoustic trauma. Exp Neurol 177:202–221

Muly SM, Gross JS, Potashner SJ (2004) Noise trauma alters D-[^{4}H]aspartate release and AMPA binding in chinchilla cochlear nucleus. J Neurosci Res 75:585–596.

Olson WO, Noffsinger D, Kurdziel S (1975) Speech discrimination in quiet and in white noise by patients with peripheral and central lesions. Acta Otolaryngol 80:375–382.

Osen KK (1970) Course and termination of the primary afferents in the cochlear nuclei of the cat. An experimental anatomical study. Arch Ital Biol 108:21–51.

Potashner SJ, Suneja SK, Benson CG (1997) Regulation of D-aspartate release and uptake in adult brain stem auditory nuclei after unilateral middle ear ossicle removal and cochlear ablation. Exp Neurol 148:222–235.

Potashner SJ, Suneja SK, Benson CG (2000) Altered glycinergic synaptic activities in guinea pig brain stem auditory nuclei after unilateral cochlear ablation. Hear Res 147:125–136.

Prosen CA, Moody DB, Sommers MS, Stebbins WC (1990) Frequency discrimination in the monkey. J Acoust Soc Am 88:2152–2158.

Rajan R, Irvine DRF, Wise LZ, Heil P (1993) Effect of unilateral partial cochlear lesions in adult cats on the representation of lesioned, and unlesioned cochleas in primary auditory cortex. J Comp Neurol 338:17–49.

Rasmussen GL (1990) Spiral ganglion lesions, notebook 1, pp. 1–58. Research Notebooks, History of Medicine Division, National Library of Medicine, ACC 653.

Rasmussen GL, RR Gacek, McCrane EP, Baker CC (1960) Model of cochlear nucleus (cat) displaying its afferent and efferent connections. Anat Rec 136:344.

Recanzone GH, Schreiner CE, Merzenich MM (1993) Plasticity in the frequency representation of primary auditory cortex following discrimination training in adult owl monkeys. J Neurosci 13:87–103.

Robertson D, Irvine DRF (1989) Plasticity of frequency organization in auditory cortex of guinea pigs with partial unilateral deafness. J Comp Neurol 282:456–471.

Salvi RJ, Powers NL, Saunders SS, Botcher FA, Clock AE (1992) Enhancement of evoked response amplitude and single unit activity after noise exposure. In: Dancer A, Henderson D, Salvi RJ, Hamernik RP (eds) Effects of noise on the auditory system. St. Louis: Mosby Year Book, pp. 156–171.

Salvi RJ, Wang J, Ding D (2000) Auditory plasticity and hyperactivity following cochlear damage. Hear Res 147:261–274.

Sato K, Shiraishi S, Nakagawa H, Kuriyama H, Altschuler RA (2002) Diversity and plasticity in amino acid receptor subunits in the rat auditory brain stem. Hear Res 147:137–144.

Saunders JC, Cohen YE, Szymko YM (1991) The structural and functional consequences of acoustic injury in the cochlea and peripheral auditory system: A five year update. J Acoust Soc Am 90:136–146.

Smith L, Gross J, Morest DK (2002) Fibroblast growth factors (FGFs) in the cochlear nucleus of the adult mouse following acoustic overstimulation. Hear Res 169:1–12.

Suneja SK, Potashner SJ (2003) ERK and SAPK signaling in auditory brainstem neurons after unilateral cochlear ablation. J Neurosci Res 73:235–245.

Suneja SK, Benson CG, Potashner SJ (1998a) Glycine receptors in adult guinea pig brain stem auditory nuclei: regulation after unilateral cochlear ablation. Exp Neurol 154:473–488.

Suneja SK, Potashner SJ, Benson CG (1998b) Plastic changes in glycine and GABA release and uptake in adult brain stem auditory nuclei after unilateral middle ear ossicle removal and cochlear ablation. Exp Neurol 151:273–288.

Suneja SK, Potashner SJ, Benson CG (2000) AMPA receptor binding in adult guinea pig brain stem auditory nuclei after unilateral cochlear ablation. Exp Neurol 165:355–369.

Suneja SK, Yan L, Potashner SJ (2005) Regulation of NT-3 and BNF levels in guinea pig auditory brain stem nuclei after unilateral cochlear ablation. J Neurosci Res 80: 381–390.

Syka J (2002) Plastic changes in the central auditory system after hearing loss, restoration of function, and during learning. Physiol Rev 82:601–636.

Wailed JF, Lu SM (1982) Noise-induced hearing loss can alter neural coding and increase excitability in the central nervous system. Science 216:1331–1334.

Webster DB, Webster M (1978) Cochlear nerve projections following organ of Corti destruction. Otolaryngology 86:342–353.

Weinberger NM (1993) Learning-induced changes of auditory receptive fields. Curr Opin Neurobiol 3:570–577.

Yan L, Suneja SK, Potashner SJ (2006) Protein kinases regulate glycine receptor binding in brain stem auditory nuclei after unilateral cochlear ablation. Brain Res doi:10.1016/j.brainres.2006. 12.013

Zhang J, Suneja SK, Potashner SJ (2002) Protein kinase C regulates D-[^{3}H]aspartate release in auditory brain stem nuclei. Exp Neurol 175:245–256.

Zhang J, Suneja SK, Potashner SJ (2003a) Protein kinase A and calcium/calmodulin-dependent protein kinase II regulate D-[^{3}H]aspartate release in auditory brain stem nuclei. J Neurosci Res 74:81–90.

Zhang J, Suneja SK, Potashner SJ (2003b) Protein kinase C regulation of glycine and γ-aminobutyric acid release in brain stem auditory nuclei. Exp Neurol 182:75–86.

Zhang J, Suneja SK, Potashner SJ (2004) Protein kinase A and calcium/calmodulin-dependent protein kinase II regulate glycine and GABA release in auditory brain stem nuclei. J Neurosci Res 75:361–370.

Zhou X, Hossain WA, Rutledge D, Baier , C, Morest DK (1996) Basic fibroblast growth factor (FGF-2) affects development of acoustico-vestibular neurons in the chick embryo brain in vitro. Hear Res 101:186–207.

Zwislocki J, Maire F, Feldman AS, Rubin H (1958) On the effect of practice and motivation on the threshold of audibility. J Acoust Soc Am 30:254–262.

10
Cell Death and Cochlear Protection

STEVEN H. GREEN, RICHARD A. ALTSCHULER, AND JOSEF M. MILLER

1. Introduction

The sensorineural cells of the inner ear may be subjected to a variety of stresses—genetic, aging-related, environmental, or disease (covered in other chapters in this volume)—that result in cell death and hearing loss. Protective strategies and interventions are now becoming possible based on an increased understanding of the mechanisms of cell death as well as an increased understanding of the intrinsic cellular mechanisms for protection. There are three alternative outcomes to cellular stress or trauma. A mild sublethal stress induces protective coping mechanisms that protect not only against the inducing stress but also, for a period of time, against subsequent stresses or traumata. In this case, the inducing stress *preconditions* the cell (Niu and Canlon 2002 and Section 4.2). If the cell cannot cope with the stress, then a second outcome is possible in which the cell engages mechanisms for an orderly death or apoptosis. This is also the outcome for neurons deprived of neurotrophic support. If the stress is sudden and very severe, both protective and apoptotic mechanisms are overwhelmed, leading to a third outcome: necrotic cell death. Study of the first outcome reveals the cells' intrinsic mechanisms for identifying the most deleterious consequences of stress on the molecular level and the intrinsic mechanisms for coping with them. This can give important insights for developing protective therapies. In the second outcome, therapeutic strategies aimed at preventing cell death are likely to be based on prevention of apoptosis, coupled with cellular protection. The inclusion of protective therapy is crucial because apoptosis is a physiological, not pathological, mechanism recruited to eliminate cells that are functionally compromised. Merely blocking apoptosis could result in a cell that is alive but dysfunctional and, possibly, detrimental to the organ. In the third outcome, necrotic cell death, the most effective method of protection for the cochlea is reducing or eliminating, before injury, the exposure or circumstances that cause cell death.

The cellular decision to commit to apoptosis or to self-protection involves a delicate balance between proapoptotic and prosurvival/protective regulatory intracellular signaling pathways—a balance affected by the severity of the stress. This is similarly reflected in the homeostatic mechanisms discussed by Wangemann (Chapter 3). Therapeutic interventions must consider this balance in enhancing prosurvival and reducing cell death pathways. The fundamental assumption of this chapter is that this can be accomplished best by thoroughly understanding the molecular machinery of cell death and its regulation and by understanding the intracellular and intercellular signaling that controls the commitment to cell death.

This chapter begins by reviewing these subjects and then proceeds to discuss how the knowledge can be used to enhance protection in the auditory nerve, the organ of Corti, and in other cochlear elements. Typically, it is the hair cells in the organ of Corti that primarily die as a consequence of stress (discussed in detail by Henderson and Hu, Chapter 7 and by Rybak, Talaska, and Schacht, Chapter 8). The death of spiral ganglion neurons (SGNs) is most often encountered secondary to the loss of hair cells, presumably due to loss of hair cell–derived neurotrophic support. In some cases, SGNs may be compromised directly by a toxic stress, notably by excitotoxicity (Section 6). Therefore, discussion of protection of hair cells focuses on consequences and mitigation of the direct effects of stress and discussion of protection of SGNs focuses on consequences and mitigation of the loss of hair cell–deprived neurotrophic support. Indeed, the only effective treatment for complete sensorineural hearing loss currently is the cochlear implant, the function of which depends entirely on the survival and integrity of the spiral ganglion neurons.

2. General Principles of Cell Death and Apoptosis

Neurons or non-neuronal cells that have been subjected to severe stress such as ototoxins, pH changes, temperature extremes, oxygen or nutrient deprivation, extreme osmotic shock, or other stresses will rapidly die. In these cases, the cell membrane ruptures, either as a direct result of the stress or as a consequence of metabolic failure causing shutdown of membrane pumps and swelling of the cell and organelles. The affected region becomes necrotic, damaging adjacent cells and tissue because of leakage of intracellular contents and inflammation. Cells subjected to a trauma sufficient to kill them, but not so severe that they die immediately, engage the mechanism for programmed cell death or "cell suicide," termed apoptosis. This averts the adverse consequences of necrotic death to surrounding cells. In the process of apoptosis, which has been extensively reviewed (Hengartner 2000), the nucleus, cytoskeletal, organelle, and cytoplasmic components are disassembled and condensed prior to disruption of the membrane. The apoptotic cell also signals to adjacent cells and to phagocytic cells such as macrophages or microglia to alert them to remove the cellular "corpse" rapidly. Apoptosis is also recruited for tidy removal of cells in other circumstances where cell death is required, e.g., for cancer or virally infected cells, for supernumerary cells in development, and, most relevant for this chapter, for neurons that die as a consequence of loss of neurotrophic support.

2.1 Apoptotic Mechanisms: Caspases

In apoptosis, disassembly of the cell is initiated mainly by proteases of the caspase family (Hengartner 2000). Caspases are activated in response to extrinsic "death signals" (e.g., tumor necrosis factor), cellular trauma or stress (e.g., oxidative stress, excitotoxicity, toxic agents), or insufficient trophic support. In the case of extrinsic proapoptotic signals, these bind to cell surface receptors that directly activate associated caspases. Intrinsic proapoptotic stimuli cause caspase activation through a complex mechanism that involves the mitochondria, and is thus termed the *mitochondrial* pathway. The mitochondrial pathway is particularly relevant to the death of neurons that have lost synaptic partners, to the death of neurons or sensory cells following neurotoxic stress or trauma, and to neuronal loss in neurodegenerative disorders. Adult neurons are much more resistant than embryonic neurons to cell death after loss of target-derived neurotrophic factors. This is, in some cases, correlated with increased expression of inhibitor of apoptosis protein (IAP) (Perrelet et al. 2002). IAPs bind and antagonize caspases (Robertson et al. 2000).

2.2 Apoptotic Mechanisms: The Mitochondrial Pathway and Bcl-2 Family Proteins

Caspases are activated by proteins released into the cytosol from the mitochondria. These include cytochrome c , which activates caspases by forming a complex (the "apoptosome") with the caspase-binding protein Apaf-1 (Hengartner 2000), and Smac/DIABLO, which inactivates IAPs (Fesik and Shi 2001). The initiation of apoptosis occurs on the outer mitochondrial membrane (OMM) where a pore must be made to allow cytochrome c and other proteins to emerge (Kroemer and Reed 2000). Formation of the pore depends crucially on proteins of the Bcl-2 family of apoptotic regulatory proteins, which includes both pro- and antiapoptotic members (Reed 1998; Hengartner 2000). A simple current hypothesis is that assembly of the pore depends on multidomain *proapoptotic* Bcl-2 family members, e.g., Bax and Bak. Formation of the apoptotic pore is prevented by multidomain *prosurvival* Bcl-2 family members, e.g., Bcl-2 and Bcl-X , that associate with the structurally similar multidomain proapoptotic Bcl-2 family members. In this way, the prosurvival Bcl-2 family proteins prevent apoptosis. This prevention of apoptosis is, in turn, antagonized by BH3-only proapoptotic Bcl-2 family members. (The Bcl-2 Homology 3 or BH3 domain is a protein–protein interaction domain present in Bcl-2 family proteins.) When appropriately triggered, BH3-only proteins translocate to the OMM, associate with the multidomain Bcl-2 prosurvival family members, and disrupt their association with multidomain proapoptic proteins, allowing the latter to form an apoptotic pore and apoptosis to be initiated. There are a large number of different BH3-only proteins and they are the targets of different regulatory and signaling pathways. Different BH3-only proteins may be activated/inactivated by transcriptional regulation, by proteolytic cleavage, or by phosphorylation. This accounts for much of the diversity of signals that can control apoptosis (Huang and Strasser 2000).

The proposed mechanism implies that the ratio of proapoptotic to antiapoptotic regulators present on the OMM determines whether apoptosis is initiated. In the normal viable cell there is a functional preponderance of antiapoptotic regulators. When the ratio shifts to favor proapoptotic regulators, apoptosis is initiated (Reed 1998; Hengartner 2000). The ratio of pro- to antiapoptotic regulators on the OMM is controlled by posttranslational mechanisms in the cytoplasm and by transcriptional regulation in the nucleus. Different trophic stimuli use different intracellular signaling pathways to accomplish these two goals in parallel. Protective agents that may be clinically useful likewise target one or more of these signaling pathways to reduce proapoptotic signaling or increase prosurvival signaling or both.

Expression of pro- and antiapoptotic Bcl-2 family members is developmentally regulated and contributes to the relative independence of many mature neurons from target-derived neurotrophic factors (unlike neurons in embryos or neonates, neurons in mature animals do not die or die only very slowly after loss of their synaptic targets.) Thus, the ratio of the antiapoptotic regulator Bcl-X to proapoptotic Bax increases concomitantly with the decline in neurotrophic factor dependence of maturing dorsal root ganglion sensory neurons (Vogelbaum et al. 1998). (As noted in Section 2.1, decline in neurotrophic factor dependence is also correlated with other molecular changes, e.g., increased IAP expression.) Bcl-2, Bcl-X , and Bax transcripts are present in the rat spiral ganglion by postnatal day 1 (P1) and their expression is maintained thereafter (Ishii et al. 1996).

3. Transcriptional Regulation of Apoptosis and Cell Survival

Intracellular signaling pathways control transcription of genes encoding regulators of cell survival and apoptosis and also control the activity of these regulators by posttranslational modification (generally phosphorylation and dephosphorylation). Prosurvival intracellular signaling—which is activated by neurotrophic stimuli—and proapoptotic intracellular signaling—which is activated by trauma, stress, or withdrawal of neurotrophic stimuli— influence cell survival or death and they do so by interacting with the core apoptotic regulatory machinery summarized in the preceding text. This happens, in parallel, at two levels of regulation: posttranslational and transcriptional. Posttranslational modification of proteins involved in apoptosis affects the probability that the mitochondrial pathway will be initiated or that it will result in apoptosis, if initiated. In particular, posttranslational modification of Bcl-2 family apoptotic regulators affects their ability to translocate to the mitochondria and form a functional apoptotic channel. Transcriptional regulation, over a longer time scale, affects the quantitative balance between pro- and antiapoptotic regulators by regulating their synthesis.

There are two important caveats. First, this is an area of intense research activity and undoubtedly, additional regulatory mechanisms will yet be identified. Second, there is extensive crosstalk among these pathways; they constitute an integrated and intricate signaling network, not separate signaling pathways. Moreover, each of these signaling pathways is also involved in regulation of cellular processes other than cell death. The exact cellular response or outcome depends on the cellular context and the state of the network: locally, in the relevant subcellular compartment, and globally throughout the cell.

3.1 Posttranslational Regulation of Bcl-2 Family Proteins

An important component of posttranslational control of apoptosis is control over translocation of proapoptotic Bcl-2 family members from the cytosol to the OMM. Bax and BH3-only family members are typically sequestered in the cytoplasm away from the mitochondria. Diverse posttranslational mechanisms including acetylation, proteolytic cleavage, and phosphorylation/dephosphorylation of Bcl-2 family members or their cytoplasmic binding partners control the translocation of other proapoptotic proteins (e.g., Bax, Bim, Bid) to the mitochondrial surface (Reed 1998; Harris 2000).

A particularly well studied, illustrative, example (although by no means the only well studied example) is control of the proapoptotic BH3-only Bcl-2 family member Bad by phosphorylation/dephosphorylation (Downward 1999) involving prosurvival and proapoptotic protein kinases. Prosurvival protein kinases (discussed later in detail) include the extracellular signal-regulated kinase (ERK) family of MAP kinases (MAPKs), protein kinase B (PKB/Akt), and cyclic AMP-dependent protein kinase (protein kinase A, PKA) (Downward 1999). Opposing these prosurvival protein kinases are protein kinases participating in proapoptotic intracellular signaling. Cyclin-dependent protein kinase Cdc2 and c-Jun N-terminal kinase (JNK) phosphorylate Bad on a site different from those targeted by the prosurvival kinases, a phosphorylation that causes Bad activation (Donovan et al. 2002; Konishi et al. 2002). As might be expected, protein dephosphorylation by protein phosphatases also plays a significant role in these regulatory networks. Compelling evidence implicates phosphatases such as protein phosphatase 2A (PP2A) (Strack et al. 2004) and PP2B/calcineurin (Wang et al. 1999) in promoting apoptosis by dephosphorylating proapoptotic regulators such as Bad. Analogous regulation by phosphorylation/dephosphorylation involving these and other protein kinases and phosphatases occurs at many other key apoptotic decisions.

3.2 Regulation of Expression of Prosurvival Genes

3.2.1 CREB

The cAMP/Ca^{2+}-regulatory element binding (CREB) protein (Dawson and Ginty 2002) is a particularly well investigated example of transcriptional regulation. Several intracellular signaling pathways activated by survival-promoting stimuli converge on the transcription factor CREB, which is necessary, at least in part, for these stimuli to promote survival (Dawson and Ginty 2002). Activation of CREB results in increased expression of a number of genes including genes involved in prosurvival signaling. For example, the antiapoptotic regulatory protein Bcl-2 is upregulated by CREB (Wilson et al. 1996), as is brain-derived neurotrophic factor (BDNF) (Shieh et al. 1998; Tao et al. 1998). BDNF gene expression in cultured SGNs is reduced in cells transfected with dominant-negative mutant CREB (Zha et al. 2001).

3.2.2 NF-κB

The typically prosurvival transcription factor NF-κB is also activated by neurotrophic stimuli (Maggirwar et al. 1998). In the central nervous system (CNS), NF-κB activity is constitutive (Bhakar et al. 2002), and suppression of NF-κB activity leads to increased susceptibility to oxidative and excitotoxic stress (Lezoualch et al. 1998; Bhakar et al. 2002; Fridmacher et al. 2003) but, apparently, does not profoundly affect developmental programmed neuronal death. Although a role for NF-κB in neurotrophic support of SGNs after loss of hair cells has not been reported, higher levels of NF-κB activity in type II SGNs relative to type I SGNs, have been conjectured to be protective against neurotoxic insult by ouabain (Lang et al. 2005). Moreover, NF-κB-deficient mice evince an accelerated age-related hearing loss associated with increased SGN death (but not hair cell death) that has the appearance of excitotoxic death (Lang et al. 2006). Thus, NF-κB may be required for protection of SGNs against traumatic insults including excitotoxicity.

3.3 Stress Pathways

Generalized stress responses are important intracellular protective mechanisms. Crucial pathways are those involving the heat shock response and homeostatic mechanisms triggered by reactive oxygen species.

3.3.1 Heat Shock Response

An important and lethal consequence of many types of stress is misfolding of cellular proteins, which results in their dysfunction and formation of cytotoxic protein aggregates (reviewed in Morimoto et al. 1997). The classical stress response involves the induction of heat shock proteins (HSPs) by moderate stress (not restricted to heat stress), protecting cells from subsequent severe stress. Some HSPs (HSP10, HSP27, HSP40, HSP47, HSP60, HSP70, HSP90, HSP105/110, TriC) are chaperones that stabilize protein structure and prevent aggregation. Other functions for HSPs include nonlysosomal protein degradation (HSP8), free radical scavenging (HSP32/HO1), regulation of the actin cytoskeleton (HSP27, αβ-crystallin), inhibition of apoptosis (HSP27, HSP70), regulation of cell growth and differentiation (HSP27, αβ-crystallin), and signal transduction (HSP90).

Stress-associated upregulation of HSPs is directed by a group of stress-responsive transcription factors termed heat shock factors (Hsfs), with Hsf1 playing the major role (Morimoto et al. 1997). HSP upregulation has a protective function in the cochlea (Section 4.2.1).

3.3.2 Reactive Oxygen Species

After removal of neurotrophic factors, there is a rapid increase in reactive oxygen species (ROS) levels in neurons (Greenlund et al. 1995). ROS play a key role in neurodegeneration, exerting direct toxic effects through their chemical reactivity. In particular, reactive oxygen plays an important role in cell death in the cochlea: the role of reactive oxygen in noise- and drug-induced hair cell death is detailed by Henderson and Hu (Chapter 7) and by Rybak et al. (Chapter 8), respectively. Protection of hair cells by antioxidants is discussed inSections 4.1.1.1 and 4.1.2.1 and, to a limited extent, in Henderson and Hu, Chapter 7 and Rybak, Talaska, and Schacht, Chapter 8.

In the context of apoptosis or programmed cell death ROS appear to be important as intracellular signals (Deshmukh and Johnson 1998) apart from their direct cytotoxic actions. Cellular stress, including stress caused by reactive oxygen species, activates a number of proapoptotic signaling pathways, involving c-Jun, p53, and other transcription factors, thereby promoting apoptosis. Cell death is delayed by the introduction of superoxide dismutase (SOD) to sympathetic neurons after nerve growth factor (NGF) withdrawal (Greenlund et al. 1995), suggesting that ROS play a role in the initiation of apoptosis. A variety of interventions into apoptosis act at a later point in the cascade and do not affect the increase in ROS observed with NGF withdrawal. This implies that the ROS generation, like other cellular stresses, is not necessarily fatal in itself (Deshmukh and Johnson 1998) and depends on the severity of the trauma and the activation of other opposing or compensatory transcription factors, such as NF-κB (Lezoualch et al. 1998). Indeed, elevation of the prosurvival transcription factor NF-κB is associated with antioxidant protection of hair cells against aminoglycoside ototoxicity (Jiang and Schacht 2005).

3.4 Proapoptotic Gene Expression

Many neurons appear to lack "competence to die" even after triggering cytochrome *c* release (Deshmukh and Johnson 1998), possibly because proapoptotic regulatory proteins are constitutively expressed only at low levels in neurons. Thus, to carry out the apoptotic program, increased synthesis of such proteins must accompany initiation of apoptosis. This accounts for the characteristic requirement for transcription in programmed neuronal cell death (Martin et al. 1988). Consequently, transcription factors associated with proapoptotic gene expression are negatively regulated by neurotrophic stimuli. A common theme across these signaling pathways, among which there is considerable "crosstalk" and interaction, is that they are activated by various cellular stresses, including reactive oxygen, hypoglycemia, heat, osmotic stress, and others, as well as by exogenous proapoptotic stimuli (e.g., tumor necrosis factor). Conversely, these pathways are inhibited by neurotrophic stimuli. Thus, apoptosis caused by withdrawal of neurotrophic support reflects, in part, a loss of prosurvival signaling and, in part, a disinhibition of the proapoptotic pathways.

Examples of proapoptotic genes upregulated during acquisition of "competence to die" include the proapoptotic Bcl-2 family members Bim (Whitfield et al. 2001) and Bax (Miyashita and Reed 1995). Transcription factors implicated in increased synthesis of proapoptotic regulators and necessary, in part, for apoptosis include E2F (Liu and Greene 2001; Konishi and Bonni 2003; Biswas et al. 2005), B- and c-Myb (Liu and Greene 2001), the forkhead transcription factor FKHRL (Linseman et al. 2002), the transcription factor c-Jun (Ham et al. 2000), and the transcription factor p53 (Miller et al. 2000).

3.4.1 Forkhead

Phosphorylation of FKHRL by PKB/Akt prevents it from activating transcription so that withdrawal of neurotrophic factors results in reduced PKB activity and increased expression of FKHRL-regulated genes (Brunet et al. 1999). PKB is essential for support of SGN survival in vitro by neurotrophic factors (Hansen et al. 2001b).

3.4.2 JNK–Jun

JNK–Jun signaling is crucial for neuronal apoptosis in response to stress and neurotrophic factor withdrawal, and JNK signaling plays a role in the death of both hair cells (Sections 4.1.1.4 and 4.1.2.2) and SGNs (Section 5.4). The c-Jun N-terminal kinase (JNK) phosphorylates c-Jun, and which is a necessary step for Jun activity in apoptosis (Eilers et al. 2001), with Bim upregulation being an important consequence (Whitfield et al. 2001).

The JNK MAP kinase module includes the MAP kinase kinases (MKKs) that activate JNKs and the mixed lineage kinases (MLKs) and other MKK kinases that activate the MKKs (Davis 2000). MLK inhibition prevents JNK activation and prevents apoptosis in neurons that have been deprived of neurotrophic support or subjected to trauma (Wang et al. 2004). MLK and JNK inhibitors appear promising as therapeutics for neurodegeneration (Bogoyevitch et al. 2004; Wang et al. 2004).

Nevertheless, apoptosis is not an inevitable outcome of stress or "death signals" and JNK is not necessarily a proapoptotic signal (Liu et al. 1996). The response to JNK signaling depends on the cellular context. For example, activation of transcription factor ATF3 results in upregulation of proteins such as HSP27 that antagonize JNK signaling (Nakagomi et al. 2003), accounting, in part, for the protective effect of HSPs (Section 4.2.1). NF-κB-dependent transcription also inhibits JNK signaling or apoptosis (Lin 2003). Finally, in some contexts, JNK acts as a neuroprotective or differentiation signal (Waetzig and Herdegen 2003).

JNKs are multifunctional kinases that perform diverse functions throughout the cell so JNK inhibition, while preventing apoptosis, may have unwanted side effects. Notably, JNK inhibitors strongly reduce neurite growth in cultured SGNs (Bodmer et al. 2002). Such side effects may be avoidable by targeting specific isoforms as there may be functional differences among the JNK isoforms, JNK1, JNK2, and JNK3. Of these, it is JNK3 that is activated and required for neuronal cell death in sympathetic and CNS neurons (Yang et al. 1997; Waetzig 2003). In these and other studies, JNK1 and/or JNK2 could not substitute for JNK3 in apoptosis but, rather, appeared to provide basal JNK activity required for other cellular functions. JNK1/2 isoforms may play a role in neuronal apoptosis during early brain development (Kuan et al. 1999) or stress (Miao et al. 2005).

3.4.3 p53

Neuronal death after withdrawal of neurotrophic support has a requirement for, and is promoted by, the transcription factor p53, an important regulator of cell proliferation and cell death (Miller et al. 2000). The proapoptotic protein Bax is one important transcriptional target of p53-dependent gene expression (Miyashita and Reed 1995). Also, p53 and NF-κB reciprocally inhibit each other's transcriptional efficacy by competing for a common transcriptional coactivator, p300 (Culmsee et al. 2003). Pifithrin-α, a p53 inhibitor, reduces cisplatin-induced hair cell death (Zhang et al. 2003), suggesting a role for p53 in hair cell death (Cheng et al. 2005).

p53 is an important point of convergence for multiple intracellular signals in control of proapoptotic gene expression. p53 expression is itself positively regulated by JNK (Ghahremani et al. 2002) and the Cdk–Rb–E2F pathway (Hiebert et al. 1995). PKB, activated by neurotrophic signaling, promotes survival, in part, by stimulating a pathway that leads to p53 degradation (Ogawara et al. 2002).

4. Therapeutic Interventions for Protection of Hair Cells

Permanent hearing loss is most often associated with loss of hair cells, making hair cells the most important target for protective interventions. Hair cell loss can result from ototoxic drugs (Rybak, Talaska, and Schacht, Chapter 8), noise (Henderson and Hu, Chapter 7), aging (Frisina and Ohlemiller, Chapter 6), genetic mutations (Shalit and Avraham, Chapter 2), autoimmune diseases (Gopen and Harris, Chapter 5), and Ménière's disease and other disorders of cochlear homeostasis (Wangemann, Chapter 3). It is notable that many of these converge on common death effectors, e.g., overproduction of ROS in hair cells or supporting cells. Therefore, intervention strategies that prevent formation of ROS or that block proapoptotic signaling pathways downstream of ROS can be effective against multiple causes of hair cell loss. In this section, first, protection against noise, ototoxic drugs, aging, and genetic mutations are surveyed. Second, endogenous protective systems for hair cells are discussed.

4.1 Protection Against Specific Causes of Hair Cell Death

4.1.1 Noise-Induced Hearing Loss

Causes of noise-induced hearing loss (NIHL) are discussed in detail by Henderson and Hu (Chapter 7). Until a decade ago, the prevailing view was that NIHL was caused primarily by a mechanical destruction of the delicate membranes of the hair cells and supporting structures of the organ of Corti (Hamernik and Henderson 1974; Hunter-Duvar and Bredberg 1974; Bohne 1976) with an additional contribution from an effect of intense noise on blood flow to the inner ear (Scheibe et al. 1992; Miller et al. 1996). With this view, the only rational intervention to prevent NIHL is to reduce the intensity of mechanical energy reaching the inner ear, i.e., the use of ear-protectors. More recent studies have implicated ROS and Ca^{2+} ions as intracellular mediators of noise-induced hair cell death, which raises the possibility that targeting these species will be an effective protective strategy.

4.1.1.1 Reactive Oxygen Species

Lim (Lim and Melnick 1971) suggested that intense metabolic activity may contribute to inner ear pathology consequent to noise but it was not until the 1990s that free radical formation in the inner ear after noise exposure was demonstrated. Noise was shown to increase a wide array of reactive oxygen species (ROS) including superoxide radicals (Yamane et al. 1995), peroxynitrite radicals (Shi and Nuttall 2003), and hydroxyl radicals (Ohlemiller et al. 1999) in the inner ear. ROS reduce cell survival by direct cytotoxic action and proapoptotic intracellular signaling (Section 3.3.2).

Increased levels of the antioxidant glutathione in lateral wall tissues after intense noise suggest that activation of an endogenous protective system is triggered in order to reduce noise-induced damage (Yamasoba et al. 1999). Indeed, antagonizing endogenous antioxidant protective mechanisms resulted in greater NIHL (Yamasoba et al. 1998) while boosting endogenous antioxidants reduces NIHL (Ohinata et al. 2000b). Administration of any of several exogenous antioxidants also attenuates NIHL (Henderson et al. 2006). These observations not only demonstrate the role of antioxidants in preventing neuronal death but also indicate that upregulation of endogenous antioxidant systems or application of exogenous antioxidants might be useful therapeutic strategies to prevent NIHL.

As an aggravating factor, noise-induced ROS can also result in lipid peroxidation and the formation of 8-isoprostane , a potent vasoconstrictor (Ohinata et al. 2000a). Vasoconstriction can transiently reduce oxygen tension and yield a rebound in blood flow that may contribute to additional ROS formation, similar to that seen in stroke and cardiac infarctions. There is evidence that antioxidants and vasodilators can act in synergy to attenuate NIHL (see Section 8).

A question pertinent for clinical protection is how much delay can be tolerated; i.e., must therapeutic prevention be applied before noise exposure or are treatments effective even if applied after noise exposure? A relevant observation is that increased ROS and proapoptotic signaling continues up to 2 weeks after noise overstimulation (Yamashita et al. 2004, also see Henderson and Hu, Chapter 7) with a major peak a week after the noise stress. Consistent with this, several studies have found that antioxidants still provide some protection when applied several days following the noise overstimulation (van Campen et al. 2002).

4.1.1.2 Ca^{2+}

Noise exposure also results in increased cytosolic Ca^{2+} concentration ($[Ca^{2+}]_i$) in hair cells (Fridberger et al. 1998), with detrimental consequences to calcium homeostasis and cell survival. For example, the Ca^{2+}-activated phosphatase PP2B/calcineurin increases in hair cells following noise exposure, possibly triggering mitochondrial cell death pathways (Minami et al. 2004). Blocking Ca^{2+} channels before noise exposure moderates the increase in $[Ca^{2+}]_i$ and protects hair cells (Heinrich et al. 1999), as do calcineurin inhibitors (Minami et al. 2004). Thus, Ca^{2+} buffering or other means to support intracellular Ca^{2+} homeostasis are also potential targets for protective interventions.

4.1.1.3 Mitochondrial Function

Both the generation of ROS and some aspects of calcium homeostasis are influenced by the integrity of mitochondrial metabolism (see Wangemann, Chapter 3). Although the underlying mechanisms remain speculative, acetyl-L-carnitine and α-lipoic acid, both of which modulate mitochondrial function, can reduce NIHL (Seidman 2000).

4,1.1.4 JNK and Apoptotic Regulators

Proapoptotic signaling pathways discussed in Section 3 can be activated after noise exposure (Henderson and Hu, Chapter 7) and may mediate cell death downstream of ROS or Ca^{2+}. Consequently, these are potential targets for protective interventions and in vivo administration of inhibitors of JNK activation, CEP-1347 (Pirvola et al. 2000) or D-JNK-1 (Wang et al. 2003), indeed protects against NIHL. Noise exposure that results in only a temporary threshold shift upregulates expression of the antiapoptotic Bcl-2 member, Bcl-x, in the outer hair cell region; in contrast, intense noise exposure that results in permanent threshold changes and hair cell loss activates the proapoptotic Bcl-2 family members, Bak and Bad (Vicente-Torres and Schacht 2006). Thus, differential regulation of Bcl-2 family members can contribute to protective or apoptotic responses depending on the severity of the stress.

4.1.1.5 Neurotrophic Factors

Glial cell line–derived neurotrophic factor (GDNF) delivered directly to the scala tympani protected from NIHL (Ylikoski et al. 1998; Shoji et al. 2000), but a caveat is that very high concentrations of GDNF increased damage (Shoji et al. 2000). Neurotrophic factor-3 (NT-3) was also effective (Shoji et al. 2000), while BDNF, fibroblast growth factor (FGF)-1, and FGF-2 were not (Shoji et al. 2000; Yamasoba et al. 2001). The mechanism by which NT-3 and GDNF support hair cell survival remains to be elucidated. Both GDNF (Ylikoski et al. 1998) and NT-3 (Sugawara et al. 2007) are expressed in the postnatal cochlea but hair cells do not appear to express receptors for these factors. An indirect mode of action is therefore rather likely.

4.1.1.6 Medial Olivary Complex Efferents and Protection

The olivocochlear efferent pathway has its origin in the superior olivary complex and terminates in the cochlea. It can be divided into lateral and medial systems based on their origin and terminations (Warr 1992). The lateral olivary complex (LOC) is targeted mainly to SGNs and is discussed in Section 6.3. The medial olivary complex (MOC) is part of a sound-activated feedback loop targeted mainly to the outer hair cells, especially outer hair cells of the basal turns of the cochlea (Brown et al. 2003). MOC efferents reduce the sensitivity of the cochlear amplifier during loud ambient sound, via release of the neurotransmitters γ-aminobutyric acid (GABA) and acetylcholine (Dallos et al. 1997). By dampening outer hair cell responses, the MOC efferents could also serve a protective function, e.g., by reducing Ca^{2+} entry and metabolic activity in outer hair cells. Indeed, the strength of the olivocochlear response correlates with susceptibility to acoustic injury (Maison and Liberman 2000). Moreover, susceptibility to acoustic injury is reduced in transgenic mice by forced overexpression of $\alpha 9$ nicotinic acetylcholine receptor subunits in outer hair cells (Maison et al. 2002).

4.1.2 Ototoxic Drugs

As for NIHL, investigations of the mechanisms by which ototoxic drugs cause damage (see Rybak, Talaska, and Schacht, Chapter 8) has revealed potential targets for protective interventions, with ROS and apoptotic mechanisms again playing prominent roles.

4.1.2.1 Reactive Oxygen Species

Aminoglycosides such as gentamicin increase levels of ROS (Lautermann et al. 1995) and the other prominent ototoxin, cisplatin, likewise increases ROS and related oxidative radicals (Rybak et al. 2000; Kelly et al. 2003). Supporting a causal role for ROS in these pathologies, a wide variety of antioxidants have successfully reduced hair cell loss and hearing loss due to aminoglycosides (Song and Schacht 1996) or cisplatin (Rybak et al. 2007) in animal models. As the ROS generation by aminoglycosides is also iron dependent (Priuska and Schacht 1995), iron chelators are also useful protective agents (Song and Schacht 1996).

Antioxidant supplementation has promise as a clinical application. Salicylate, whose properties include antioxidant and iron chelating actions, emerged as a potential protectant from animal experimentation (Sha and Schacht 1999). Consequently, aspirin (acetyl salicylate) was tested in a randomized, placebo-controlled double-blind clinical trial in patients receiving gentamicin as an antibiotic (Sha et al. 2006). There was a significant 75% reduction in the incidence of hearing loss in patients receiving salicylate at the same time as gentamicin, relative to those receiving a placebo.

Hair cell loss and hearing loss continue after cessation of aminoglycoside treatment and can reach a peak days to weeks after treatment (Rybak, Talaska, and Schacht, Chapter 8). This may be due to persistence of the drugs in the cochlea after cessation of their administration (Forge and Schacht 2000). Alternatively, it may be due to continued elevated ROS after ototoxins are cleared akin to the situation after noise exposure. Thus, it can be speculated that some hair cell loss may be prevented even if antioxidant treatment is initiated after the end of drug treatment.

4.1.2.2 Apoptotic Regulation

As for NIHL, protective interventions into drug-induced hearing loss might be successful if targeted at proapoptotic signaling downstream of ROS. Some features of apoptosis, such as JNK activation (Pirvola et al. 2000) or mobilization of AIF and EndoG from mitochondria, appear after aminoglycoside administration in vivo. However, other features, e.g., caspase activation, cytochrome c release, and terminal uridine deoxynucleotidyl transferase dUTP nick end labeling (TUNEL), are not apparent (Jiang et al. 2006). Possibly, apoptosis of cochlear hair cells after aminoglycoside exposure involves mitochondrial pathways other than those after noise exposure. Alternatively, caspase-mediated pathways are rapid and transient and evident only briefly before death and disappearance of the hair cell and thus may be missed. Caspase inhibition does provide protection from hair cell loss and hearing loss due to cisplatin (Van de Water et al. 2004) but it is not yet clear whether inhibition of these apoptotic events is a useful protective strategy for aminoglycoside ototoxicity. With regard to JNK signaling, inhibition of JNK activation by CEP-1347 or by D-JNKI-1 prevented aminoglycoside ototoxicity in vitro and partially reduced it in vivo (Pirvola et al. 2000; Wang et al. 2003), suggesting further exploration as a therapeutic strategy.

4.1.3 Age-Related Hearing Loss

There is significant hair cell loss associated with the loss of hearing in aging humans and experimental animals (Ohlemiller and Frisina, Chapter 6). Age-related hearing loss (ARHL) may result from oxidative stress accumulated over a lifetime or, alternatively, from an age-related reduction in the capacity of endogenous protective mechanisms, such that a stress that does not cause hair cell loss in the young cochleae will damage the older cochlea (e.g., Jiang et al. 2006). Animal models in which endogenous protective mechanisms are reduced often show premature or enhanced ARHL (Ohlemiller et al. 2000; Keithley et al. 2005; Lang et al. 2006). Age-related loss of hair cells may also be associated with an accumulation of mitochondrial mutations (Pickles 2004), another expression of oxidative stress. Exogenous antioxidants can reduce the extent of mitochondrial mutations and delay ARHL (Seidman 2000), although such protection is much less effective than in hearing loss due to noise or ototoxic drugs.

4.1.4 Hearing Loss of Genetic Origin

For hearing loss of genetic origin, it is possible to design interventions with gene transfer methods, such as viral vectors, that can introduce a nonmutated copy of a gene. This strategy has been used to replace the mutated *Myo15* gene, causing hair cell death in *shaker-2* mice (Probst et al. 1998) and effectively rescue cochlear morphology and hearing (Kanzaki et al. 2006). Other methods of gene transfer, in combination with appropriate genes, may offer potential for correction of genetic deafness at later ages, particularly for later onset deafness (Raphael and Heller, Chapter 11).

4.2 Preconditioning and Endogenous Protective Mechanisms

Exposure to a mild sublethal stress is generally protective against a subsequent strong, even otherwise lethal, stress, a phenomenon known as preconditioning. Preconditioning the cochlea with sound over a period of time indeed protects from later louder noise (reviewed in Niu and Canlon 2002). Cell defense mechanisms that are invoked during preconditioning are generally effective against many types of stress. Therefore, they likely involve homeostatic protective and antiapoptotic systems discussed earlier in this chapter.

Preconditioning has been intensively investigated in brain ischemia (Gidday 2006), where the molecular basis of preconditioning has been found to include HSP70 upregulation (McLaughlin et al. 2003), activation of prosurvival effectors such as NF-κB (Blondeau et al. 2001), and CREB (Mabuchi et al. 2001), and downregulation of proapoptotic effectors such as JNK (Miao et al. 2005) and the BH3-only protein Bim (Meller et al. 2006). Interestingly, preconditioning may require caspase activation (Garnier et al. 2003; McLaughlin et al. 2003). Presumably, in the context of mild stress, caspases cleave proteins that would otherwise be deleterious.

With regard to the cochlea, sound conditioning increases antiapoptotic regulators such as Bcl-2, and antioxidant enzymes (Jacono et al. 1998; Harris et al. 2006). Three intrinsic protective mechanisms that bear special mention in this regard are heat shock proteins (HSPs), adenosine, and steroid hormones.

4.2.1 Heat Shock Proteins

Stress-induced upregulation of HSPs serves a protective function in the cochlea (Altschuler et al. 1996). Hsf1, the major transcription factor responsible for regulating HSP expression is found in hair cells, stria vascularis and SGN in the rodent cochlea (Fairfield et al. 2002). HSP70 is upregulated in the organ of Corti and stria vascularis by potentially damaging stresses including noise, heat, and ototoxins such as cisplatin (Yoshida et al. 1999; Oh et al. 2000; Yoshida and Liberman 2000). Levels of HSP27 (Leonova et al. 2002) and HSP32 (Fairfield et al. 2004) likewise increase after stress.

Following noise exposure, Hsf1$^{-/-}$ fmice, in which HSP upregulation does not occur, had greater hearing loss and loss of outer hair cells than did Hsf1$^{+/+}$ mice (Fairfield et al. 2005), indicating that HSP upregulation is necessary for the ability of moderate stress to protect against subsequent severe stress. Animals exposed to moderate noise (i.e., sufficient to induce temporary but not permanent threshold shift), to heat, or to geranylgeranylacetone, all of which cause HSP upregulation, show protection from otherwise permanently damaging noise exposure (Yoshida et al. 1999; Yoshida and Liberman 2000; Mikuriya et al. 2005).

4.2.2 ATP and Adenosine

ATP and other nucleotides (e.g., GTP, UTP) are neurotransmitters or intercellular signals that have modulatory functions in the cochlea including regulation of sensory transduction in hair cells (Thorne et al. 2004), regulation of neuro-transmission at the inner hair cell–SGN synapse (Robertson and Paki 2002), and regulation of endolymph volume and composition (Housley and Thorne 2000).

Both ionotropic (P2X) and metabotropic (P2Y) receptors are present on cells in the cochlea, including hair cells, supporting cells, cells in Reisner's membrane, and in the stria vascularis (Housley and Thorne 2000). P2X receptors increase $[Ca^{2+}]_i$ by acting as plasma membrane Ca^{2+} channels and by opening voltage-gated Ca^{2+} channels. This action is further amplified by Ca^{2+}-induced Ca^{2+} release from intracellular Ca^{2+} stores via ryanodine receptor Ca^{2+} channels (Bobbin 2002). P2Y receptors increase $[Ca^{2+}]_i$ through G-protein–dependent signal transduction via activation of phospholipase C, consequent generation of inositol trisphosphate (IP), and Ca^{2+} release from the endoplasmic reticulum by way of IP-gated Ca^{2+} channels (Harden et al. 1995).

Noise exposure increases the response of cells in the cochlea to ATP (Chen et al. 1995). Thus, the elevated $[Ca^{2+}]_i$ which contributes to noise damage (Section 4.1.1.2) may in part be due to an ATP-induced increase in $[Ca^{2+}]_i$. Specifically, a wave of elevated $[Ca^{2+}]_i$ in supporting cells is initiated by damage to an outer hair cell and propagated by ATP release, activation of P2Y4 receptors on adjacent supporting cells, $[Ca^{2+}]_i$ increase in the supporting cell, and Ca^{2+}-triggered release of ATP (Piazza et al. 2007). Blockade of ATP receptors attenuates the effects of intense sound (Bobbin 2001).

Extracellular ATP in the cochlea is rapidly hydrolyzed to adenosine by ectonucleotidases present in the perilymph (Vlajkovic et al. 2004). Adenosine is an agonist at four G-protein–coupled receptor subtypes—A1, A2A, A2B, and A3—of which the A1 and A3 subtypes are present in the cochlea (Ramkumar et al. 1994; Ford et al. 1997). Thus, ATP released from cells in the cochlea may stimulate P2Y or P2X receptors and then be broken down to adenosine that stimulates A1 and A3 receptors. The latter action may antagonize the former. Noise exposure increased cochlear A1 receptor expression (Ramkumar et al. 2004) and intracochlear infusion of an adenosine agonist, (R)-phenylisopropyladenosine, attenuated NIHL (Hu et al. 1997). Adenosine may also be involved in protection of the cochlea from cisplatin ototoxicity (Whitworth et al. 2004) and hypoxia (Bobbin and Blesoe 2005). Similarly, adenosine prevented ATP-induced $[Ca^{2+}]_i$

increase and cell death in the retina (Zang et al. 2006). The mechanisms by which adenosine protects include increasing antioxidant enzymes, SOD and glutathione peroxidase (Ford et al. 1997).

4.2.3 Glucocorticoids

Another potential mechanism by which preconditioning could protect hearing is by influencing the hypothalamic–pituitary–adrenal axis (HPA). Sound conditioning elevates plasma corticosterone and upregulates glucocorticoid receptors in the cochlea, which in turn reduces the sensitivity to acoustic trauma (Tahera et al. 2007).

In the context of hormonal influences on the cochlea, higher levels of serum aldosterone were correlated with reduced ARHL (Tadros et al. 2005). Conversely, progestin treatment—as part of hormone replacement therapy in women—resulted in poorer hearing (Ohlemiller and Frisina, Chapter 6). Thus, elevating aldosterone or its agonists might protect from ARHL.

5. Neurotrophic Support of Spiral Ganglion Neurons

Neurotrophic factors have been briefly discussed for their potential role in protecting hair cells. Summarized below are the intracellular signaling pathways recruited by peptide neurotrophic factors and by neural activity to promote neuronal survival, specifically SGN survival. The mechanisms by which peptide neurotrophic factors prevent apoptosis, although originally investigated and identified in other neuronal types or in neuronal cell lines, have been verified in SGNs. The mechanisms by which neural activity prevent apoptosis have depended significantly on studies of SGNs, which have proved an ideal system for such investigation.

5.1 Hair Cells are Necessary for Spiral Ganglion Neuron Survival

5.1.1 SGNs Die in the Absence of Hair Cells

Hair cells provide the principal excitatory input to SGNs and, after the loss of hair cells or hair cell function, SGNs have no or greatly reduced electrical activity levels. Further, hair cells are also an essential source of neurotrophic support to the SGNs: after hair cell loss SGNs soon lose the peripheral axon that projects to the organ of Corti and eventually die (Spoendlin 1975; Webster and Webster 1981; Koitchev et al. 1982; Bichler et al. 1983). In most of these studies, hair cells were rapidly and selectively killed with aminoglycoside antibiotics. The rate at which SGNs die after hair cell loss is slow and differs among species. In the rat, greater than 90% of the SGNs die within 3 months (Bichler et al. 1983). In the guinea pig, ≈15% of the SGNs remain alive 110 days after hair cell loss, with half of that fraction remaining a year after hair cell loss (Webster and Webster 1981; Koitchev et al. 1982). In the chinchilla, SGN death is complete in 2–4 months (McFadden et al. 2004) while in the cat, SGN death occurs over a period of years (Leake and Hradek 1988). Studies of human postmortem tissue suggest that SGNs are capable of surviving for many years in the absence of hair cells (Leake and Hradek 1988; Fayad et al. 1991; Nadol 1997).

Type I SGNs convey auditory information from the inner hair cells to the brain and comprise approximately 95% of all SGNs. In humans, both type I and type II SGNs die at comparable rates after hair cell loss (Zimmermann et al. 1995), although earlier experimental studies on animal models had suggested that type I SGNs might be more sensitive to hair cell loss or die at a faster rate (Spoendlin 1975; Leake and Hradek 1988).

5.1.2 Support of SGN Survival by Hair Cells Can Involve Electrical Stimulation and/or Peptide Neurotrophic Factors

Membrane depolarization (Hegarty et al. 1997) as well as peptide neurotrophic factors (Hegarty et al. 1997) support the survival of cultured SGNs. Moreover, peptide neurotrophic factors and membrane depolarization are additive in their ability to promote SGN survival (Hegarty et al. 1997), suggesting that SGN survival may depend on their receiving multiple neurotrophic stimuli simultaneously. This is supported by experimental studies of SGN survival in vivo in the absence of hair cells (see later). In the following sections, the mechanisms by which neurotrophic factors and membrane depolarization prevent apoptosis are discussed.

5.2 Support of SGN Survival by Neurotrophic Factors

5.2.1 Neurotrophic Factors in the Cochlea

The expression and function of neurotrophic factors with respect to survival and neurite growth in the cochlea have been extensively reviewed recently (Fritzsch et al. 2004, 2005) so are only briefly summarized here. SGNs express at least two receptors for neurotrophins: TrkC, which binds NT-3 with high affinity, and TrkB, which binds BDNF with high and NT-3 with lower affinity. BDNF and NT-3 are two of the four members of the neurotrophin family of neurotrophic factors. Both BDNF and NT-3 do promote SGN survival, evidenced by experiments showing that SGN survival in vitro is supported by either BDNF or NT-3 (Lefebvre et al. 1994; Hegarty et al. 1997).

BDNF is expressed in mammalian cochlear and vestibular hair cells early in development and appears to play an important, although not exclusive, role in guiding sensory neuronal peripheral projections, particularly for vestibular neurons (Fritzsch et al. 2005). In the cochlea, BDNF expression gradually declines through embryogenesis and in the early postnatal period, in a base to apex gradient, and is not expressed in the mature mammalian organ of Corti (Fritzsch et al. 2004). In contrast, expression of BDNF in the spiral ganglion itself may persist in the mature cochlea (Wiechers et al. 1999; Hansen et al. 2001b; Zha et al. 2001).

NT-3 is initially expressed throughout the organ of Corti sensory epithelium in mammals but during embryonic development becomes more restricted. Expression of NT-3 early in development is absolutely required for survival of SGNs: most SGNs die during embryogenesis in knockout mice lacking NT-3 or TrkC (Fritzsch et al. 2004). In the postnatal rodent organ of Corti, NT-3 is expressed only in the inner hair cells and adjacent supporting cells (Sugawara et al. 2007).

In addition to the neurotrophins, there are other families of peptide factors in which some or all members are neurotrophic. One such peptide neurotrophic factor, GDNF, appears to be synthesized in the organ of Corti, starting in the second postnatal week, and can support SGN survival in vitro and in vivo (Ylikoski et al. 1998; Yagi et al. 2001).

There has been no test of the postnatal requirement for hair cell–derived neurotrophic factors, such as NT-3 and GDNF, for SGN survival. Mice lacking these factors either lose their SGNs before birth in the case of NT-3 or die before birth in the case of GDNF. It is possible that given physiological electrical stimulation via normal afferent input, SGNs may be able to survive in the absence of hair cell–derived neurotrophic factors. Also, it should be noted that neurotrophic factors have roles other than survival, including synaptic maintenance and function and control of the mature neuronal phenotype. For example, BDNF and NT-3 have been implicated in inducing, respectively, the characteristically basal and apical physiological properties of SGNs including ion channel composition and firing pattern (Adamson et al. 2002; Zhou et al. 2005). Targeted and conditional gene deletion will be necessary to resolve the role of neurotrophic factors in SGN survival and function in the postnatal cochlea.

5.2.2 Support of SGNs by Multiple Neurotrophic Factors In Vitro

As noted earlier, BDNF, NT-3, or GDNF can each support survival of SGNs in culture (Lefebvre et al. 1994; Hegarty et al. 1997; Ylikoski et al. 1998). Combining neurotrophins BDNF and NT-3 results in increased SGN survival relative to individual neurotrophins (Lefebvre et al. 1994; Hegarty et al. 1997). Moreover, combining neurotrophins with other peptide neurotrophic factors, FGF, Leukemia Inhibitory Factor (LIF), transforming growth factor-β (TGF-β), or ciliary neurotrophic factor (CNTF) also increases survival relative to neurotrophin alone (Hartnick et al. 1996; Marzella et al. 1997, 1998). In addition, these factors promote neurite growth from cultured SGNs (Lefebvre et al. 1994; Hegarty et al. 1997).

5.2.3 Receptors for Peptide Neurotrophic Factors

5.2.3.1 Neurotrophins and Trks

Receptors for most peptide neurotrophic factors are either receptor protein-tyrosine kinases or associate with and activate protein-tyrosine kinases. Of the neurotrophic factors, the most widespread in the nervous system and best studied are the four members of the neurotrophin family, NGF, BDNF, NT-3, and NT-4 (Huang and Reichardt 2001). The cognate family of receptor protein-tyrosine kinases is the tropomyosin-related kinase (Trk) family which consists of three members, TrkA, TrkB, and TrkC. NGF binds and signals via TrkA receptors, BDNF and NT-4 via TrkB receptors; NT-3 acts principally via TrkC receptors but NT-3 can bind and signal via both TrkA and TrkB receptors (Huang and Reichardt 2003).

5.2.3.2 Neurotrophin Receptor p75^{NTR}

In addition to binding and signaling via Trk receptors, all four neurotrophins, and their unprocessed precursors, bind and signal through an unrelated receptor, the neurotrophin receptor p75^{NTR} (Huang and Reichardt 2003; Gentry et al. 2004). p75^{NTR} is not a protein-tyrosine kinase but does interact with Trk receptors, with the consequence that these receptors reciprocally modulate each other's signaling. p75^{NTR} also initiates several pathways that activate NF-κB, and, importantly, proapoptotic signaling, including JNK. p75^{NTR} is expressed in the developing cochlea in SGNs and in non-neuronal cells (von Bartheld et al. 1991; Gestwa et al. 1999). In the mature cochlea, p75^{NTR} is at low or undetectable levels but is strongly upregulated in the spiral ganglion following hair cell loss (Tan and Shepherd 2006) and so may contribute to SGN degeneration. Mice with mutant p75^{NTR} have apparently normal cochlear development and function but lose hair cells and SGNs at an accelerated rate as they age (Sato et al. 2006). It remains to be established whether the loss of the sensory and neural elements is due directly to loss of p75^{NTR} function in these cells or is an indirect result of loss of p75^{NTR} function in non-neuronal cells.

5.2.3.3 GDNF Family Receptors

The GDNF subfamily of the TGF-β family of growth factors, GDNF, neurturin, artemin, and persephin (Baloh et al. 2000), all bind and signal via the Ret receptor protein-tyrosine kinase or by binding to the neuronal cell adhesion molecule (NCAM; Paratcha et al. 2003). For high-affinity binding an additional "coreceptor" is required: a member of the GDNF family receptorα(GFRα) family. This family has four members, GFRα1–4, which are the cognate coreceptors for the four GDNF family ligands. As noted in the preceding text, SGNs express GDNF receptors and GDNF is a neurotrophic factor for SGNs.

5.2.4 Neurotrophic Factor Signal Transduction and Prosurvival Intracellular Signaling Pathways

Trks and Ret, receptors for neurotrophins and GDNF-family ligands, respectively, are receptor protein-tyrosine kinases. For neurotrophins, activation of the receptor protein-tyrosine kinase appears to be sufficient for initiation of intracellular signaling. Signal transduction for GDNF-family ligands is more complex. Whether the ligand binds to Ret or to NCAM, recruitment of other protein-tyrosine kinases by the receptor is additionally required (Encinas et al. 2001;

Paratcha et al. 2003). In all cases, understanding receptor protein-tyrosine kinase signal transduction is fundamental to understanding how neurotrophins, GDNF, and other peptide neurotrophic factors can initiate prosurvival intracellular signaling.

The signal transduction mechanisms for receptor protein-tyrosine kinases and, in particular, for Trks, have been extensively reviewed (Huang and Reichardt 2003). In brief, binding of ligand to the receptor results in dimerization, cross-phosphorylation of the dimerized protein-tyrosine kinases, and assembly of protein complexes containing adaptor and signal transducer proteins that initiate the various pathways.

One intracellular signaling pathway is initiated by activation of the lipid-modifying enzyme phosphatidylinositol 3-OH kinase (PI3K), which leads to recruitment and activation at the plasma membrane of the protein kinase PKB/Akt, crucial prosurvival kinase for promotion of survival by peptide neurotrophic factors (reviewed in Datta et al. 1999). Inhibition of PKB abolishes the survival-promoting effect of BDNF or NT-3 on SGNs in vitro (Hansen et al. 2001b). PKB promotes survival via phosphorylation of multiple targets (Datta et al. 1999). As noted earlier, PKB inhibits apoptosis by phosphory-lating and inactivating forkhead transcription factors (Brunet et al. 1999) and by phosphorylating and inactivating the proapoptotic Bcl-2 family protein Bad (Downward 1999).

Another neurotrophic factor–activated intracellular pathway important for survival is the ERK subfamily of the MAPKs. ERKs promote survival by phosphorylating diverse targets. ERK appears to directly phosphorylate Bad, inactivating this proapoptotic effector (Downward 1999). Another outcome of ERK signaling is activation of the prosurvival transcription factor CREB in the nucleus (Dawson and Ginty 2002).

5.3 Support of SGN Survival by Electrical Activity/Membrane Depolarization

Electrical activity, specifically depolarization, is a survival-promoting stimulus for neurons, principally due to increased cytosolic Ca^{2+} (Franklin and Johnson 1994; Green 2000). In particular, for SGNs, membrane electrical activity provides a prosurvival stimulus, evidenced by experiments on chroni-cally depolarized SGNs in vitro (Hegarty et al. 1997) and electrically stimulated SGNs in vivo (reviewed in Miller 2001). This raises the possibility that electrical stimulation provided by a cochlear implant, alone or in combination with other therapy, in supporting survival of SGNs, further discussed later.

5.3.1 Membrane Depolarization Supports Neuronal Survival Via Ca^{2+} Signaling

Multiple intracellular signaling pathways link membrane depolarization to survival (Hansen et al. 2001b), although there may be variability among neurons with regard to the relative importance of different pathways. A common element among these intracellular signaling pathways is that they are initiated by entry of Ca^{2+} ions into the cytosol via voltage-gated Ca^{2+} channels (VGCCs). In neurons chronically depolarized by elevating extracellular K^+, Ca^{2+} entry via L-type VGCCs is required for neuronal survival (Franklin et al. 1995; Galli et al. 1995). It has been suggested that Ca^{2+} entry through other VGCCs, e.g., N-type, may be more important when neurons are stimulated by patterned electrical activity (Brosenitsch and Katz 2001) in vitro; nevertheless, support of SGNs in vivo by patterned electrical stimulation is inhibited by blockade of L-type Ca^{2+} channels (Miller et al. 2003), similar to what was observed with SGNs chronically depolarized in vitro (Hegarty et al. 1997).

It may seem paradoxical that cytosolic Ca^{2+} acts as a prosurvival signal, given that it has been long known that high $[Ca^{2+}]_i$ results in excitotoxic cell death (see Section 6). The resolution of this paradox lies in the concentration dependence of the consequences of cytosolic Ca^{2+}. Where $[Ca^{2+}]_i$ has been experimentally raised by depolarization from the resting level of 50–100 nM, it has been noted that a moderate increase to 150–250 mM in sympathetic neurons (Franklin et al. 1995) or to 200–400 nM in SGNs (Hegarty et al. 1997) results in survival; higher $[Ca^{2+}]_i$ levels result in excitotoxicity (see Section 6 and Wangemann, Chapter 3). As further evidence of protection by elevated cytosolic Ca^{2+}, increased levels of cytosolic Ca^{2+} (Tong et al. 1996) and increased levels of L-type voltage-gated Ca^{2+} channels (Koike and Tanaka 1991) have been implicated in the reduced dependence of mature neurons on peptide neurotrophic factors.

5.3.2 Ca^{2+} Signaling Pathways That Support SGN Survival

Ca^{2+} is an important second messenger (Berridge 1998); after Ca^{2+} entry via VGCCs, the increased level of cytosolic Ca^{2+} causes the activation of several Ca^{2+}-dependent signaling systems. Most studies have focused on those in which the Ca^{2+}-binding regulatory protein calmodulin (CaM) is involved. In SGNs, depolarization primarily promotes survival via at least three Ca^{2+}/CaM-dependent signaling pathways: Ca^{2+}/CaM-dependent protein kinase II (CaMKII), Ca^{2+}/CaM-dependent protein kinase IV (CaMKIV), and protein kinase A (PKA) (Hansen et al. 2001b, 2003; Bok et al. 2003). These act independently and additively: activation of any one of these pathways promotes SGN survival, although not as effectively as depolarization (Hansen et al. 2001b, 2003) while, conversely, inhibition of any one of these pathways partially inhibits the ability of depolarization to support SGN survival (Hansen et al. 2001b, 2003; Bok et al. 2007). These Ca^{2+}-dependent signaling pathways act in distinct subcellular compartments on distinct substrates, directed toward either cytoplasmic effectors or the cell nucleus. Thus, depolarization promotes SGN survival by inhibiting apoptotic signaling coordinately on the transcriptional and posttranslational levels.

CaMKIV, which is primarily nuclear in SGNs, targets the transcription factor CREB and so constitutes a nuclear pathway—consistent with observations of other neurons (See et al. 2001). Genes regulated by CREB, such as Bcl-2 (Section 3.1), may therefore be important targets of depolarization-dependent signaling in SGNs.

CaMKIIα and CaMKIIβ isoforms are expressed in SGNs (Bok et al. 2007). However, in contrast to CaMKIV, CaMKII promotion of survival is unaffected by CREB inhibition; possible nuclear targets for CaMKII include the kinesin family motor protein KIF4, which releases the prosurvival factor PARP-1 in a CaMKII-dependent manner to promote neuronal survival (Midorikawa et al. 2006). With regard to cytoplasmic CaMKII targets, CaMKII has been shown to activate the prosurvival transcriptional regulator NF-κB in neurons (Meffert et al. 2003). Also, CaMKII in SGNs functionally inactivates the proapoptotic regulator Bad (Bok et al. 2007) and the proapoptotic protein kinase JNK (J. Huang and S.H. Green, unpublished observations).

The third Ca^{2+}-dependent signaling pathway involves Ca^{2+}/CaM-sensitive adenylyl cyclases that link depolarization to cAMP synthesis and activation of PKA. The second messenger cAMP is a prosurvival signal for neurons in general (Rydel and Greene 1988; Galli et al. 1995) and SGNs in particular (Hegarty et al. 1997). While cAMP, like CaMKs, appears to be a prosurvival signal in neurons, the mechanism may be different in different types of neurons. In CNS neurons, cAMP appears to promote survival by increasing responsiveness to neurotrophins (Meyer-Franke et al. 1998) but, in SGNs, cAMP promotes survival independently of neurotrophins (Hansen et al. 2001b). In SGNs (Bok et al. 2003) and other cells (Harada et al. 1999), the mechanism appears to involve phosphorylation and functional inactivation of the proapoptotic effector Bad by PKA, facilitated by the location of PKA action on the outer mitochondrial membrane (Harada et al. 1999; Affaitati 2003; Y.S. Cho, J. Huang, and S.H. Green, unpublished observations}. Although PKA can enter the nucleus and phosphorylate and activate CREB (De Cesare and Sassone-Corsi 2000), this appears to be irrelevant to support of SGN survival by cAMP. Rather, cytoplasmic PKA activity is necessary and sufficient for SGN survival promoted by cAMP signaling and nuclear activity is dispensable (Bok et al. 2003). Dominant-negative CREB mutants that blocked the ability of CaMKIV to promote survival had no effect on PKA (Bok et al. 2003).

Protein kinase C (PKC) is a CaM-independent Ca^{2+}-activated protein kinase. PKC appears to be necessary, in part, for support of sympathetic neurons by NGF (Pierchala et al. 2004). Activation of PKC allows SGN survival in a MEK- and PKB-dependent manner (Lallemend et al. 2003, 2005) but it is not known whether PKC activity is necessary for support of SGNs by depolarization. In contrast, PKC inhibition *prevents* programmed cell death in cerebellar Purkinje cells (Ghoumari et al. 2002), indicating that the consequences of particular Ca^{2+}-dependent signals can be very different depending on the cellular context.

5.4 Suppression of JNK Signaling by Neurotrophic Stimuli

Jun phosphorylation and JNK activation can be detected in neurotrophic factor–deprived cultured SGNs and in SGNs in vivo after hair cell loss (Alam et al. 2007), implying a role in the death of deafferented SGNs. JNK inhibitors reduce death of neurotrophic factor–deprived cultured SGNs (Pirvola et al. 2000) and of SGNs in vivo after oxidative stress (Scarpidis et al. 2003). As noted in Section 4.1.1.4, MLKs are important upstream activators of JNK and MLK inhibitors prevent JNK activation after neurotrophic factor withdrawal. Peptide neurotrophic factors suppress JNK activation by preventing MLK activation. At least two distinct pathways are involved. Neurotrophic factor receptors inhibit the small GTPases Rac and Cdc42, which are MLK activators (Xu et al. 2001). Also, the protein-tyrosine kinase effector PKB can directly phosphorylate and inactivate MLKs (Barthwal et al. 2003).

Preliminary data on cultured SGNs indicates that, like peptide neurotrophic factors, depolarization suppresses JNK activation (J. Huang and S.H. Green, unpublished observations) and does so in a CaMKII-dependent manner.

5.5 Support of SGN Survival by Cells Other Than Hair Cells

The slow death of deafferented SGNs in the postnatal cochlea is strikingly different from the very rapid death of SGNs in NT-3$^{-/-}$ mouse embryos (Fritzsch et al. 2004). This may reflect the general loss of dependence on target-derived neurotrophic factors that occurs as neurons mature, which is due to multiple cellular changes. Another (not necessarily exclusive) explanation is that, in the mature cochlea, SGNs may receive survival-promoting stimuli from cells other than hair cells, so loss of the hair cells would not mean a complete lack of neurotrophic support, accounting for survival of some SGNs even long after hair cell loss. What could be the sources of such neurotrophic support? Neurons typically receive neurotrophic support from postsynaptic cells and it is likely that SGNs receive such support from the cochlear nucleus, because cutting the VIIIth nerve results in SGN death (Spoendlin 1971). It is also likely that SGNs can be supported by neurotrophic factors derived from paracrine/autocrine sources. Cell cultures, as well as freshly dissected spiral ganglia, contain BDNF and NT-3 (Hansen et al. 2001a, b; Zha et al. 2001). Survival of SGNs in cultures containing only cells from spiral ganglia (neurons and glia) is reduced when NT-3 or BDNF signaling is blocked (Hansen et al. 2001a, b). Last but not least, expression of NT-3 is not restricted to the inner hair cells, even in the mature cochlea, but is also present in supporting cells (inner pillar cell, inner phalangeal cell) (Sugawara et al. 2007), and long-term survival of SGNs after hair cell loss can be correlated with persistence of these supporting cells (Sugawara et al. 2005). Supporting cells in the mature cochlea play an active role in providing trophic support to SGN through the neuregulin-ErbB signaling pathway and genetic ablation of supporting cells therefore compromises SGN survival (Stankovic et al. 2004).

6. Excitotoxicity

Excitotoxicity is a neuronal trauma that can cause cell death or degeneration and may involve molecular mechanisms other than those participating in apoptosis due to loss of neurotrophic support (Mattson 2003). Excitotoxic death is principally due to an excessively high level of cytosolic Ca^{2+}, generally the result of neural exposure to unusually high levels of excitatory neurotransmitter for prolonged periods (Mattson 2003). The principal excitatory neurotransmitter in the CNS (and in the inner ear) being glutamate, Ca^{2+} enters via Ca^{2+}-permeable glutamate receptors—NMDA-type and some AMPA-type—and via voltage-gated Ca^{2+} channels that open as a result of the depolarization.

6.1 Excitotoxicity in the Cochlea

Excitotoxicity is seen in the cochlea mainly as a rapid destruction of the peripheral endings of type I SGNs after noise exposure. This is a result of the increased glutamate release from the inner hair cells, and these terminals can be effectively protected by intracochlear infusion of glutamate receptor antagonists (Puel et al. 1998). The SGNs themselves typically survive and the terminals regenerate. The ability of the SGNs to survive this insult is likely due to the fact that just a single postsynaptic site, well removed from the soma, is exposed to the glutamate and so that there may be only little Ca^{2+} entry in the soma. Consequently, it has been assumed that there is no long-term deleterious effect of noise-induced excitotoxicity on the SGNs themselves. However, recent work (Kujawa and Liebermann 2006) has shown that noise exposure in young mice causes an accelerated age-related hearing loss. This hearing loss is particularly unusual in that it is associated with a primary degeneration of the neurons rather than a primary loss of hair cells followed by secondary loss of neurons. Thus, while the neurons appear to recover from the noise, they may be compromised in a way that increases the probability of degeneration in the older mouse. Accelerated age-related hearing loss in individuals exposed to loud noise when young has also been observed in human epidemiological surveys (Gates 2006).

Excitotoxicity may be relevant to another inner ear pathology: Ménière's disease involves the periodic exposure of the hair cells and spiral ganglion neurons to elevated extracellular K^+ ([K^+]) because rupture of the membranous labyrinth allows high K^+ endolymph to contaminate the perilymph (Schuknecht 1993). High [K^+] is directly toxic to hair cells (Zenner 1986) and to spiral ganglion neurons (Hegarty et al. 1997). In the latter case, the toxicity is correlated with elevated cytosolic Ca^{2+} entering through voltage-gated Ca^{2+} channels (Hegarty et al. 1997), suggesting that this SGN death is excitotoxic. In extreme cases of Ménière's disease, loss of hair cells or spiral ganglion neurons or both occurs, exacerbating the hearing loss (Schuknecht 1993).

6.2 Intracellular Mediators of Excitotoxicity

Excitotoxic cell death has some of the hallmarks of apoptosis, including proteolytic cleavage of key cellular proteins, but *calpains*—Ca^{2+}-activated proteases—rather than caspases, appear to be the crucial proteases in excitotoxic death (Lankiewicz et al. 2000; Stefanis 2005) (although see Adamec et al. 1998). In agreement with this notion, genetic inhibition of caspases in transgenic mice in vivo by expression of the caspase inhibitor protein p35 does not prevent excitotoxic neuronal death (Higuchi et al. 2005). Analogous in vivo inhibition of calpains in transgenic mice by expression of the inhibitor protein calpastatin does inhibit excitotoxic neuronal death (Higuchi et al. 2005). In addition to calpain, the Ca^{2+}-activated phosphatase *calcineurin* has been implicated in cell death acting by dephosphorylating and activating the proapoptotic effector Bad (Wang et al. 1999).

It should be noted that glutamate receptors and Ca^{2+} modulate other intracellular signals and some of these, notably nitric oxide (NO), appear to contribute to excitotoxicity (Araujo and Carvalho 2005). Also, NF-κB deficiency increases the susceptibility of SGNs to excitotoxicity and accelerates age-related hearing loss due to SGN death (Lang et al. 2006).

6.3 LOC Efferents and Protection from Excitotoxicity

LOC efferents originate in the lateral superior olive and terminate on peripheral processes of SGNs. LOC efferents suppress auditory nerve activity in noisy conditions, presumably by inhibition at the postsynaptic bouton (Zheng et al. 1999; Groff and Liberman 2003). As discussed for the medial olivocochlear system in Section 4.1.1.6, this suppression of activity could also have a protective role; in this case, protection of SGNs. Three lines of evidence support such a role for the LOC. First, animals in which the entire olivocochlear projection is cut are much more vulnerable to acoustic injury than are animals in which only the crossed MOC projections were cut (Kujawa and Liberman 1997). Second, perfusing the dopamine (one of the neurotransmitters used by the LOC) receptor agonist piribedil into the cochlea protects against excitotoxic damage caused by noise or ischemia (d'Aldin et al. 1995). Third, dopamine in the LOC may be upregulated during sound conditioning (Niu and Canlon 2002). This would increase the efficacy of this system and contribute to the increased protection of preconditioning (Section 4.2).

7. Therapeutic Interventions to Support SGN Survival After Loss of Hair Cells

Interest in SGN survival mechanisms derives in large part from the need to prevent SGN death in cochlear implant users. Death of all SGNs in the absence of hair cells would render the implant ineffective and make it impossible in the future to restore hearing via hair cell regeneration. Even if SGNs survive in the absence of hair cells, they lose their peripheral processes (Leake and Hradek 1988; Fayad et al. 1991; Nadol 1997). This presumably raises the threshold required to stimulate the neuron electrically, limiting the effectiveness and precision of stimulation by a cochlear implant. Strategies to support SGN survival in vivo in the absence of hair cells or to maintain or regrow the peripheral processes, are based on an understanding of SGN death and trophic support of SGNs in the normal cochlea. These are summarized in the text that follows and have also been reviewed elsewhere (Roehm and Hansen 2005).

While it might be expected that SGN survival plays a significant role in determining the efficacy of cochlear implants, the somewhat counterintuitive observation is that this does not appear to be the case. Rather, speech perception by cochlear implant users is not positively correlated with SGN number, provided that at least $\approx 10\%$ of the neurons are present (e.g., Nadol and Eddington 2006). However, this may simply indicate that central rather than peripheral processing is the limiting factor for *current* technology: even a small number of SGNs is sufficient to convey to the CNS the limited amount of information provided by current cochlear implants and coding strategies. As the technology of cochlear implants improves, poor SGN survival or lack of the peripheral process may well become a limiting factor in their efficacy.

7.1 Protection of SGNs by Neurotrophic Factors In Vivo

Delivery of neurotrophic factors to rescue SGNs in vivo is a subject that has also been recently reviewed (Bianchi and Raz 2004; Roehm and Hansen 2005). BDNF, NT-3, and GDNF each will enhance SGN survival after hair cell loss. Two strategies have generally been used for delivery. In one, the factors are directly infused into the scala tympani via a microcannulation–osmotic pump system (Ernfors et al. 1996; Staecker et al. 1996; Miller et al. 1997; Keithley et al. 1998; Ylikoski et al. 1998). The second is a virally mediated gene therapy strategy in which a genetically modified virus is injected into the scala tympani, infecting cells lining the membranous labyrinth and causing them to produce and secrete neurotrophic factors (Staecker et al. 1998; Yagi et al. 2001; Kanzaki et al. 2002; Lalwani et al. 2002). These data support the efficacy of promoting SGN survival in vivo in the absence of hair cells by supplying peptide neurotrophic factors to which the neurons respond because they are likely to be exposed to them in vivo in the normal cochlea.

As noted in the preceding text, in vitro studies (Hartnick et al. 1996; Hegarty et al. 1997; Marzella et al. 1997, 1998) have shown that combining peptide neurotrophic factors results in an approximate additive increase in SGN survival. Similar results have been obtained in vivo: BDNF combined with FGF-1 was shown to be more effective than either factor alone (Altschuler et al. 1999). Increased SGN survival was seen even if administration started after SGN degeneration had begun (Yamagata et al. 2004). Moreover, the enhanced SGN survival, induced by BDNF with CNTF, was associated with improved electrical excitability of the auditory nerve (Shinohara et al. 2002).

For translation and human application, there are a number of outstanding issues that require resolution, having to do with technical aspects of drug delivery and side effects of neurotrophic factors. For gene therapy, the primary issues relate to safety and efficacy of the vector as well as safety of the drug. Current data suggest that neurotrophic factor therapy, once initiated, must be maintained continuously. After cessation of BDNF infusion, SGNs rapidly degenerated so that, within 2 weeks after cessation, SGN density had fallen to the same level as without intervention (Gillespie et al. 2003). While direct drug infusion may be possible in conjunction with cochlear implantation, long-term presence of a cannula in the cochlea poses a significant risk of infection and is unlikely to be practical. Rather, gene therapy approaches based on those used in animal studies appear more promising.

7.2 SGN Survival in Response to Electrical Stimulation In Vivo

SGN death after loss of hair cells is significantly reduced if SGNs are stimulated by an implanted electrode (reviewed in Miller 2001). These results, obtained from studies of cats and guinea pigs, imply that deaf humans with cochlear implants should also experience reduced loss of SGNs.

SGN rescue may depend on place and intensity of the electrical stimulation (ES; Leake et al. 1991). The effect was greatest in the vicinity of the stimulating electrode. In kitten, lower intensity of ES (near-threshold for excitation) appears more effective than higher levels (Leake et al. 1995), although threshold sensitivity may have been changing throughout the course of the stimulation in these studies (Miller 2001). In the mature guinea pig, higher levels of ES appear most effective (Miller and Altschuler 1995; Mitchell et al. 1997). When stimulation was delayed for 2 or 4 weeks after deafening, in the guinea pig, SGN protection was reduced and the threshold for ES-induced protections was elevated (Miller and Altschuler 1995; Mitchell et al. 1997). The reduction in protection may be a consequence of a smaller surviving population of SGNs through the delay in initiating chronic ES.

Some studies of electrical stimulation in vivo did not find increased SGN survival after hair cell destruction in kitten (Araki et al. 1998, 2000) and in mature guinea pig (Li et al. 1999). Recently Shepherd and colleagues (Shepherd et al. 2005), while corroborating their earlier finding of no ES-induced SGN rescue, did report an ES-induced enhancement in SGN size, consistent with earlier findings of Leake and colleagues (Leake et al. 1999). Differences in experimental design and other technical factors might also account for this discrepancy (Miller 2001). An important difference may be the density of surviving SGNs at the time when ES is initiated; that is, higher SGN density is synergistic with ES in promotion of SGN survival, and a "critical mass" of cells is necessary for ES to be successful. Deafening procedures that cause a rapid initial loss of SGNs, e.g., a single dose of aminoglycoside in combination with a loop diuretic, result in a lower density of SGNs at the time when ES is initiated. In contrast, repeated daily administrations of aminoglycoside alone presumably do not cause an immediate loss of all hair cells. The resulting death of SGNs would then be more gradual, even in the initial period, and SGN density higher at the initiation of ES.

Although there is no direct evidence to support the hypothesis that increased cell density in the spiral ganglion synergizes with ES to increase SGN survival, it is consistent with the presence of neurotrophin expression in the spiral ganglion in vivo (Zha et al. 2001) and autocrine and/or paracrine neurotrophic support of SGNs in vitro (Hansen et al. 2001a). Glia may also contribute to paracrine neurotrophic support of SGNs: as noted in the preceding text, spiral ganglion glia produce neurotrophic factors (Hansen et al. 2001a).

7.3 Combining Electrical Stimulation and Peptide Neurotrophic Factors Enhances SGN Survival over Either Alone

Given that neurotrophins and depolarization are additive in promoting survival of cultured SGNs (Hegarty et al. 1997; Hansen et al. 2001b), therapy involving a combination of these factors should be more effective than either alone in enhancing SGN survival after hair cell loss in vivo. Indeed, ES potentiated BDNF-dependent survival of deafferented SGNs in vivo in cats even when ES alone had no effect on SGN survival (Shepherd et al. 2005). Also, combination of ES with GDNF, the latter delivered either by a viral vector (Kanzaki et al. 2002) or by intracochlear perfusion, resulted in enhancement of SGN survival in vivo over GDNF alone. An important remaining question is whether ES alone will maintain SGN survival following withdrawal of neurotrophic factors. If so, this will permit maintenance of SGNs after hair cell loss without a need for continued long-term application of neurotrophic factors.

7.4 Protection of SGNs In Vivo Using Small-Molecule Therapeutics

To date little in vivo work has been done testing the protective capacity of small molecules in the cochlea. However, on the basis of work in other systems and the initial work in the auditory system, at least three strategies—enhancement of neurotrophic signaling, inhibition of apoptosis, and antioxidants—hold promise, along with many challenges.

7.4.1 Enhancement of Neurotrophic Signaling with Small Molecules

Neurotrophin signaling is enhanced by gangliosides such as GM1 (Ferrari et al. 1995). Treatment with GM1 moderately reduces SGN death in vivo after hair cell loss, although severe shrinkage of the surviving SGNs was noted after termination of the GM1 treatment (Leake et al. 2007). However, a promising result of this study is that the trophic effects of GM1 were additive with those of ES and were maintained by continued ES after termination of the GM1.

7.4.2 Inhibition of Apoptosis by Small Molecules

Because apoptosis depends on caspases, caspase inhibition is a potential approach for therapy. The endogenous caspase inhibitors, IAPs, are thought to be particularly promising therapeutic targets and have been investigated in this regard in CNS trauma and neurodegenerative disease (Robertson et al. 2000). Targeting of caspases via IAPs has not yet been used as a protective mechanism for spiral ganglion neurons but inhibition of caspases by this or by the use of cell membrane–permeable inhibitors may be of therapeutic value in SGN protection (Liu et al. 1998). While more associated with excitotoxic cell death, calpains may play a role in the death of SGNs caused by other traumata including loss of neurotrophic support, suggesting that calpain inhibition may have a therapeutic role in SGN protection (Ding et al. 2002).

Inhibitors of JNK or its upstream activators appear promising as therapeutics for neurodegeneration (Bogoyevitch et al. 2004), and JNK inhibitors have been used to protect SGNs (Section 5.4). Since neurotrophic factor deprivation leads to a change in the oxidative state of deafferented neurons, antioxidant treatment may provide another approach to protect SGN from degeneration after hair cell loss.

7.4.3 Antioxidants

Antioxidants (Trolox and acorbic acid) administered either locally (intrascalar) or systemically after ototoxic deafening may reduce degeneration of SGNs and maintain the sensitivity of the auditory nerve to electrical excitation (Miller et al. 2002). Moreover, this efficacy was observed to continue for at least 2 weeks after cessation of systemic administration.

7.5 Regrowth of the Peripheral Process

Studies of humans and experimental animal models have shown that the SGN peripheral process that normally projects to the organ of Corti degenerates quickly after loss of hair cells and is absent even in surviving SGNs (Leake and Hradek 1988; Fayad et al. 1991; Nadol 1997). As a consequence, stimulation by a cochlear implant requires superthreshold depolarization at the SGN cell body or axon within the modiolus, necessitating high stimulating currents that broadly stimulate SGNs. To achieve maximal benefit, therapy to promote SGN survival should also promote regrowth of the peripheral process to allow more focal stimulation of SGNs near an electrode. Neurotrophins promote neurite growth from SGNs in vitro (Lefebvre et al. 1994; Hegarty et al. 1997), and some studies suggest that neurotrophic factor therapy can promote in vivo growth of the SGN peripheral process in guinea pigs: neurotrophins or GDNF induce a regrowth of peripheral processes lost within days following hair cell death (Altschuler et al. 1999; Wise et al. 2005), with BDNF-induced regrowth of peripheral processes enhanced by FGF or electrical stimulation (Altschuler et al. 1999).

Depolarization reduces axon growth in cultured SGNs (Hegarty et al. 1997), suggesting that electrical stimulation might have adverse effects on peripheral process growth, maintenance, of physiology in vivo. The particular depolarization-initiated intracellular signals responsible for inhibiting SGN neurite growth are being identified (e.g., Hansen et al. 2003) and may be able to be specifically inhibited as part of therapy. However, patterned electrical activity, as opposed to chronic depolarization, appears to promote rather than inhibit neurite growth in other neurons (Goldberg et al. 2002). This may account for recent observations that have demonstrated positive effects of chronic ES on SGN peripheral processes (Altschuler et al. 1999; Altschuler et al., unpublished observations): ES shortly following hair cell death, and before complete peripheral process degeneration is capable of maintaining these processes; ES initiated weeks after hair cell death, after complete process degeneration, is able to induce a regrowth of processes through the habenula perforata and into the scarred remnants of organ of Corti.

8. Cochlear Blood Flow and Protection

Inner ear blood flow is reduced by intense noise as shown by laser-Doppler measurements (Thorne and Nuttall 1987). This may be a consequence of a noise-induced increase in levels of the vasoconstrictor 8-isoprostane , a lipid proxidation product, in the cochlea (Ohinata et al. 2000a). Direct administration of 8-isoprostane to the anterior inferior cerebellar artery (main blood supply to the inner ear) resulted in a reduction in blood flow. This was blocked by a specific 8-isoprostane antagonist, SQ29548, which also blocked the noise-induced reduction in CBF (Miller et al. 2003). ROS are presumably involved because glutathione-mono-ethyl ester, a scavenger of oxidative free radicals (Miller et al. 2003) also blocked the noise-induced reduction in CBF. Antioxidants reduce 8-isoprostane formation in the lateral wall and provide protection from noise-induced hearing loss (Ohinata et al. 2003). These data implicate ROS and one of its products, the vasoconstrictor 8-isoprostane , in noise-induced reduction in CBF.

Magnesium, which increases cochlear blood flow, reduces NIHL in guinea pigs (Scheibe et al. 2002) and humans (Attias et al. 2004). This not only indicates that noise-induced reduction in CBF can exacerbate NIHL but also suggests that interventions including vasodilators can be protective against acoustic trauma.

Combination of antioxidants (vitamins A, C, E) with the vasodilator magnesium provides greater attenuation of NIHL than either alone (Le Prell et al. 2007).

9. Protection of Other Cochlear Elements

Damage from noise and ototoxins is not limited to the organ of Corti or spiral ganglion. Other cochlear elements, including the stria vascularis and fibrocytes of the lateral wall, are also affected (Hirose and Liberman 2003; Imamura and Adams 2003; see also Henderson and Hu, Chapter 7; Rybak, Talaska, and Schacht, Chapter 8). These latter effects can contribute directly to hearing loss, for example, by altering the endocochlear potential (Hirose and Liberman 2003), or can have indirect effects on hair cells and SGNs by affecting cochlear homeostasis. Therefore, protection of lateral wall structures might reduce hearing loss caused by noise or other stresses.

10. Summary

The increasing understanding of the basic biology of cell death, protection, and neurotrophic mechanisms has suggested means to reduce the effects of trauma to the inner ear. Generally, traumata that cause permanent hearing loss—e.g., noise, ototoxins, aging—directly affect the hair cells, with spiral ganglion neuron (SGN) death being secondary to the loss of hair cells. However, SGNs are directly susceptible to excitotoxic damage. Known intrinsic protective systems in hair cells include heat shock proteins, antioxidants, and Ca^{2+} homeostasis; known exogenous systems include the olivocochlear efferents. If these protective systems are overwhelmed by the trauma, hair cells die, typically via known apoptotic pathways but not necessarily in the same fashion in all types of stress. Current strategies for protection of hair cells target enhancing intrinsic protective mechanisms—e.g., treatment with antioxidants—or blocking apoptotic pathways—e.g., inhibition of JNK. In the case of SGNs, protective strategies are based on current understanding of the neurotrophic support SGNs receive from hair cells, the prosurvival intracellular signaling pathways that the neurotrophic stimuli activate and the proapoptotic pathways that they inhibit. Current strategies for protecting SGNs involve enhancing neurotrophic support, e.g., intracochlear application of peptide neurotrophic factors or electrical stimulation by an implanted electrode. In vitro and in vivo studies emphasize the value of combinatorial approaches. One challenge is the identification of the most effective approach(es). A possibly more daunting challenge is the identification of effective means to deliver protective molecules to their intended cellular or subcellular targets, particularly means that will allow translation from animal models to clinical application.

Acknowledgments. Richard A. Altschuler and Josef M. Miller would like to acknowledge the contributions of Drs. Margaret Lomax, Amy Miller, Annieliese Shrott Fischer, and Mats Ulfendahl and SHG, and the contributions of all the members of the Green lab and the University of Iowa Auditory Neuroscience Group, especially Dr. Marlan Hansen. We thank Drs. Richard Bobbin and, especially, Jochen Schacht for their considerable and helpful contributions to the text. Studies of Drs. Miller and Altschuler were supported by NIH/NIDCD grants R01 DC003820 and P30 DC005188. Studies in the Green lab were supported by NIH/NIDCD grant R01 DC002961 and by the American Hearing Research Foundation.

References

Adamec E, Beermann ML, Nixon RA (1998) Calpain I activation in rat hippocampal neurons in culture is NMDA receptor selective and not essential for excitotoxic cell death. Brain Res Mol Brain Res 54:35–48.

Adamson CL, Reid MA, Davis RL (2002) Opposite actions of brain-derived neurotrophic factor and neurotrophin-3 on firing features and ion channel composition of murine spiral ganglion neurons. J Neurosci 22:1385–1396.

Affaitati A,Cardone L, de Cristofaro T, Carlucci A, Ginsberg MD, Varrone S, Gottesman ME, Avvedimento EV, Feliciello A (2003) Essential role of A-kinase anchor protein 121 for cAMP signaling to mitochondria. J Biol Chem 278:4286–4294.

Alam S, Robinson BK, Huang J, Green SH (2007) Prosurvival and proapoptotic intracellular signaling in rat spiral ganglion neurons in vivo after the loss of hair cells. J Comp Neurol 503:832–852.

Altschuler RA, Lim HH, Ditto J, Dolan D, Raphael Y (1996) Protective mechanisms in the cochlea: heat shock proteins. In:Salvi RJ, Henderson D, Fiorino F, Colletti V (eds) Auditory Plasticity and Regeneration. New York: Tieman Med Publishers, pp. 202–212.

Altschuler RA, Cho Y, Ylikoski J, Pirvola U, Magal E, Miller JM (1999) Rescue and regrowth of sensory nerves following deafferentation by neurotrophic factors. Ann NY Acad Sci 884:305–311.

Araki S, Kawano A, Seldon L, Shepherd RK, Funasaka S, Clark GM (1998) Effects of chronic electrical stimulation on spiral ganglion neuron survival and size in deafened kittens. Laryngoscope 108:687–695.

Araki S, Kawano A, Seldon HL, Shepherd RK, Funasaka S, Clark GM (2000) Effects of intracochlear factors on spiral ganglion cells and auditory brain stem response after long-term electrical stimulation in deafened kittens. Otolaryngol Head Neck Surg 122:425–433.

Araujo IM, Carvalho CM (2005) Role of nitric oxide and calpain activation in neuronal death and survival. Curr Drug Targets CNS Neurol Disord 4:319–324.

Attias J, Sapir S, Bresloff I, Reshef-Haran I, Ising H (2004) Reduction in noise-induced temporary threshold shift in humans following oral magnesium intake. Clin Otolaryngol Allied Sci 29:635–641.

Baloh RH, Enomoto H, Johnson EM, Jr., Milbrandt J (2000) The GDNF family ligands and receptors—implications for neural development. Curr Opin Neurobiol 10:103–110.

Barthwal MK, Sathyanarayana P, Kundu CN, Rana B, Pradeep A, Sharma C, Woodgett JR, Rana A (2003) Negative regulation of mixed lineage kinase 3 by protein kinase B/AKT leads to cell survival. J Biol Chem 278:3897–3902.

Berridge MJ (1998) Neuronal calcium signaling. Neuron 21:13–26.

Bhakar AL, Tannis LL, Zeindler C, Russo MP, Jobin C, Park DS, MacPherson S, Barker PA (2002) Constitutive NF-κB activity is required for central neuron survival. J Neurosci 22:8466–8475.

Bianchi LM, Raz Y (2004) Methods for providing therapeutic agents to treat damaged spiral ganglion neurons. Curr Drug Targets CNS Neurol Disord 3:195–199.

Bichler E, Spoendlin H, Rauchegger H (1983) Degeneration of cochlear neurons after amikacin intoxication in the rat. Arch Otorhinolaryngol 237:201–208.

Biswas SC, Liu DX, Greene LA (2005) Bim is a direct target of a neuronal E2F-dependent apoptotic pathway. J Neurosci 25:8349–8358.

Blondeau N, Widmann C, Lazdunski M, Heurteaux C (2001) Activation of the nuclear factor-κB is a key event in brain tolerance. J Neurosci 21:4668–4677.

Bobbin RP (2001) PPADS, an ATP antagonist, attenuates the effects of a moderately intense sound on cochlear mechanics. Hear Res 156:10–16.

Bobbin RP (2002) Caffeine and ryanodine demonstrate a role for the ryanodine receptor in the organ of Corti. Hear Res 174:172–182.

Bobbin RP, Blesoe SJ (2005) Asphyxia and depolarization increase adenosine levels in perilymph. Hear Res 205:110–114.

Bodmer D, Gloddek B, Ryan AF, Huverstuhl J, Brors D (2002) Inhibition of the c-Jun N-terminal kinase signaling pathway influences neurite outgrowth of spiral ganglion neurons *in vitro*. Laryngoscope 112:2057–2061.

Bogoyevitch MA, Boehm I, Oakley A, Ketterman AJ, Barr RK (2004) Targeting the JNK MAPK cascade for inhibition: basic science and therapeutic potential. Biochim Biophys Acta 1697:89–101.

Bohne BA (1976) Safe level for noise exposure? Ann Otol Rhinol Laryngol 85:711–724.

Bok J, Huang J, Wang Q, Green SH (2007) CaMKII and CaMKIV mediate divergent prosurvival signaling pathways in response to depolarization in neurons. Mol Cell Neurosci 36:13–26.

Bok J, Zha XM, Cho YS, Green SH (2003) An extranuclear locus of cAMP-dependent protein kinase action is necessary and sufficient for promotion of spiral ganglion neuronal survival by cAMP. J Neurosci 23:777–787.

Brosenitsch TA, Katz DM (2001) Physiological patterns of electrical stimulation can induce neuronal gene expression by activating N-type calcium channels. J Neurosci 21:2571–2579.

Brown MC, deV Venecia RK, Guinan JJ (2003) Responses of medial olivocochlear neurons. Specifying the central pathways of the medial olivocochlear reflex. Exp Brain Res 153:491–498.

Brunet A, Bonni A, Zigmond MJ, Lin MZ, Juo P, Hu LS, Anderson MJ, Arden KC, Blenis J, Greenberg ME (1999) Akt promotes cell survival by phosphorylating and inhibiting a forkhead transcription factor. Cell 96:857–868.

Chen C, Nenov A, Bobbin RP (1995) Noise exposure induced change in outer hair cell response to ATP. Hear Res 88:215–221.

Cheng AG, Cunningham LL, Rubel EW (2005) Mechanisms of hair cell death and protection. Curr Opin Otolaryngol Head Neck Surg 13:343–348.

Culmsee C, Siewe J, Junker V, Retiounskaia M, Schwarz S, Camandola S, El-Metainy S, Behnke H, Mattson MP, Krieglstein J (2003) Reciprocal inhibition of p53 and nuclear factor-kB transcriptional activities determines cell survival or death in neurons. J Neurosci 23:8586–8595.

d'Aldin C, Puel JL, Leducq R, Crambes O, Eybalin M, Pujol R (1995) Effects of a dopaminergic agonist in the guinea pig cochlea. Hear Res 90:202–211.

Dallos P, He DZ, Lin X, Sziklai I, Mehta S, Evans BN (1997) Acetylcholine, outer hair cell electromotility, and the cochlear amplifier. J Neurosci 17:2212–2226.

Datta SR, Brunet A, Greenberg ME (1999) Cellular survival: a play in three Akts. Genes Dev 13:2905–2927.

Davis RJ (2000) Signal transduction by the JNK group of MAP kinases. Cell 103:239–252.

Dawson TM, Ginty DD (2002) CREB family transcription factors inhibit neuronal suicide. Nat Med 8:450–451.

De Cesare D, Sassone-Corsi P (2000) Transcriptional regulation by cyclic AMP-responsive factors. Prog Nucleic Acid Res Mol Biol 64:343–369.

Deshmukh M, Johnson EM Jr (1998) Evidence of a novel event during neuronal death: development of competence-to-die in response to cytoplasmic cytochrome C. Neuron 21:695–705.

Ding L, McFadden SL, Salvi RJ (2002) Calpain immunoreactivity and morphological damage in chinchilla inner ears after carboplatin. J Assoc Res Otolaryngol 3:68–79.

Donovan N, Becker EB, Konishi Y, Bonni A (2002) JNK phosphorylation and activation of BAD couples the stress-activated signaling pathway to the cell death machinery. J Biol Chem 277:40944–40949.

Downward J (1999) How BAD phosphorylation is good for survival. Nat Cell Biol 1:E33–35.

Eilers A, Whitfield J, Shah B, Spadoni C, Desmond H, Ham J (2001) Direct inhibition of c-Jun N-terminal kinase in sympathetic neurons prevents c-jun promoter activation and NGF withdrawal-induced death. J Neurochem 76:1439–1454.

Encinas M, Tansey MG, Tsui-Pierchala BA, Comella JX, Milbrandt J, Johnson EM (2001) c-Src is required for glial cell line-derived neurotrophic factor (GDNF) family ligand-mediated neuronal survival via a phosphatidylinositol-3 kinase (PI-3K)-dependent pathway. J Neurosci 21:1464–1472.

Ernfors P, Duan ML, ElShamy WM, Canlon B (1996) Protection of auditory neurons from aminoglycoside toxicity by neurotrophin-3. Nature Med 2:463–467.

Fairfield DA, Kanicki AC, Lomax MI, Altchuler RA (2002) Expression and localization of heat shock factor (HSF) 1 in the rodent cochlea. Hear Res 173:109–118.

Fairfield DA, Kanicki AC, Lomax MI, Altchuler RA (2004) Induction of heat shock protein 32 (Hsp32) in the rat cochlea following hyperthermia. Hear Res 188:1–11.

Fairfield DA, Lomax MI, Dootz GA, Chen S, Galecki TA, Benjamin IJ, Dolan DF, Altchuler RA (2005) Heat shock factor 1 (Hsf1) deficient mice exhibit decreased recovery of hearing following noise overstimulation. J Neurosci Res 81:589–596.

Fayad J, Linthicum FH, Jr., Otto SR, Galey FR, House WF (1991) Cochlear implants: histopathologic findings related to performance in 16 human temporal bones. Ann Otol Rhinol Laryngol 100:807–811.

Ferrari G, Anderson BL, Stephens RM, Kaplan DR, Green LA (1995) Prevention of apoptotic neuronal death by GM1 ganglioside. Involvement of Trk neurotrophin receptors. J Biol Chem 270:3074–3080.

Fesik SW, Shi Y (2001) Controlling the caspases. Science 294:1477–1478.

Ford MS, Maggirwar SB, Rybak LP, Whitworth C, Ramkumar V (1997) Expression and function of adenosine receptors in the chinchilla cochlea. Hear Res 105:130–140.

Forge A, Schacht J (2000) Aminoglycoside antibiotics. Audiol Neurootol 5:3–22.

Franklin JL, Johnson EM Jr (1994) Block of neuronal apoptosis by a sustained increase of steady-state free Ca^{2+} concentration. Philos Trans R Soc Lond B Biol Sci 345:251–256.

Franklin JL, Sanz-Rodriguez C, Juhasz A, Deckwerth TL, Johnson EM Jr (1995) Chronic depolarization prevents programmed death of sympathetic neurons *in vitro* but does not support growth: requirement for Ca^{2+} influx but not Trk activation. J Neurosci 15:643–664.

Fridberger A, Flóck A, Ulfendahl M, Flóck B (1998) Acoustic overstimulation increases outer hair cell Ca^{2+} concentrations and causes dynamic contractions of the hearing organ. Proc Natl Acad Sci USA 95:7127–7132.

Fridmacher V, Kaltschmidt B, Goudeau B, Ndiaye D, Rossi FM, Pfeiffer J, Kaltschmidt C, Israel A, Memet S (2003) Forebrain-specific neuronal inhibition of nuclear factor-κB activity leads to loss of neuroprotection. J Neurosci 23:9403–9408.

Fritzsch B, Tessarollo L, Coppola E, Reichardt LF (2004) Neurotrophins in the ear: their roles in sensory neuron survival and fiber guidance. Prog Brain Res 146:265–278.

Fritzsch B, Pauley S, Matei V, Katz DM, Xiang M, Tessarollo L (2005) Mutant mice reveal the molecular and cellular basis for specific sensory connections to inner ear epithelia and primary nuclei of the brain. Hear Res 206:52–63.

Galli C, Meucci O, Scorziello A, Werge TM, Calissano P, Schettini G (1995) Apoptosis in cerebellar granule cells is blocked by high KCl, forskolin, and IGF-1 through distinct mechanisms of action: the involvement of intracellular calcium and RNA synthesis. J Neurosci 15:1172–1179.

Garnier P, Ying W, Swanson RA (2003) Ischemic preconditioning by caspase cleavage of poly(ADP-ribose) polymerase-1. J Neurosci 23:7967–7973.

Gates GA (2006) The effect of noise on cochlear aging. Ear Hear 27:91.

Gentry JJ, Barker PA, Carter BD (2004) The p75 neurotrophin receptor: multiple inter-actors and numerous functions. Prog Brain Res 146:25–39.

Gestwa G, Wiechers B, Zimmermann U, Praetorius M, Rohbock K, Kopschall I, Zenner HP, Knipper M (1999) Differential expression of trkB.T1 and trkB.T2, truncated trkC, and p75[NGFR] in the cochlea prior to hearing function. J Comp Neurol 414:33–49.

Ghahremani MH, Keramaris E, Shree T, Xia Z, Davis RJ, Flavell R, Slack RS, Park DS (2002) Interaction of the c-Jun/JNK pathway and cyclin-dependent kinases in death of embryonic cortical neurons evoked by DNA damage. J Biol Chem 277:35586–35596.

Ghoumari AM, Wehrle R, De Zeeuw CI, Sotelo C, Dusart I (2002) Inhibition of protein kinase C prevents Purkinje cell death but does not affect axonal regeneration. J Neurosci 22:3531–3542.

Gidday JM (2006) Cerebral preconditioning and ischaemic tolerance. Nat Rev Neurosci 7:437–448.

Gillespie LN, Clark GM, Bartlett PF, Marzella PL (2003) BDNF-induced survival of auditory neurons in vivo: cessation of treatment leads to accelerated loss of survival effects. J Neurosci Res 71:785–790.

Goldberg JL, Espinosa JS, Xu Y, Davidson N, Kovacs GT, Barres BA (2002) Retinal ganglion cells do not extend axons by default. Promotion by neurotrophic signaling and electrical activity. Neuron 33:689–702.

Green SH (2000) Neurotrophic signaling by membrane electrical activity in spiral ganglion neurons. In: Lim DJ (ed) Cell and Molecular Biology of the Ear. New York: Kluwer Academic/Plenum, pp. 165–182.

Greenlund LJS, Deckwerth TL, Johnson EM Jr (1995) Superoxide dismutase delays neuronal apoptosis: a role for reactive oxygen species in programmed neuronal death. Neuron 14:303–315.

Groff JA, Liberman MC (2003) Modulation of cochlear afferent response by the lateral olivocochlear system: activation via electrical stimulation of the inferior colliculus. J Neurophysiol 90:3178–3200.

Ham J, Eilers A, Whitfield J, Neame SJ, Shah B (2000) c-Jun and the transcriptional control of neuronal apoptosis. Biochem Pharmacol 60:1015–1021.

Hamernik RP, Henderson D (1974) Impulse noise trauma. A study of histological suscep-tibility. Arch Otorhinolaryngol 99:118–121.

Hansen MR, Vijapurkar U, Koland JG, Green SH (2001a) Reciprocal signaling between spiral ganglion neurons and Schwann cells involves neuregulin and neurotrophins. Hear Res 161:87–98.

Hansen MR, Zha X-M, Bok J, Green SH (2001b) Multiple distinct signal pathways, including an autocrine neurotrophic mechanism, contribute to the survival-promoting effect of depolarization on spiral ganglion neurons. J Neurosci 21:2256–2267.

Hansen MR, Devaiah AK, Bok J, Zha X, Green SH (2003) Ca^{2+}/calmodulin-dependent protein kinases II and IV function similarly in neurotrophic signaling but differ in their effects on neurite growth in spiral ganglion neurons. J, Neurosci Res 72:169–184.

Harada H, Becknell B, Wilm M, Mann M, Huang LJ, Taylor SS, Scott JD, Korsmeyer SJ (1999) Phosphorylation and inactivation of BAD by mitochondria-anchored protein kinase A. Mol Cell 3:413–422.

Harden TK, Boyer JL, Nicholas RA (1995) P2-purinergic receptors: subtype-associated signaling responses and structure. Annu Rev Pharmacol Toxicol 35:541–579.

Harris KC, Bielefeld E, Hu BH, Henderson D (2006) Increased resistance to free radical damage induced by low-level sound conditioning. Hear Res 213:118–129.

Harris MH, Thompson CB (2000) The role of the Bcl-2 family in the regulation of outer mitochondrial membrane permeability. Cell Death Differ 7:1182–1191.

Hartnick C, Staecker H, Malgrange B, Lefebvre P, Liu W, Moonen G, van de Water TR (1996) Neurotrophic effects of BDNF and CNTF, alone and in combination, on postnatal day 5 rat acoustic ganglion neurons. J Neurobiol 30:246–254.

Hegarty JL, Kay AR, Green SH (1997) Trophic support of cultured spiral ganglion neurons by depolarization exceeds and is additive with that by neurotrophins or cyclic AMP, and requires elevation of $[Ca^{2+}]_i$ within a set range. J Neurosci 17:1959–1970.

Heinrich UR, Maurer J, Mann W (1999) Ultrastructural evidence for protection of the outer hair cells of the inner ear during intense noise exposure by application of the organic calcium channel blocker diltiazem. ORL J Otolaryngolg Relat Spec 61:321–327.

Henderson D, Bielefeld EC, Harris KC, Hu BH (2006) The role of oxidative stress in noise-induced hearing loss. Ear Hear 27:1–19.

Hengartner MO (2000) The biochemistry of apoptosis. Nature 407:770 –776.

Hiebert SW, Packham G, Strom DK, Haffner R, Oren M, Zambetti G, Cleveland JL (1995) E2F-1:DP-1 induces p53 and overrides survival factors to trigger apoptosis. Mol Cell Biol 15:6864–6874.

Higuchi M, Tomioka M, Takano J, Shirotani K, Iwata N, Masumoto H, Maki M, Itohara S, Saido TC (2005) Distinct mechanistic roles of calpain and caspase activation in neurodegeneration as revealed in mice overexpressing their specific inhibitors. J Biol Chem 280:15229–15237.

Hirose K, Liberman MC (2003) Lateral wall histopathology and endocochlear potential in the noise-damaged mouse cochlea. J Assoc Res Otolaryngol 4:339–352.

Housley GD, Thorne PR (2000) Purinergic signalling: an experimental perspective. J Auton Nerv Syst 81:139–145.

Hu BH, Zheng XY, McFadden SL, Kopke RD, Henderson D (1997) R-phenylisopro-pyladenosine attenuates noise-induced hearing loss in the chinchilla. Hear Res 113: 198–206.

Huang DC, Strasser A (2000) BH3-only proteins—essential initiators of apoptotic cell death. Cell 103:839–842.

Huang EJ, Reichardt LF (2001) Neurotrophins: roles in neuronal development and function. Annu Rev Neurosci 24:677–736.

Huang EJ, Reichardt LF (2003) Trk receptors: roles in neuronal signal transduction. Annu Rev Biochem 72:609–642.

Hunter-Duvar IM, Bredberg G (1974) Effects of intense auditory stimulation: hearing losses and inner ear changes in the chinchilla. J Acoust Soc Am 55:795–801.

Imamura S, Adams JC (2003) Changes in cytochemistry of sensory and nonsensory cells in gentamicin-treated cochleas. J Assoc Res Otolaryngol 4:196–218.

Ishii N, Wanaka A, Ohno K, Matsumoto K, Eguchi Y, Mori T, Tsujimoto Y, Tohyama M (1996) Localization of bcl-2, bax, and bcl-x mRNAs in the developing inner ear of the mouse. Brain Res 726:123–128.

Jacono AA, Hu B, Kopke RD, Henderson D, Van de Water TR, Steinman HM (1998) Changes in cochlear antioxidant enzyme activity after sound conditioning and noise exposure in the chinchilla. Hear Res 117:31–38.

Jiang H, Schacht J (2005) NF-kappaB pathway protects cochlear hair cells from aminoglycoside-induced ototoxicity. J Neurosci Res 79:644–651.

Jiang H, Sha S-H, Forge A, Schacht J (2006) Caspase-independent pathways of hair cell death induced by kanamycin in vivo. Cell Death Differ 13:20–30.

Kanzaki S, Stover T, Kawamoto K, Prieskorn DM, Altschuler RA, Miller JM, Raphael Y (2002) Glial cell line-derived neurotrophic factor and chronic electrical stimulation prevent VIII cranial nerve degeneration following denervation. J Comp Neurol 454:350–360.

Kanzaki S, Beyer L, Karolyi IJD, D.F., Fang Q, Probst FJ, Camper SA, Raphael Y (2006) Transgene correction maintains normal cochlear structure and function in 6–month-old Myo15a mutant mice. Hear Res 214:37–44.

Keithley EM, Ma CL, Ryan AF, Louis JC, Magal E (1998) GDNF protects the cochlea against noise damage. NeuroReport 9:2183–2187.

Keithley EM, Canto C, Zheng QY, Wang X, Fischel-Ghodsian N, Johnson KR (2005) Cu/Zn superoxide dismutase and age-related hearing loss. Hear Res 209:76–85.

Kelly TC, Whitworth CA, Husain K, Rybak LP (2003) Aminoguanidine reduces cisplatin ototoxicity. Hear Res 186:10–16.

Koike T, Tanaka S (1991) Evidence that nerve growth factor dependence of sympathetic neurons for survival in vitro may be determined by levels of cytoplasmic free Ca^{2+}. Proc Natl Acad Sci USA 88:3892–3896.

Koitchev K, Guilhaume A, Cazals Y, Aran J-M (1982) Spiral ganglion changes after massive aminoglycoside treatment in the guinea pig. Counts and ultrastructure. Acta Otolaryngol 94:431–438.

Konishi Y, Bonni A (2003) The E2F-Cdc2 cell-cycle pathway specifically mediates activity deprivation-induced apoptosis of postmitotic neurons. J Neurosci 23: 1649–1658.

Konishi Y, Lehtinen M, Donovan N, Bonni A (2002) Cdc2 phosphorylation of BAD links the cell cycle to the cell death machinery. Mol Cell 9:1005–1016.

Kroemer G, Reed JC (2000) Mitochondrial control of cell death. Nat Med 6:513–519.

Kuan CY, Yang DD, Samanta Roy DR, Davis RJ, Rakic P, Flavell RA (1999) The Jnk1 and Jnk2 protein kinases are required for regional specific apoptosis during early brain development. Neuron 22:667–676.

Kujawa SG, Liberman MC (1997) Conditioning-related protection from acoustic injury: effects of chronic de-efferentation and sham surgery. J Neurophysio 78:3095–3106.

Kujawa SG, Liebermann MC (2006) Acceleration of age-related hearing loss by early noise exposure: evidence of a misspent youth. J Neurosci 26:2115–2123.

Lallemend F, Lefebvre PP, Hans G, Rigo JM, Van de Water TR, Moonen G, Malgrange B (2003) Substance P protects spiral ganglion neurons from apoptosis via PKC-Ca^{2+}-MAPK/ERK pathways. J Neurochem 87:508–521.

Lallemend F, Hadjab S, Hans G, Moonen G, Lefebvre PP, Malgrange B (2005) Activation of protein kinase CbI constitutes a new neurotrophic pathway for deafferented spiral ganglion neurons. J Cell Sci 118:4511–4525.

Lalwani AK, Han JJ, Castelein CM, Carvalho GJ, Mhatre AN (2002) In vitro and in vivo assessment of the ability of adeno-associated virus-brain-derived neurotrophic factor to enhance spiral ganglion cell survival following ototoxic insult. Laryngoscope 112:1325–1334.

Lang H, Schulte BA, Schmiedt RA (2005) Ouabain induces apoptotic cell death in type I spiral ganglion neurons, but not type II neurons. J Assoc Res Otolaryngol 6:63–74.

Lang H, Schulte BA, Zhou D, Smythe N, Spicer SS, Schmiedt RA (2006) Nuclear factor kB deficiency is associated with auditory nerve degeneration and increased noise-induced hearing loss. J Neurosci 26:3541–3550.

Lankiewicz S, Marc Luetjens C, Truc Bui N, Krohn AJ, Poppe M, Cole GM, Saido TC, Prehn JH (2000) Activation of calpain I converts excitotoxic neuron death into a caspase-independent cell death. J Biol Chem 275:17064–17071.

Lautermann J, McLaren J, Schacht J (1995) Glutathione protection against gentamicin ototoxicity depends on nutritional status. Hear Res 86:15–24.

Leake PA, Hradek GT (1988) Cochlear pathology of long term neomycin induced deafness in cats. Hear Res 33:11–33.

Leake PA, Hradek GT, Rebscher SJ, Snyder RL (1991) Chronic intracochlear electrical stimulation induces selective survival of spiral ganglion neurons in neonatally deafened cats. Hear Res 54:251–271.

Leake PA, Snyder RL, Hradek GT, Rebscher SJ (1995) Consequences of chronic extra-cochlear electrical stimulation in neonatally deafened cats. Hear Res 82:65–80.

Leake PA, Hradek GT, Snyder RL (1999) Chronic electrical stimulation by a cochlear implant promotes survival of spiral ganglion neurons after neonatal deafness. J Comp Neurol 412:543–562.

Leake PA, Hradek GT, Vollmer M, Rebscher SJ (2007) Neurotrophic effects of GM1 ganglioside and electrical stimulation on cochlear spiral ganglion neurons in cats deafened as neonates. J Comp Neurol 501:837–853.

Lefebvre PP, Malgrange B, Staecker H, Moghadass M, Van De Water TR, Moonen G (1994) Neurotrophins affect survival and neuritogenesis by adult injured auditory neurons in vitro. NeuroReports 5:865–868.

Leonova EV, Fairfield DA, Lomax MI, Altschuler RA (2002) Constitutive expression of HSP 27 in the rat cochlea. Hear Res 163:61–70.

Le Prell CG, Hughes LF, Miller JM (2007) Free radical scavengers vitamins A, C, and E plus magnesium reduce noise trauma. Free Radic Biol Med 42:1454–1463.

Lezoualch F, Sagara Y, Holsboer F, Behl C (1998) High constitutive NF-κB activity mediates resistance to oxidative stress in neuronal cells. J Neurosci 18:3224–3232.

Li L, Parkins CW, Webster DB (1999) Does electrical stimulation of deaf cochleae prevent spiral ganglion degeneration? Hear Res 133:27–39.

Lim DJ, Melnick W (1971) A scanning and transmission electron microscopic observation. Arch Otorhinolaryngol 94:294–305.

Lin A (2003) Activation of the JNK signaling pathway: breaking the brake on apoptosis. Bioessays 25:17–24.

Linseman DA, Phelps RA, Bouchard RJ, Le SS, Laessig TA, McClure ML, Heidenreich KA (2002) Insulin-like growth factor-I blocks bcl-2 interacting mediator of cell death (bim) induction and intrinsic death signaling in cerebellar granule neurons. J Neurosci 22:9287–9297.

Liu DX, Greene LA (2001) Regulation of neuronal survival and death by E2F-dependent gene repression and derepression. Neuron 32:425–438.

Liu W, Staecker H, Stupak H, Malgrange B, Lefebvre P, Van De Water TR (1998) Caspase inhibitors prevent cisplatin-induced apoptosis of auditory sensory cells. NeuroReport 9:2609–2614.

Liu ZG, Hsu H, Goeddel DV, Karin M (1996) Dissection of TNF receptor 1 effector functions: JNK activation is not linked to apoptosis while NF-κB activation prevents cell death. Cell 87:565–576.

Mabuchi T, Kitagawa K, Kuwabara K, Takasawa K, Ohtsuki T, Xia Z, Storm D, Yanagihara T, Hori M, Matsumoto M (2001) Phosphorylation of cAMP response element-binding protein in hippocampal neurons as a protective response after exposure to glutamate in vitro and ischemia in vivo. J Neurosci 21:9204–9213.

Maggirwar SB, Sarmiere PD, Dewhurst S, Freeman RS (1998) Nerve growth factor-dependent activation of NF-kB contributes to survival of sympathetic neurons. J Neurosci 18:10356–10365.

Maison SF, Liberman MC (2000) Predicting vulnerability to acoustic injury with a noninvasive assay of olivocochlear reflex strength. J Neurosci 20:4701–4707.

Maison SF, Luebke AE, Liberman MC, Zuo J (2002) Efferent protection from acoustic injury is mediated via alpha9 nicotinic acetylcholine receptors on outer hair cells. J Neurosci 22:10838–10846.

Martin DP, Schmidt RE, DiStefano PS, Lowry OH, Carter JG, Johnson EM, Jr. (1988) Inhibitors of protein synthesis and RNA synthesis prevent neuronal death caused by nerve growth factor deprivation. J Cell Biol 106:829–844.

Marzella PL, Clark GM, Shepherd RK, Bartlett PF, Kilpatrick TJ (1997) LIF potentiates the NT-3–mediated survival of spiral ganglia neurones *in vitro*. NeuroReport 8: 1641–1644.

Marzella PL, Clark GM, Shepherd RK, Bartlett PF, Kilpatrick TJ (1998) Synergy between TGF-b3 and NT-3 to promote the survival of spiral ganglia neurones *in vitro*. Neurosci Lett 240:77–80.

Mattson MP (2003) Excitotoxic and excitoprotective mechanisms: abundant targets for the prevention and treatment of neurodegenerative disorders. Neuromol Med 3:65–94.

McFadden SL, Ding D, Jiang H, Salvi RJ (2004) Time course of efferent fiber and spiral ganglion cell degeneration following complete hair cell loss in the chinchilla. Brain Res 997:40–51.

McLaughlin B, Hartnett KA, Erhardt JA, Legos JJ, White RF, Barone FC, Aizenman E (2003) Caspase 3 activation is essential for neuroprotection in preconditioning. Proc Natl Acad Sci USA 100:715–720.

Meffert MK, Chang JM, Wiltgen BJ, Fanselow MS, Baltimore D (2003) NF-κB functions in synaptic signaling and behavior. Nat Neurosci 6:1072–1078.

Meller R, Cameron JA, Torrey DJ, Clayton CE, Ordonez AN, Henshall DC, Minami M, Schindler CK, Saugstad JA, Simon RP (20006) Rapid degradation of Bim by the ubiquitin-proteasome pathway mediates short-term ischemic tolerance in cultured neurons. J Biol Chem 281:7429–7436.

Meyer-Franke A, Wilkinson GA, Kruttgen A, Hu M, Munro E, Hanson MG Jr, Reichardt LF, Barres BA (1998) Depolarization and cAMP elevation rapidly recruit TrkB to the plasma membrane of CNS neurons. Neuron 21:681–693.

Miao B, Yin XM, Pei DS, Zhang QG, Zhang GY (2005) Neuroprotective effects of preconditioning ischemia on ishemic brain injury through down-regulating activation of JNK1/2 via N-methyl-D-aspartate receptor-mediated Akt1 activation. J Biol Chem 280:21693–21699.

Midorikawa R, Takei Y, Hirokawa N (2006) KIF4 motor regulates activity-dependent neuronal survival by suppressing PARP-1 enzymatic activity. Cell 125:371–383.

Mikuriya T, Sugahara K, Takemoto T, Tanaka K, Takeno K, Shimogori H, Nakai A, Yamashita H (2005) Geranylgeranylacetone, a heat shock protein inducer, prevents acoustic injury in the guinea pig. Brain Res 1065:107–114.

Miller AL (2001) Effects of chronic stimulation on auditory nerve survival in ototoxically deafened animals. Hear Res 151:1–14.

Miller AL, Prieskorn D, Altschuler RA, Miller JM (2003) Mechanism of electrical stimulation-induced neuroprotection: effects of verapamil on protection of primary auditory afferents. Brain Res 966:218–230.

Miller FD, Pozniak CD, Walsh GS (2000) Neuronal life and death: an essential role for the p53 family. Cell Death Differ 7:880–888.

Miller JM, Altschuler RA (1995) Effectiveness of different electrical stimulation conditions in preservation of spiral ganglion neurons following deafness. Ann Otol Rhinol Laryngol Suppl 166 104:57–60.

Miller JM, Ren TY, Dengerink HA, Nuttall AL (1996) Cochlear blood flow changes with short sound stimulation. In: Axelsson A, Borchgrevink H, Hamernik RP, Hellstrom P, Henderson D, Salvi RJ (eds) Scientific Basis of Noise-Induced Hearing Loss. New York: Thieme Medical Publishers, pp. 95–109.

Miller JM, Chi DH, O'Keeffe LJ, Kruszka P, Raphael Y, Altschuler RA (1997) Neurotrophins can enhance spiral ganglion cell survival after inner hair cell loss. Int J Dev Neurosci 15:631–643.

Miller JM, Miller AL, Yamagata T, Bredberg G, Altschuler RA (2002) Protection and regrowth of the auditory nerve after deafness: neurotrophins, antioxidants and depolarization are effective in vivo. Audiol Neurootol 7:175–179.

Miller JM, Brown JN, Schacht J (2003) 8-Iso-prostaglandin F2a, a product of noise exposure, reduces inner ear flow. Audiol Neurootol 8:207–221.

Minami SB, Yamashita D, Schacht J, Miller JM (2004) Calcineuron activation contributes to noise-induced hearing loss. J Neurosci Res 78:383–392.

Mitchell A, Miller JM, Finger PA, Heller JW, Raphael Y, Altschuler RA (1997) Effects of chronic high-rate electrical stimulation on the cochlea and eighth nerve in the deafened guinea pig. Hear Res 105:30–43.

Miyashita T, Reed JC (1995) Tumor suppressor p53 is a direct transcriptional activator of the human bax gene. Cell 80:293–299.

Morimoto RI, Kline MP, Bimston DN, Cotto JJ (1997) The heat-shock response: regulation and function of heat-shock proteins and molecular chaperones. Essays Biochem 32:17–29.

Nadol JB Jr (1997) Patterns of neural degeneration in the human cochlea and auditory nerve: implications for cochlear implantation. Otolaryngol Head Neck Surg 117: 220–228.

Nadol JB Jr, Eddington DK (2006) Histopathology of the inner ear relevant to cochlear implantation. Adv Otorhinolaryngol 64:31–49.

Nakagomi S, Suzuki Y, Namikawa K, Kiryu-Seo S, Kiyama H (2003) Expression of the activating transcription factor 3 prevents c-Jun N-terminal kinase-induced neuronal death by promoting heat shock protein 27 expression and Akt activation. J Neurosci 23:5187–5196.

Niu X, Canlon B (2002) Protective mechanisms of sound conditioning. Adv Otorhinolaryngol 59:96–105.

Ogawara Y, Kishishita S, Obata T, Isazawa Y, Suzuki T, Tanaka K, Masuyama N, Gotoh Y (2002) Akt enhances Mdm2–mediated ubiquitination and degradation of p53. J Biol Chem 277:21843–21850.

Oh SH, Yu WS, Song BH, Lim D, Koo JW, Chang SO, Kim CS (2000) Expression of heat shock protein 72 in rat cochlea with cisplatin-induced acute ototoxicity. Acta Otolaryngol 120:146–150.

Ohinata Y, Miller JM, Altschuler RA, Schacht J (2000a) Intense noise induces formation of vasoactive lipid peroxidation products in the cochlea. Brain Res 878: 163–173.

Ohinata Y, Yamasoba T, Schacht J, Miller JM (2000b) Glutathione limits noise-induced hearing loss. Hear Res 146:28–34.

Ohinata Y, Miller JM, Schacht J (2003) Protection from noise-induced lipid peroxidation and hair cell loss in the cochlea. Brain Res 966:265–273.

Ohlemiller KK, Wright JS, Dugan LL (1999) Early elevation of cochlear reactive oxygen species following noise exposure. Audiol Neurootol 4:229–236.

Ohlemiller KK, McFadden SL, Ding D, Lear PM, Ho YS (2000) Targeted mutation of the gene for cellular glutathione peroxidaase (Gpx1) increases noise-induced hearing loss in mice. J Assoc Res Otolaryngol 1:243–254.

Paratcha G, Ledda F, Ibáñez CF (2003) The neural cell adhesion molecule NCAM is an alternative signaling receptor for GDNF family ligands. Cell 113:867–879.

Perrelet D, Ferri A, Liston P, Muzzin P, Korneluk RG, Kato AC (2002) IAPs are essential for GDNF-mediated neuroprotective effects in injured motor neurons in vivo. Nat Cell Biol 4:175–179.

Piazza V, Ciubotaru CD, Gale JE, Mammano F (2007) Purinergic signalling and intercellular Ca^{2+} wave propagation in the organ of Corti. Cell Calcium 41:77–86.

Pickles JO (2004) Mutation in mitochondrial DNA as a cause of presbyacusis. Audiol Neurootol 9:23–33.

Pierchala BA, Ahrens RC, Paden AJ, Johnson EM Jr (2004) Nerve growth factor promotes the survival of sympathetic neurons through the cooperative function of the protein kinase C and phosphatidylinositol 3-kinase pathways. J Biol Chem 279: 27986–27993.

Pirvola U, Liang XQ, Virkkala J, Saarma M, Murakata C, Marie Camoratto AM, Walton KM, Ylikoski J (2000) Rescue of hearing, auditory hair cells, and neurons by CEP-1347/KT7515, an inhibitor of c-Jun N-terminal kinase activation. J Neurosci 20:43–50.

Priuska E, Schacht J (1995) Formation of free radicals by gentamicin and iron and evidence for an iron/gentamicin complex. Biochm Pharmacol 50:1749–1752.

Probst FJ, Fridell RA, Raphael Y, Saunders TL, Wang A, Liang Y, Morell RJ, Touchman JW, Lyons RH (1998) Corrrection of deafness in shaker-2 mice by an unconventional myosin in a BAC transgene. Science 280:1444–1447.

Puel JL, Ruel J, Gervais d'Aldin C, Pujol R (1998) Excitotoxicity and repair of cochlear synapses after noise-trauma induced hearing loss. NeuroReport 9:2109–2114.

Ramkumar V, Ravi R, Wilson MC, Gettys TW, Whitworth C, Rybak LP (1994) Identification of A1 adenosine receptors in rat cochlea coupled to inhibition of adenylyl cyclase. Am J Physiol 267:C731–737.

Ramkumar V, Whitworth CA, Pingle SC, Hughes LF, Rybak LP (2004) Noise induces A1 adenosine receptor expression in the chinchilla cochlea. Hear Res 188:47–56.

Reed JC (1998) Bcl-2 family proteins. Oncogene 17:3225–3236.

Robertson D, Paki B (2002) Role of L-type Ca^{2+} channels in transmitter release from mammalian inner hair cells. II. Single-neuron activity. J Neurophys 87:2734–2740.

Robertson GS, Crocker SJ, Nicholson DW, Schulz JB (2000) Neuroprotection by the inhibition of apoptosis. Brain Pathol 10:283–292.

Roehm PC, Hansen MR (2005) Strategies to preserve or regenerate spiral ganglion neurons. Curr Opin Otolaryngol Head Neck Surg 13:294–300.

Rybak LP, Husain K, Morris C, Whitworth C, Somani S (2000) Effect of protective agents against cisplatin ototoxicity. Am J Otol 186:513–520.

Rybak LP, Whitworth CA, Mukherjea D, Ramkumar V (2007) Mechanisms of cisplatin-induced ototoxicity and prevention. Hearing research 226:157–167.

Rydel RE, Greene LA (1988) cAMP analogs promote survival and neurite outgrowth in cultures of rat sympathetic and sensory neurons independently of nerve growth factor. Proc Natl Acad Sci USA 85:1257–1261.

Sato T, Doi K, Taniguchi M, Yamashita T, Kubo T, Tohyama M (2006) Progressive hearing loss in mice carrying a mutation in the p75 gene. Brain Res 1091:224–234.

Scarpidis U, Madnani D, Shoemaker C, Fletcher CH, Kojima K, Eshraghi AA, Staecker H, Lefebvre P, Malgrange B, Balkany TJ, Van De Water TR (2003) Arrest of apoptosis in auditory neurons: implications for sensorineural preservation in cochlear implantation. Otol Neurotol 24:409–417.

Scheibe F, Haupt H, Ludwig C (1992) Intensity-dependent changes in oxygenation of cochlear perilymph during acoustic exposure. Hear Res 63:19–25.

Scheibe F, Haupt H, Ising HC L (2002) Therapeutic effect of parenteral magnesium on noise-induced hearing loss in the guinea pig. Magnes Res 15:27–36.

Schuknecht HF (1993) Pathology of the Ear. Philadelphia: Lea & Febiger.

See V, Boutillier AL, Bito H, Loeffler JP (2001) Calcium/calmodulin-dependent protein kinase type IV (CaMKIV) inhibits apoptosis induced by potassium deprivation in cerebellar granule neurons. FASEB J 15:134–144.

Seidman MD (2000) Effects of dietary restriction and antioxidants on presbyacusis. Laryngoscope 110:727–738.

Sha S-H, Schacht J (1999) Salicylate attenuates gentamicin-induced ototoxicity. Lab Invest 79:807–813.

Sha S-H, Qiu J-H, Schacht J (2006) Aspirin to prevent gentamicin-induced hearing loss. N Engl J Med 354:1856–1857.

Shepherd RK, Coco A, Epp SB, Crook JM (2005) Chronic depolarization enhances the trophic effects of brain-derived neurotrophic factor in rescuing auditory neurons following a sensorineural hearing loss. J Comp Neurol 486:145–158.

Shi X, Nuttall AL (2003) Upregulated iNOS and oxidative damage to the cochlear stria vascularis due to noise stress. Brain Res 967:1–10.

Shieh PB, Hu SC, Bobb K, Timmusk T, Ghosh A (1998) Identification of a signaling pathway involved in calcium regulation of BDNF expression. Neuron 20:727–740.

Shinohara T, Bredberg G, Ulfendahl M, Pyykko I, Olivius NP, Kaksonen R, Lindstrom B, Altschuler R, Miller JM (2002) Neurotrophic factor intervention restores auditory function in deafened animals. Proc Natl Acad Sci USA 99:1657–1660.

Shoji F, Miller AL, Mitchell A, Yamasoba T, Altschuler RA, Miller JM (2000a) Differential protective effects of neurotrophins in the attenuation of noise-induced hair cell loss. Hear Res 146:134–142.

Shoji F, Yamasoba T, Magal E, Dolan DF, Altschuler RA, Miller JM (2000b) Glial cell line-derived neurotrophic factor has a dose dependent influence on noise-induced hearing loss in the guinea pig cochlea. Hearing Res 142:41–55.

Song B-B, Schacht J (1996) Variable efficacy of radical scavengers and iron chelators to attenuate gentamicin ototoxicity in guinea pig in vivo. Hear Res 94:87–93.

Spoendlin H (1971) Degeneration behavior of the cochlear nerve. Arch Klin Exp Ohren Nasen Kehlkopfheilk 200:275–291.

Spoendlin H (1975) Retrograde degeneration of the cochlear nerve. Acta Otolaryngol 79:266–275.

Staecker H, Kopke R, Malgrange B, Lefebvre P, Van de Water TR (1996) NT-3 and/or BDNF therapy prevents loss of auditory neurons following loss of hair cells. NeuroReport 7:889–894.

Staecker H, Gabaizadeh R, Federoff H, Van De Water TR (1998) Brain-derived neurotrophic factor gene therapy prevents spiral ganglion degeneration after hair cell loss. Otolaryngol Head Neck Surg 119:7–13.

Stankovic K, Rio C, Xia A, Sugawara M, Adams JC, Liberman MC, Corfas G (2004) Survival of adult spiral ganglion neurons requires erbB receptor signaling in the inner ear. J Neurosci 24:8651–8661.

Stefanis L (2005) Caspase-dependent and -independent neuronal death: two distinct pathways to neuronal injury. Neuroscientist 11:50–62.

Strack S, Cribbs JT, Gomez L (2004) Critical role for protein phosphatase 2A heterotrimers in mammalian cell survival. J Biol Chem 279:47732–47739.

Sugawara M, Corfas G, Liberman MC (2005) Influence of supporting cells on neuronal degeneration after hair cell loss. J Assoc Res Otolaryngol 6:136–147.

Sugawara M, Murtie JC, Stankovic KM, Liberman MC, Corfas G (2007) Dynamic patterns of neurotrophin 3 expression in the postnatal mouse inner ear. J Comp Neurol 501:30–37.

Tadros SF, S.T. F, Mapes F, Frisina DR, Frisina RD (2005) Higher serum aldosterone correlates with lower hearing thresholds: a possible protective hormone against presbycusis. Hear Res 209:10–18.

Tahera Y, Meltser I, Johansson P, Salman H, Canlon B (2007) Sound conditioning protects hearing by activating the hypothalamic-pituitary-adrenal axis. Neurobiol Dis 25:189–197.

Tan J, Shepherd RK (2006) Aminoglycoside-induced degeneration of adult spiral ganglion neurons involves differential modulation of tyrosine kinase B and p75 neurotrophin receptor signaling. Am J Pathol 169:528–543.

Tao X, Finkbeiner S, Arnold DB, Shaywitz AJ, Greenberg ME (1998) Ca^{2+} influx regulates BDNF transcription by a CREB family transcription factor-dependent mechanism. Neuron 20:709–726.

Thorne PR, Nuttall AL (1987) Laser Doppler measurements of cochlear blood flow during loud sound exposure in the guinea pig. Hear Res 27:1–10.

Thorne PR, Munoz DJ, Housley GD (2004) Purinergic modulation of cochlear partition resistance and its effect on the endocochlear potential in the Guinea pig. J Assoc Res Otolaryngol 5:58–65.

Tong JX, Eichler ME, Rich KM (1996) Intracellular calcium levels influence apoptosis in mature sensory neurons after trophic factor deprivation. Exp Neurol 138:45–52.

van Campen LE, Murphy WJ, Franks JR, Mathias PI, Torraason MA (2002) Oxidative DNA damage is associated with intense noise exposure in the rat. Hear Res 164:29–38.

Van de Water TR, Lallemand F, Eshraghi AA, Ahsan S, He J, Guzman J, Polak M, Malgrange B, Lefebvre PP, Staecker H, Balkany TJ (2004) Caspases, the enemy within, and their role in oxidative stress-induced apoptosis of inner ear sensory cells. Otol Neurotol 25:627–632.

Vicente-Torres MA, Schacht J (2006) A BAD link to mitochondrial cell death in the cochlea of mice with noise-induced hearing loss. J Neurosci Res 83:1564–1572.

Vlajkovic SM, Thorne PR, Munoz DJB, Housley GD (2004) Ectonucleotidase activity in the perilymphatic compartment of the guinea pig cochlea. Hear Res 99:31–37.

Vogelbaum MA, Tong JX, Rich KM (1998) Developmental regulation of apoptosis in dorsal root ganglion neurons. J Neurosci 18:8928–8935.

von Bartheld CS, Patterson SL, Heuer JG, Wheeler EF, Bothwell M, Rubel EW (1991) Expression of nerve growth factor (NGF) receptors in the developing inner ear of chick and rat. Development 113:455–470.

Waetzig V, Herdegen T (2003) A single c-Jun N-terminal kinase isoform (JNK3–p54) is an effector in both neuronal differentiation and cell death. J Biol Chem 278:567–572.

Wang HG, Pathan N, Ethell IM, Krajewski S, Yamaguchi Y, Shibasaki F, McKeon F, Bobo T, Franke TF, Reed JC (1999) Ca^{2+}-induced apoptosis through calcineurin dephosphorylation of BAD. Science 284:339 343.

Wang J, Van De Water TR, Bonny C, de Ribaupierre F, Puel JL, Zine A (2003) A peptide inhibitor of c-Jun N-terminal kinase protects against both aminoglycoside and acoustic trauma-induced auditory hair cell death and hearing loss. J Neurosci 23:8596–8607.

Wang LH, Besirli CG, Johnson EM Jr (2004) Mixed-lineage kinases: a target for the prevention of neurodegeneration. Annu Rev Pharmacol Toxicol 44:451–474.

Warr WB (1992) Organization of olivocochlear efferent systems in mammals. In: Webster DB, Popper AN, Fay RR (eds) Mammalian Auditory Pathway: Neuroanatomy. New York: Springer-Verlag, pp. 410–448.

Webster M, Webster DB (1981) Spiral ganglion neuron loss following organ of Corti loss: a quantitative study. Brain Res 212:17–30.

Whitfield J, Neame SJ, Paquet L, Bernard O, Ham J (2001) Dominant-negative c-Jun promotes neuronal survival by reducing BIM expression and inhibiting mitochondrial cytochrome c release. Neuron 29:629–643.

Whitworth CS, Ramkumar V, Jones B, Tsukasaki N, Rybak LP (2004) Protection against cisplatin ototoxicity by adenosine agoinists. Biochem Pharmacol 67:1801–1807.

Wiechers B, Gestwa G, Mack A, Carroll P, Zenner HP, Knipper M (1999) A changing pattern of brain-derived neurotrophic factor expression correlates with the rearrangement of fibers during cochlear development of rats and mice. J Neurosci 19:3033–3042.

Wilson BE, Mochon E, Boxer LM (1996) Induction of bcl-2 expression by phosphorylated CREB proteins during B-cell activation and rescue from apoptosis. Mol Cell Biol 16:5546–5556.

Wise AK, Richardson R, Hardman J, Clark G, O'Leary S (2005) Resprouting and survival of guinea pig cochlear neurons in response to the administration of the neurotrophins brain-derived neurotrophic factor and neurotrophin-3. J Comp Neurol 487:147–165.

Xu Z, Maroney AC, Dobrzanski P, Kukekov NV, Greene LA (2001) The MLK family mediates c-Jun N-terminal kinase activation in neuronal apoptosis. Mol Cell Biol 21:4713–4724.

Yagi M, Kanzaki S, Kawamoto K, Shin B, Shah PP, Magal E, Sheng J, Raphael Y (2001) Spiral ganglion neurons are protected from degeneration by GDNF gene therapy. J Assoc Res Otolaryngol 4:315–325.

Yamagata T, Miller JM, Ulfendahl M, Olivius NP, Altchuler RA, Pyykko I, Bredberg G (2004) Delayed neurotrophic treatment preserves nerve survival and electrophysiological responsiveness in neomycin-deafened guina pigs. J Neurosci Res 78:75–86.

Yamane H, Nakai Y, Takayama M, Iguchi H, Nakawawa T, Kojima A (1995) Appearance of free radicals in the guinea pig inner ear after noise-induced acoustic trauma. Arch Otorhinolaryngol 252:504–508.

Yamashita D, Miller JM, Jiang H-Y, Minami SB, Schacht J (2004) AIF and EndoG in noise-induced hearing loss. Neuro Report 15:2719–2722.

Yamasoba T, Harris CA, Shoji F, Lee RJ, Nuttall AL, Miller JM (1998) Influence of intense sound exposure on glutathione synthesis in the cochlea. Brain Res 804:72–78.

Yamasoba T, Schacht J, Shoji F, Miller JM (1999) Attenuation of cochlear damage from noise trauma by an iron chelator, a free radical scavenger and glial cell line-derived neurotrophic factor *in vivo*. Brain Res 815:317–325.

Yamasoba T, Altchuler RA, Raphael Y, Miller AL, Shoji F, Miller JM (2001) Absence of hair cell protection by exogenous FGF-1 and FGF-2 delivered to guinea pig cochlea in vivo. Noise Health 3:65–78.

Yang DD, Kuan CY, Whitmarsh AJ, Rincon M, Zheng TS, Davis RJ, Rakic P, Flavell RA (1997) Absence of excitotoxicity-induced apoptosis in the hippocampus of mice lacking the JNK3 gene. Nature 389:865–870.

Ylikoski J, Pirvola U, Virkkala J, Suvanto P, Liang XQ, Magal E, Altschuler R, Miller JM, Saarma M (1998) Guinea pig auditory neurons are protected by GDNF from degeneration after noise trauma. Hear Res 124:17–26.

Yoshida N, Liberman MC (2000) Sound conditioning reduces noise-induced permanent threshold shift in mice. Hear Res 148:213–219.

Yoshida N, Kristiansen A, Liberman MC (1999) Heat stress and protection from permanent acoustic injury in mice. J Neurosci 19:10116–10124.

Zang X, Zhang M, Laties AM, Mitchell CH (2006) Balance of purines may determine life or death of retinal ganglion cells as A3 adenosine receptors prevent loss following P2X7 receptor stimulation. J Neurochem 98:566–575.

Zenner HP (1986) K^+-induced motility and depolarization of cochlear hair cells. Direct evidence for a new pathphysiological mechanism in Ménière's disease. Arch Otorhinolaryngol 243:108–111.

Zha X-M, Bishop JF, Hansen MR, Victoria L, Abbas PJ, Mouradian MM, Green SH (2001) BDNF synthesis in spiral gangion neurons is constitutive and CREB-dependent. Hear Res 156:53–68.

Zhang M, Liu W, Ding D, Salvi R (2003) Pifithrin-α suppresses p53 and protects cochlear and vestibular hair cells from cisplatin-induced apoptosis. Neuroscience 120:191–205.

Zheng XY, Henderson D, McFadden SL, Ding DL, Salvi RJ (1999) Auditory nerve fiber responses following chronic cochlear de-efferentation. J Comp Neurol 406:72–78.

Zhou Z, Liu Q, Davis RL (2005) Complex regulation of spiral ganglion neuron firing patterns by neurotrophin-3. J Neurosci 25:7558–7566.

Zimmermann CE, Burgess BJ, Nadol JB Jr (1995) Patterns of degeneration in the human cochlear nerve. Hear Res 90:192–201.

11
Emerging Strategies for Restoring the Cochlea

Stefan Heller and Yehoash Raphael

1. Hair Cell Loss

1.1 Background

Throughout history, people with hearing loss have been confronted with enormous challenges. Excluded from the circles of the hearing for millennia, they were routed from society, prosecuted, even murdered. Institutionalization and cruel torture was a common fate for misdiagnosed hard of hearing people during the Middle Ages and beyond, even the first half of the 20th century. Life improved for hard of hearing people with the advent of modern hearing aids and other technological advances. These developments were paralleled by the organization of a deaf community, which offers choices of integrating with the hearing world and/or use of sign language and other alternative means of communication.

One challenge of developing technology is the constant addition of sound sources that are hazardous to hearing. Mammalian ears have evolved to respond to sound with great selectivity and sensitivity; however, they have not evolved to tolerate some of the man-made sounds that have unnatural physical characteristics. The rise time, peak energy, and sustained duration of some of the modern-age sounds are detrimental to hearing. Sounds generated by explosives, industrial noise, power tools, electric (and even acoustic) music instruments, and abused headphones often lead to permanent hearing deficits. In addition, some medications have ototoxic effects (see Henderson et al., Chapter 7 and Rybak et al., Chapter 8, which deal with ear pathologies caused by overstimulation and ototoxic drugs). The activities of daily living contain many hazards to human hearing.

In parallel, however, all progress, whether technological, biomedical, or sociological, is providing hard of hearing people with more options to live their lives the way they would like and not as society dictates. Although advancements have been made, biomedical researchers would like to find permanent solutions for the underlying causes of human hearing loss. Providing patients with the option to repair their sensory disability completely is one of the fundamental goals of hearing research. This chapter summarizes the latest studies in this direction and aims to provide a realistic view of the remaining challenges.

1.2 The Causes of Hearing Loss

The causes of sensorineural hearing loss can be divided into two general categories. In the first group, hair cell loss is the underlying cause. In the second group, hearing loss is caused by atrophic changes in other cochlear cell types, including the spiral ganglion, stria vascularis, nonsensory epithelial cells of the organ of Corti, connective tissue components, and/or other cochlear elements. The borders between these groups are not strictly defined and some hearing losses are certainly caused by hair cell loss that is primarily caused by the dysfunction of supporting cells, vasculature, or the stria vascularis. Degenerative changes in the cochlea that involve hair cell or neuron loss cannot be regenerated because most epithelial and neural cell types in the adult cochlea, including hair cells, are strictly postmitotic. Loss of cochlear hair cells in the mammalian cochlea at any age is irreversible and leads to hearing impairment, which usually corresponds in severity to the degree of hair cell loss. Hair cell loss is the most common cause of sensorineural hearing loss; therefore, most of the research on protection, repair, and regeneration in the inner ear has focused on hair cells. Nevertheless, other cell types are often the primary target of the insult and also need to be studied in depth.

2. Regenerating Versus Nonregenerating Hair Cell Systems

2.1 Hair Cell Regeneration in Nonmammalian Vertebrates

The cells that line the membranous labyrinth are epithelial derivates of the otocyst. Most epithelial cells in the body are capable of regeneration, and in most cases they turnover (replace themselves) constantly (e.g., the gut brush-border epithelium replaces itself rapidly, and the corneal epithelium turns over and is capable of repairing lesions efficiently). Turnover and regenerative responses in epithelial cells are usually dependent on the basal cell population. Basal cells are nondifferentiated cells that can proliferate (constantly, and/or in response to a lesion) and produce new cells, some of which differentiate while others remain as basal cells ready for a new round of proliferation. In the cochlear sensory epithelium, all cells appear to be differentiated (supporting cells and hair cells), and nondifferentiated basal cells are absent. The absence of basal cells is one way to explain the lack of hair cell regeneration following cochlear hair cell loss.

Hair cell epithelia in vertebrates other than mammals vary in their degree of growth, turnover, and ability to regenerate. Yet, in most cases, regeneration is possible. Some vertebrates, such as sharks, bony fish, and amphibians, grow throughout life and add new hair cells to their ears as they grow (Corwin 1981, 1983, 1985). In the avian basilar papilla (the avian equivalent of the organ of Corti), the addition of new hair cells in the mature ear is negligible, but regeneration of hair cells after an experimentally induced lesion is rapid, effective, and impressive (Corwin and Cotanche 1988; Ryals and Rubel 1988). New hair cells can be generated in the avian ear after different types of induced lesions such as those caused by overstimulation, ototoxic drugs, and genetic disease (Corwin and Cotanche 1988; Ryals and Rubel 1988; Tucci and Rubel 1990; Wilkins et al. 2001). Along with the regeneration of hair cells, birds can restore their auditory function (Adler et al. 1993; Niemiec et al. 1994; Marean et al. 1995; Dooling et al. 1997). Interestingly, hair cell regeneration in the avian ear can occur more than once (Niemiec et al. 1994).

The basilar papilla does not contain a clearly identifiable population of basal cells. Nevertheless, after a lesion to the sensory epithelium, the supporting cells of the basilar papilla replicate their DNA and divide near the luminal surface (Raphael 1992, 1993; Hashino and Salvi 1993), then some of the new cells become new hair cells (Cotanche et al. 1994; Stone and Cotanche 1994). When avian supporting cells proliferate, they dedifferentiate, undergo S-phase near the basal lamina, then round up near the luminal surface and divide (Raphael et al. 1994). The importance of this finding is in the fact that supporting cells are differentiated cells with specialized function and structure. As such, their ability to dedifferentiate and produce new hair cells is among the very few examples of transdifferentiation (phenotypic conversion). The transdifferentiation of supporting cells into new hair cells in the avian basilar papilla can be even more drastic when it occurs without mitosis (Adler and Raphael 1996; Roberson et al. 1996).

In addition to the transdifferentiation of supporting cells into new hair cells, there may be additional contributions from surrounding tissues to the repair of the traumatized basilar papilla. Hyalin and cuboidal cells, neighboring the basilar papilla, have been shown to proliferate following lesions to the hair cells (Girod et al. 1989). These cells have also been shown to migrate into the basilar papilla after a severe lesion that depletes hair cells along with the original supporting cells (Cotanche et al. 1995).

2.2 The Molecular Control of Avian Hair Cell Regeneration

Research on hair cell regeneration in the avian basilar papilla has provided invaluable information about the morphological and physiological outcome of the regenerative process and identified the differentiated supporting cells as the main cellular contributors to the process. More recently, some aspects of the molecular regulation of hair cell regeneration in the basilar papilla have been elucidated (Bermingham-McDonogh et al. 2001; Witte et al. 2001; Warchol 2002; Hawkins et al. 2003; Matsui et al. 2004; Stone et al. 2004). Data obtained from gene expression assays (Hawkins et al. 2003, 2006) combined with other methods will likely facilitate the identification of the signals that initiate the regenerative activity in the supporting cells of the basilar papilla. Given that basal cells (or stem cells) are apparently absent in both the avian and the mammalian ear, it is not clear why birds can and do regenerate hair cells and humans (along with other mammals) do not.

In searching for the reasons that account for the inability of mammals to regenerate hair cells, it is tempting to consider mostly the nonsensory components of the epithelium, as the nonsensory cells give rise to new hair cells. Nevertheless, important clues can also be gained from studying hair cells, considering that on their demise, these cells might be able to provide a signal that initiates regenerative activity in the supporting cells.

2.3 Limited Regenerative Capability in Mammalian Vestibular Sensory Epithelia

The morphology of vestibular hair cells and the organization of the balance sense organs have some features that are similar to those of nonmammalian hair cell systems. One of these features is the ability of the vestibular epithelium to generate new hair cells. After a severe ototoxic lesion to the mammalian vestibular epithelium, much like in the cochlea, hair cell regeneration could not be identified (Lindeman 1969; Hawkins and Preston 1975; Meiteles and Raphael 1994). However, after less severe lesions to the vestibular sensory epithelium, a limited number of new hair cells were observed (Forge et al. 1993; Lopez et al. 1998). The mechanism of vestibular hair cell regeneration is based on the transdifferentiation of supporting cells, and the extent of supporting cell proliferation may vary between species (Warchol et al. 1993; Lopez et al. 1998).

While the finding that a small number of stem cells are present in the vestibular organs may account for the regenerative capability in this epithelium (Li et al. 2003a), the signals for regeneration are not well characterized. The vestibular sensory epithelium in mammals could potentially be a useful model for understanding the signals that mediate and regulate the regeneration thanks to the abundance of markers, reagents, and advanced status of whole genome mapping in mammals. However, the low level of regeneration impairs the ability to use the plethora of available resources in a fruitful manner. When the signals initiating and regulating the regeneration of hair cells are identified (in mammals, birds, or other vertebrates), they may be applicable for enhancing vestibular regeneration. This would be extremely beneficial for treating balance disorders due to hair cell loss, for which no treatment is currently available. Clinically, hearing loss can be treated with amplification and/or cochlear implant, while balance disorders still await treatment.

3. Reactivation of Developmental Programs in the Damaged Organ of Corti

3.1 Transdifferentiation: Induced Phenotypic Conversion

There are two main feasible options to introduce new hair cells to replace lost cochlear sensory cells (Fig. 11.1). One option is to tap into the population of nonsensory cells that remain in the damaged cochlea and use them as a

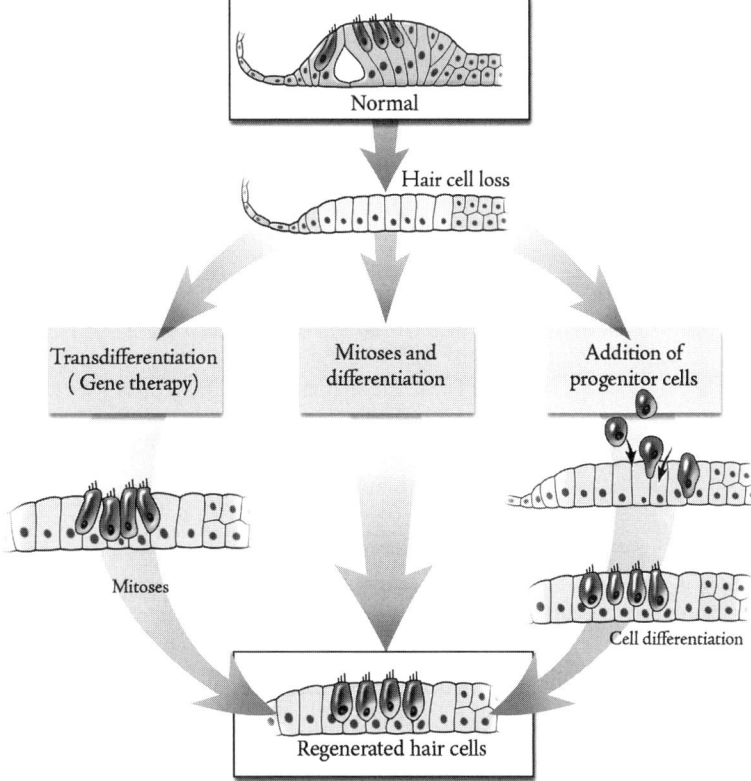

FIGURE 11.1. After hair cells (top box) degenerate, only supporting cells remain in the organ of Corti. The three options to generate new hair cells are by transdifferentiation of supporting cells (left), inducing proliferation (middle), or adding progenitor cells such as stem cells (right). Combining proliferation with transdifferentiation may enhance the regenerative response.

source for generating new hair cells. The other option is to introduce cells from external sources, such as stem cells. Both technologies have advantages and disadvantages. Approaches for using the intrinsic cells in the cochlea as a source for generating new hair cells are discussed in this section and the use of stem cells is discussed in Section 4.

The theory and logic behind the attempt to use cochlear nonsensory cells to generate new hair cells are in the similarity of this principle to what birds do spontaneously. One logical and practical way to accomplish such trans-differentiation is by reactivation of the deve- lopmental program that regulates formation of hair cells. Many of the genes that are involved in the develop-mental of the sensory epithelium have been identified and characterized (Torres and Giraldez 1998; Chen et al. 2002; Fekete and Wu 2002; Kelley 2003; Montcouquiol and Kelley 2003; Barald and Kelley 2004; Woods et al. 2004;

Fritzsch et al. 2006). The spatial and temporal sequence of gene expression has been linked to cell–cell and cell–matrix interactions. Cell cycle regulation in the developing and mature sensory epithelium is of interest because manipulating the genes involved in this regulation may help produce new cells in mature tissues. Hair cells and supporting cells share common progenitors. Therefore, the developmental stages during which these two cell types undergo fate commitment and differentiation are of immediate relevance to designing strategies to induce the transdifferentiation of nonsensory cells into new hair cells.

Because hair cells and supporting cells are clonally related (Fekete 2000; Fekete and Wu 2002), they have both gone through similar stages of developmental signaling and gene expression and are therefore responsive to similar signals. During development, the *bHLH* gene *Atoh1* (formerly *Math1*) signals cells to choose the fate of hair cells, rather than supporting cells (Bermingham et al. 1999; Zine 2003; Woods et al. 2004). As such, *Atoh1* is an excellent candidate gene to attempt transdifferentiation of nonsensory cells into new hair cells. Initial experiments on the overexpression of *Atoh1* were done in tissue cultures. Induced expression of this gene in explants of the cochlea resulted in the transdifferentiation of some of the nonsensory cells into extranumerary hair cells (Zheng and Gao 2000; Shou et al. 2003). When *Atoh1* was expressed in the normal guinea pig cochlea in vivo, ectopic hair cells were detected (Kawamoto et al. 2003b). Using neuronal-specific staining, it was determined that neurons can find new hair cells even in ectopic locations (Kawamoto et al. 2003b). These results provide evidence for the principle that mature nonsensory cells in the mammalian cochlea retain their competence to respond to gene expression of a hair cell–specific gene (*Atoh1*) and transdifferentiate into the hair cell phenotype. These experiments paved the way for testing the outcome of *Atoh1* gene expression in the mature deafened cochlea.

The technological ability to introduce genes into supporting cells in vivo (discussed later) depended on inoculation of the adenovirus vectors with transgene inserts into the endolymph (Ishimoto et al. 2002). This technology was used to insert *Atoh1* into the nonsensory cells of mature deafened guinea pigs in vivo. The deafening was accomplished by using a high concentration of an aminoglycoside antibiotic in combination with a potent diuretic, which resulted in the elimination of most or all of the hair cells in the cochlea. After the bilateral elimination of the hair cells, the *Atoh1* expressive adenovirus was inoculated into the left ear. Transgenic expression of *Atoh1* in the nonsensory cells that remained in these ears was efficient. In animals tested two months after the inoculation, Ad.*Atoh1* induced the generation of a significant number of new hair cells in the organ of Corti and hearing thresholds improved (Izumikawa et al. 2005). The improvement in threshold does not attest to the qualitative features of hearing in these *Atoh1*-treated animals. Based on the morphology of these cochleae, hearing is likely to be distorted and abnormal.

These studies, along with findings on the role of *Atoh1* during inner ear development (Bermingham et al. 1999; Chen et al. 2002; Woods et al. 2004), suggest that *Atoh1* functions as a master regulatory gene, which is both necessary and sufficient for hair cell development or regeneration. The data provide the proof for the principle that reactivation of developmental programs may be utilized as therapeutic means in mature tissues of the inner ear and elsewhere. Nevertheless, this type of therapy will need to be improved and characterized along several different avenues before it can be applied clinically.

Atoh1 and other genes that induce phenotypic change are not expected to directly enhance proliferation in the auditory epithelium. It remains to be determined if cell proliferation occurs spontaneously as a secondary effect of the induced transdifferentiation. Even if it does, there will be many cases in which a need to increase the population of nonsensory cells will be the first task in the induced regenerative process. Moreover, it is important to test whether inducing cell division will result in some of the progeny differentiating into new hair cells. Therefore, studies looking at the regulation of the cell cycle in the auditory epithelium and attempts to manipulate the cell cycle are important and exciting.

3.2 Inducement of Proliferation in the Inner Ear Epithelium

The first experiments that showed generation of new cells in the organ of Corti past the normal end of mitosis involved work on the $p27^{Kip1}$ (*p27*) gene. p27 inhibits cyclin-dependent kinase-2 (cdk-2) (Sherr and Roberts 1999) and therefore has an antimitogenic role. In the organ of Corti, p27 appears to be responsible for the quiescence of the supporting cells and is selectively expressed in these cells (Löwenheim et al. 1999). During development of the organ of Corti, *p27* expression is induced around E13, when cell division of hair cell progenitors stops (Chen and Segil 1999). *p27* expression persists at high levels in differentiated supporting cells of the mature organ of Corti. As such, the traumatized ear, in which no hair cells remain, may be an attractive target for blocking *p27* (expression or function) in order to induce mitosis in supporting cells. The inner ears of *p27* knockout mice display continued cell division into the postnatal period, as well as supernumerary hair cells (Chen and Segil 1999; Löwenheim et al. 1999). In parallel to the continued mitosis and excessive number of hair cells, the *p27* knockout mice are severely hearing impaired (Chen and Segil 1999; Kanzaki et al. 2006). The reason for the deafness in these mice is unknown at present.

While p27 is restricted to mature supporting cells, it appears that *Ink4* is expressed in hair cells and prevents their reentry into the cell cycle. *Ink4d-/-* animals display continued mitosis and cell death in the hair cell population (Chen et al. 2003). Like *p27* mutations, *Ink4d* mutations also lead to hearing loss.

Another gene involved in cell cycle regulation in the organ of Corti is the retinoblastoma tumor suppressor gene *Rb1*. Mutations in *Rb1* cause tumors and loss of cell cycle regulation in multiple tissues (Lohmann and Gallie 2004). Disruption of *Rb1* in the inner ear was accomplished by crossing floxed *Rb1* mice with collagen1A1 (Col1A1)-Cre mice (Sage et al. 2005). The resulting mutant mice lacked inner ear expression of *Rb1* and exhibited a large number of supernumerary hair cells in the cochlea and the vestibular epithelium. One of the novel and important findings in this study was that hair cells themselves might be able to replicate their DNA and divide in the absence of functional *Rb1*. The supernumerary hair cells had several phenotypic features of hair cells. Future improvement in the technology to regulate the control of cell proliferation genes via somatic cell–specific methods will enable the use of the knowledge of cell cycle regulation to design clinical approaches to regenerate hair cells.

3.3 Technological Needs for Inducing Therapeutic Transdifferentiation and Proliferation

To design clinically applicable therapies based on the scientific knowledge generated in the laboratory, it is necessary to develop safe and efficient methods for specific inner ear application of such therapies. Much of the practical application of knowledge at the genetic level will depend on the ability to regulate gene expression in a time- and place-specific manner and at the optimal level. Regulating gene expression can involve overexpression of a gene or blocking gene expression. Overexpression can sometimes be accomplished by the use of ligands that bind to extracellular receptors and initiate gene expression cascades. Thus, genes encoding secreted proteins can be introduced to cells in the vicinity of the target cell and the secreted gene product will act in a paracrine fashion as shown for several growth factors genes (Staecker et al. 1998; Yagi et al. 1999; Chen et al. 2001; Kawamoto ct al. 2003a).

In other cases, it is necessary to introduce the gene itself into the cell, along with a promoter that can function in the specific cell type. The use of a variety of vehicles and techniques to influence gene expression in the cells of the inner ear has recently been reviewed (Avraham and Raphael 2003; Patel et al. 2004; Crumling and Raphael 2006). One major challenge in manipulating gene expression in the inner ear is the need to deliver therapeutic agents into cells or fluid spaces. This task involves the risk of disrupting the membranous labyrinth. The ideal route of delivery would be oral or systemic via intravenous injection, but the final concentration in the cochlea would be impractically low. In many cases, the blood–ear barrier would prevent the entry of the delivered reagents into the inner ear. Delivery of genes will probably continue to require direct inoculation into the cochlear fluid. Vector inoculation into the perilymph is a minimally invasive method to penetrate the round window using a micropipette (Stöver et al. 1999). Alternatively, cochleostomy can be used. However, if the vector solution needs to be introduced into the scala media, the procedure is likely to result in damage to the cochlear tissues (Ishimoto et al. 2002). With the present adenovirus vectors, it is necessary to inoculate into the endolymph to achieve transgene expression in nonsensory cells of the cochlea (Ishimoto et al. 2002). This procedure is technically complicated, leads to excessive variability in the results, and is not easily applicable to clinical use. The future may bring alternative or improved vectors that will accomplish viral delivery of nonsensory epithelial cells via the perilymph. To utilize cell cycle regulatory genes for hair cell regeneration therapy, it will be necessary to develop methods to promote temporally and spatially regulated gene expression.

4. Inner Ear Stem Cells

4.1 Progenitor and Stem Cells from Vestibular and Cochlear Tissues

The advent of stem cell biology research has recently provided new insights and technology that allows the direct testing of the hypothesis that inner ear sensory epithelia with regenerative capacity contain a population of progenitor or stem cells with high proliferative capacity. The inner ear sensory epithelia obviously do not contain undifferentiated basal cells, but some residing differentiated supporting cells may have the ability to dedifferentiate and to proliferate. The defining feature of a stem cell is its capacity for long-term self-renewal without losing the ability to spawn differentiating cells (McKay 1997). Particularly, the ability to proliferate without elaborate stimulation has been used routinely to isolate various stem cell populations from different organs. Today, the terms "progenitor cell" and "stem cell" are used very loosely. Progenitor cells are defined here as not fully differentiated cells, which are able to undergo a limited number of mitoses; stem cells are defined as multipotent or pluripotent cells with the capability of long-term self-renewal.

A first indication for the possible existence of inner ear stem cells in the postnatal mammalian inner ear arose from the observation that dissociated cells from the early postnatal rat organ of Corti developed into floating spherical colonies (Malgrange et al. 2002). Hair cells were detectable within these floating colonies after a 2-week culture period. Malgrange and colleagues further demonstrated that these new hair cells arose from dividing progenitors that incorporated the thymidine analog bromo-deoxyuridine during S-phase. They did not demonstrate, however, that individual cells are capable of generating floating colonies and that these spheres can be propagated. In a related study using dissociated cells from adult vestibular sensory epithelia, Li and colleagues (Li et al. 2003a) also found floating spherical colonies. A series of tests demonstrated that these spheres arose from single cells with high proliferative capacity and that it is possible to propagate and to maintain the spheres over many generations; ergo, the sphere-forming cells within adult vestibular epithelia are stem cells. Grafting of spheres derived from murine inner ear vestibular stem cells into the inner ears of chicken embryos showed that murine cells were able to differentiate into hair cells. When grafted at earlier stages, inner ear vestibular stem cells gave rise to many different cell types in organs derived from all three germ layers. Hence, inner ear vestibular stem cells are pluripotent (Li et al. 2003a).

Propagation of inner ear–derived spheres revealed a second characteristic feature: the majority of the 50–100 cells that make up an individual sphere are differentiating progenitors and only 1–3 cells are stem cells, which are able to reform new spheres (Li et al. 2003a). This is a problem because with a very limited capacity for expansion, it is difficult to obtain sufficient numbers of stem cells for extensive experiments. Nevertheless, improvements for growing and expanding other stem cell populations have been developed (Svendsen et al. 1998; Sen et al. 2002), and future improvements of inner ear stem cell passaging will lead to the greater availability of these cells.

The initial observation of sphere formation from dissociated early postnatal rat organ of Corti (Malgrange et al. 2002) raised the question of whether these spheres are, like the vestibular spheres, the manifestation of proliferating stem cells. This is indeed the case, as spheres derived from dissociated postnatal mouse organ of Corti can also self-renew for many generations (Oshima et al. 2007). Interestingly, during the second and third postnatal weeks in mice, the organ of Corti loses about 99% of its sphere-forming capacity. It is unlikely that several hundred stem cells disappear from the organ of Corti during this postnatal maturation period. It has been speculated that the loss or reduction of proliferative capacity goes hand-in-hand with the final maturation of supporting cells or of differentiation of greater epithelial ridge cells to inner sulcus cells. This observation is encouraging for potential hair cell regeneration in the adult mammalian organ of Corti because the prospective hair cell progenitor cells have not vanished but are possibly unable to respond to mitogenic stimulation. It is unclear, however, whether any mitogenic substances are increased, or conversely, whether any antimitotic factors are decreased in the mammalian cochlea as a consequence of hair cell loss (Tsue et al. 1994).

Within the cochlea, progenitor cells or stem cells are not limited to the organ of Corti. Rask-Anderson and colleagues (Rask-Andersen et al. 2005) have recently described the isolation and propagation of sphere-forming cells from adult human and guinea pig spiral ganglion. These adult spiral ganglion stem cells display similar characteristics to adult neural stem cells and can differentiate into neurons and glial cell types. It appears that the 6-week-old murine spiral ganglion harbors only a few stem cells with sphere-forming characteristics, and sphere-forming stem cells from mice older than 6 weeks occur only occasionally, which is too rare to be reliably quantifiable without processing large numbers of inner ears (Oshima et al. 2007). When both studies are taken into account, one can nevertheless hypothesize that spiral ganglion stem cells exist in older mammals and that they are more readily detectable in individual spiral ganglia specimens from guinea pigs or humans because the larger ganglia in these mammals contain more total cells than the murine ganglia. Other important cochlear tissues are located in the stria vascularis. It appears that cells with proliferative capacity are also detectable in the postnatal stria of young postnatal mice (Oshima et al. 2007). Nevertheless, all stem cell populations of the murine cochlea seem to diminish substantially during the initial postnatal period, which is in stark contrast to the vestibular stem cell populations, which appear to be maintained, albeit at low numbers, throughout life.

4.2 Capacity of Inner Ear Stem Cells to Differentiate into Hair Cells

Sphere-forming stem cells can be propagated as important constituents of floating colonies in serum-free medium containing growth factors (Li et al. 2003a). Withdrawal of the growth factors and attachment of the spheres leads to patches of differentiating cells. Cells that express multiple hair cell markers become obvious after 10–14 days of differentiation, but only in populations derived from either vestibular or organ of Corti sensory epithelium–derived spheres (Li et al. 2003a; Oshima et al. 2007). Differentiating cells from spheres derived from the stria vascularis and from the spiral ganglion do not normally contain hair cell marker–positive cells. Neurons and glial cell types, on the other hand, can be found not only in populations derived from spiral ganglion spheres, but also in cells derived from vestibular cells and organ of Corti sensory epithelia.

These observations suggest that there are substantial differences among different populations of sphere-forming stem cells and that hair cells spontaneously differentiate in vitro only from sphere populations derived from inner ear sensory epithelia. This potential for generating hair cells in vivo became observable after grafting of spheres derived from murine vestibular sensory epithelia into the developing inner ears of chicken embryos at a developmental point before formation of sensory epithelia. After development continued for several days, hair cell marker–positive cells were found integrated into the maturing sensory epithelia (Li et al. 2003a). It remains to be demonstrated whether inner ear stem cells derived from vestibular sensory epithelia or the organ of Corti have the capacity to replace lost hair cells in mammalian cochleae.

4.3 Capacity of Other Non–inner Ear Stem Cells to Differentiate into Inner Ear Cell Types

Sphere-forming neural stem cells, isolated from the embryonic mouse brain, have been used for grafting experiments in the neomycin-treated cochleae of 4-week-old mice. Although the majority of the grafted cells differentiated into neurons and glia, a few hair cell marker–positive cells were found in vestibular sensory epithelia (Tateya et al. 2003). While it appears plausible that stem cells isolated from ectodermally derived organs are best suited to replace lost hair cells, other stem cell types may have features that make them uniquely suitable for use in a therapeutic situation. Bone marrow–derived stem cells, for example, have been found to survive after transplantation into cochleae and to differentiate into neuronal and glial marker–expressing cells (Naito et al. 2004), which suggests that autologous bone marrow grafts could potentially be used to replace lost spiral ganglion neurons. In vitro guidance of mesenchymal stem cells from the bone marrow with a combination of Sonic hedgehog and retinoic acid appears to enhance greatly the expression of neuronal markers and enables the bone marrow–derived cells to grow neurites toward hair cells in coculture experiments (Kondo et al. 2005). Expression of *Atoh1*, a key transcription factor for hair cell

TABLE 11.1. Challenges to effective cochlear treatment.

Challenge	Description	Outlook
Cell delivery	Gaining access to the cochlea without causing additional damage. This is particularly difficult in small animal models such as the mouse.	Refinement of surgical skills. Systematic comparison of different routes of cell administration.
Cell homing	Attachment and integration of the grafted cells into the damaged organ of Corti, not at random locations.	The correct progenitor cell type may have the intrinsic capacity for homing. Otherwise, engineering of cells with appropriate surface receptors.
Functional integration/reinnervation	The cells have to take the place of lost hair cells and become afferently and efferently innervated. Stimulation of the replacement cells has to evoke action potentials in the auditory nerve.	There are indications that ectopically placed hair cells in the cochlea are innervated (Kawamoto et al. 2003b)
Generation of different hair cell subtypes	Generation of "generic" hair cells with the ability to attract and to synaptically connect with afferent nerve fibers is the primary goal. Outer hair cell equivalents are probably necessary to restore functionality completely.	The factors that control development of different hair cell subtypes are largely unknown. There is some speculation that the local environment is able to influence the subtype of replacement hair cells.
Functional integration/placement	Proper orientation of stereociliary bundles of replacement hair cells and positioning in context with the tectorial membrane. Physical connection of the stereociliary bundles of outer hair cell equivalents with the tectorial membrane.	The grafted cells may be able to receive guidance cues from the local environment. Refinement of the functional integration is perhaps a secondary goal and not necessary for initial, but incomplete, functional recovery.
Long-term survival	Regenerated hair cells need to survive for an extended period.	No experimental data yet. Need for autologous transplants or suppression of the immune response.
Mitotic quiescence	Terminal differentiation of all grafted cells.	Cell sorting before transplantation may be required.

development, in murine mesenchymal bone marrow stem cells, increases the expression of hair cell markers in vitro (Jeon et al. 2006), raising the question of whether autologous grafts of modified bone marrow stem cells are capable of replacing lost hair cells in vivo.

Embryonic stem cells are the most powerful stem cell type because these cells can give rise to all cell types, even a complete animal, a feature that is widely used to generate knockout mice. Stepwise coaxing of murine embryonic stem cells into cells with features of inner ear progenitors has shown that it is possible to use these embryonic stem cell–derived progenitors to generate hair cell marker–positive cells in vitro and in vivo, after grafting, into embryonic chicken otocysts (Li et al. 2003b). Human embryonic stem cells have been widely proposed as a source for replacement cell types. The generation of human hair cells from embryonic stem cells will certainly be an important milestone toward future therapeutic applications.

Despite the active research in this area, the functional replacement of lost hair cells with stem cell–derived cells still faces many challenges and it is difficult to predict the future development of specific avenues for treatment (Table 11.1). Nevertheless, the field of regenerative medicine and stem cell therapy is rapidly evolving and it is certain that insights gained from other organ systems will be applied to the inner ear whenever feasible. Ultimately, a perfect experiment should lead to substantial and sustained functional improvement of hearing in an animal model, which would be the only acceptable basis for development of future human therapy.

5. The Future

5.1 Multiple Starting Points

Ten years ago, the main focus of cochlear hair cell regeneration research was to identify the growth factors that were capable of inducing hair cell regeneration in the damaged organ of Corti. Since then, the repertoire of tools and technology has increased considerably. For example, researchers are now able to manipulate gene expression and to convert existing cells into hair cells via viral infection of the damaged organ of Corti with an expression vector for *Atoh1*. It is now common knowledge that knocking out cell cycle inhibitor genes, which are normally expressed in the mature organ of Corti, leads to mitotic cell production. A logic amalgamation of this particular application would be to combine a short suppression of cell cycle inhibitors with a subsequent and transient expression of *Atoh1* via local infection with a gene therapy virus or other vehicles, transfection reagents, or even nanoparticles.

Likewise, highly efficient but transient transfection of bone marrow or other stem cell–derived cells with an *Atoh1* expression vector before grafting may increase the chance of proper integration and differentiation of replacement hair cells in the damaged organ of Corti. Progenitor cells that can be grafted autologously or cells that do not generate an immune response are probably good choices, but whether, for example, bone marrow–derived stem cells are indeed capable of engendering massive cochlear hair cell replacement needs to be determined.

Inner ear progenitors, derived from embryonic stem cells, may be able to serve as a unique tool to test growth factors or drug candidates for their ability to stimulate hair cell differentiation. Compounds identified in such a stem cell–based assay could be tested for efficacy to regenerate damaged human hair cells. Human embryonic stem cells are probably the most promising cell type that has not been explored in detail for its capacity to generate hair cells. It is already obvious that existing human embryonic stem cell lines need slightly different protocols than mouse embryonic stem cells to be converted into inner ear progenitors (Rivolta et al. 2006). Nevertheless, it is only a matter of time until the first stem cell–derived human hair cells will be generated either from embryonic stem cells or other stem cells (e.g., adult inner ear stem cells).

5.2 Science Fiction

It is understandable that the various types of hearing loss will require different therapeutic initiation points. For example, it will not be sufficient to stimulate hair cell regeneration in a patient with an underlying genetic defect that causes hair cell loss either directly or indirectly. Potential treatment scenarios for these cases could be the introduction of the wild-type version of the mutated gene in combination with stimulants of hair cell regeneration or stem cell–based transplantation therapy with cells that do not carry the mutation. Complex conditions such as connexin mutations may be treatable with a combination of gene therapy to restore the defective gap junction apparatus in the supporting cells with hair cell regeneration using stem cells.

Hair cell loss in the aging cochlea is a societal challenge. As long as the physiological effects of aging on the cochlea are not fully understood, no long-term regenerative solution for lost hair cells will be readily forthcoming unless a way is found to replenish lost hair cells constantly or to generate highly robust replacement hair cells. It is not inconceivable, however, that transgenic activation of antiapoptotic mechanisms will provide future "designer hair cells" with natural resistance to daily insults. Only time will tell whether the thoughts and speculations introduced in this paragraph are valid solutions for the treatment of hearing loss. Research and proof-of-principle experiments should be done using a variety of approaches to provide adequate choices for designs that may be selected for clinical trials aimed to cure hearing loss.

Acknowledgment. The research of Dr. Raphael has been supported by The R. Jamison and Betty Williams Professorship; Berte and Alan Hirschfield; the CHD, DRF, NOHR, and RNID; and several NIH NIDCD grants. The research of Dr. Heller has been supported by the DRF, the March of Dimes, NOHR, the McKnight Foundation, and several NIH-NIDCD grants. We thank Chris Gralapp for help with Fig. 11.1.

References

Adler HJ, Raphael Y (1996) New hair cells arise from supporting cell conversion in the acoustically damaged chick inner ear [published erratum appears in Neurosci Lett 205:17–20.

Adler HJ, Poje CP, Saunders JC (1993) Recovery of auditory function and structure in the chick after two intense pure tone exposures. Hear Res 71:214–224.

Avraham KB, Raphael Y (2003) Prospects for gene therapy in hearing loss. J Basic Clin Physiol Pharmacol 14:77–83.

Barald KF, Kelley MW (2004) From placode to polarization: new tunes in inner ear development. Development 131:4119–4130.

Bermingham NA, Hassan BA, Price SD, Vollrath MA, Ben-Arie N, Eatock RA, Bellen HJ, Lysakowski A, Zoghbi HY (1999) Math1: an essential gene for the generation of inner ear hair cells. Science 284:1837–1841.

Bermingham-McDonogh O, Stone JS, Reh TA, Rubel EW (2001) FGFR3 expression during development and regeneration of the chick inner ear sensory epithelia. Dev Biol 238:247–259.

Chen P, Segil N (1999) p27(Kip1) links cell proliferation to morphogenesis in the developing organ of Corti. Development 126:1581–1590.

Chen P, Johnson JE, Zoghbi HY, Segil N (2002) The role of Math1 in inner ear development: uncoupling the establishment of the sensory primordium from hair cell fate determination. Development 129: 2495–2505.

Chen P, Zindy F, Abdala C, Liu F, Li X, Roussel MF, Segil N (2003) Progressive hearing loss in mice lacking the cyclin-dependent kinase inhibitor Ink4d. Nat Cell Biol 5:422–426.

Chen X, Frisina RD, Bowers WJ, Frisina DR, Federoff HJ (2001) HSV amplicon-mediated neurotrophin-3 expression protects murine spiral ganglion neurons from cisplatin induced damage. Mol Ther 3:958–963.

Corwin JT (1981) Postembryonic production and aging in inner ear hair cells in sharks. J Comp Neurol 201:541–553.

Corwin JT (1983) Postembryonic growth of the macula neglecta auditory detector in the ray, Raja clavata: continual increases in hair cell number, neural convergence, and physiological sensitivity. J Comp Neurol 217:345–356.

Corwin JT (1985) Perpetual production of hair cells and maturational changes in hair cell ultrastructure accompany postembryonic growth in an amphibian ear. Proc Natl Acad Sci USA 82:3911–3915.

Corwin JT, Cotanche DA (1988) Rege- neration of sensory hair cells after acoustic trauma. Science 240:1772–1774.

Cotanche DA, Lee KH, Stone JS, Picard DA (1994) Hair cell regeneration in the bird cochlea following noise damage or ototoxic drug damage. Anat Embryol 189:1–18.

Cotanche DA, Messana EP, Ofsie MS (1995) Migration of hyaline cells into the chick basilar papilla during severe noise damage. Hear Res 91:148–159.

Crumling MA, Raphael Y (2006) Manipulating gene expression in the mature inner ear. Brain Res 1091:265–269.

Dooling RJ, Ryals BM, Manabe K (1997) Recovery of hearing and vocal behavior after hair-cell regeneration. Proc Natl Acad Sci USA 94:14206–14210.

Fekete DM (2000) Making sense of making hair cells. Trends Neurosci 23:386.

Fekete DM, Wu DK (2002) Revisiting cell fate specification in the inner ear. Curr Opin Neurobiol 12:35–42.

Forge A, Li L, Corwin JT, Nevill G (1993) Ultrastructural evidence for hair cell regeneration in the mammalian inner ear. Science 259:1616–1619.

Fritzsch B, Pauley S, Beisel KW (2006) Cells, molecules and morphogenesis: the making of the vertebrate ear. Brain Res 1091:151–171.

Girod DA, Duckert LG, Rubel EW (1989) Possible precursors of regenerated hair cells in the avian cochlea following acoustic trauma. Hear Res 42:175–194.

Hashino E, Salvi RJ (1993) Changing spatial patterns of DNA replication in the noise-damaged chick cochlea. J Cell Sci 105:23–31.

Hawkins JE Jr, Preston RE (1975) Vestibular ototoxicity. In: Naunton RF (ed) The Vestibular System. New York: Academic Press, pp. 321–349.

Hawkins RD, Bashiardes S, Helms CA, Hu L, Saccone NL, Warchol ME, Lovett M (2003) Gene expression differences in quiescent versus regenerating hair cells of avian sensory epithelia: implications for human hearing and balance disorders. Hum Mol Genet 12:1261–1272.

Hawkins RD, Helms CA, Winston JB, Warchol ME, Lovett M (2006) Applying genomics to the avian inner ear: development of subtractive cDNA resources for exploring sensory function and hair cell regeneration. Genomics 87:801–808.

Ishimoto S, Kawamoto K, Kanzaki S, Raphael Y (2002) Gene transfer into supporting cells of the organ of Corti. Hear Res 173:187–197.

Izumikawa M, Minoda R, Kawamoto K, Abrashkin KA, Swiderski DL, Dolan DF, Brough DE, Raphael Y (2005) Auditory hair cell replacement and hearing improvement by *Atoh1* gene therapy in deaf mammals. Nat Med 11:271–276.

Jeon SJ, Oshima K, Heller S, Edge AS (2006) Bone marrow mesenchymal stem cells are progenitors in vitro for inner ear hair cells. Mol Cell Neurosci.

Kanzaki S, Beyer LA, Swiderski DL, Izumikawa M, Stöver T, Kawamoto K, Raphael Y (2006) p27(Kip1) deficiency causes organ of Corti pathology and hearing loss. Hear Res 214:28–36.

Kawamoto K, Yagi M, Stöver T, Kanzaki S, Raphael Y (2003a) Hearing and hair cells are protected by adenoviral gene therapy with TGF-beta1 and GDNF. Mol Ther 7:484–492.

Kawamoto K, Ishimoto S, Minoda R, Brough DE, Raphael Y (2003b) Math1 gene transfer generates new cochlear hair cells in mature guinea pigs in vivo. J Neurosci 23:4395–4400.

Kelley MW (2003) Cell adhesion molecules during inner ear and hair cell development, including notch and its ligands. Curr Top Dev Biol 57:321–356.

Kondo T, Johnson SA, Yoder MC, Romand R, Hashino E (2005) Sonic hedgehog and retinoic acid synergistically promote sensory fate specification from bone marrow-derived pluripotent stem cells. Proc Natl Acad Sci USA 102:4789–4794.

Li H, Liu H, Heller S (2003a) Pluripotent stem cells from the adult mouse inner ear. Nat Med 9:1293–1299.

Li H, Roblin G, Liu H, Heller S (2003b) Generation of hair cells by stepwise differentiation of embryonic stem cells. Proc Natl Acad Sci USA 100:13495–13500.

Lindeman HH (1969) Regional differences in sensitivity of the vestibular sensory epithelia to ototoxic antibiotics. Acta Otolaryngol 67:177–189.

Lohmann DR, Gallie BL (2004) Retinoblastoma: revisiting the model prototype of inherited cancer. Am J Med Genet C Semin Med Genet 129:23–28.

Lopez I, Honrubia V, Lee SC, Li G, Beykirch K (1998) Hair cell recovery in the chinchilla crista ampullaris after gentamicin treatment: a quantitative approach. Otolaryngol Head Neck Surg 119:255–262.

Löwenheim H, Furness DN, Kil J, Zinn C, Gultig K, Fero ML, Frost D, Gummer AW, Roberts JM, Rubel EW, Hackney CM, Zenner HP (1999) Gene disruption of p27(Kip1) allows cell proliferation in the postnatal and adult organ of corti. Proc Natl Acad Sci USA 96:4084–4088.

Malgrange B, Belachew S, Thiry M, Nguyen L, Rogister B, Alvarez ML, Rigo JM, Van De Water TR, Moonen G, Lefebvre PP (2002) Proliferative generation of mammalian auditory hair cells in culture. Mech Dev 112:79–88.

Marean GC, Cunningham D, Burt JM, Beecher MD, Rubel EW (1995) Regenerated hair cells in the European starling: are they more resistant to kanamycin ototoxicity than original hair cells? Hear Res 82:267–276.

Matsui JI, Gale JE, Warchol ME (2004) Critical signaling events during the aminoglycoside-induced death of sensory hair cells in vitro. J Neurobiol 61:250–266.

McKay R (1997) Stem cells in the central nervous system. Science 276:66–71.

Meiteles LZ, Raphael Y (1994) Scar formation in the vestibular sensory epithelium after aminoglycoside toxicity. Hear Res 79:26–38.

Montcouquiol M, Kelley MW (2003) Planar and vertical signals control cellular differentiation and patterning in the mammalian cochlea. J Neurosci 23:9469–9478.

Naito Y, Nakamura T, Nakagawa T, Iguchi F, Endo T, Fujino K, Kim TS, Hiratsuka Y, Tamura T, Kanemaru S, Shimizu Y, Ito J (2004) Transplantation of bone marrow stromal cells into the cochlea of chinchillas. NeuroReport 15:1–4.

Niemiec AJ, Raphael Y, Moody DB (1994) Return of auditory function following structural regeneration after acoustic trauma: behavioral measures from quail. Hear Res 79:1–16.

Oshima, K, Corrales, CE, Grimm, C, Senn, P, Martinez Monedero, R, Geleoc, GS, Edge, A, Holt, JC, and Heller, S (2007). Differential distribution of stem cells in the auditory and vestibular organs of the inner ear. J Assoc Res Otolaryngol 8: 18–31.

Patel NP, Mhatre AN, Lalwani AK (2004) Biological therapy for the inner ear. Expert Opin Biol Ther 4:1811–1819.

Raphael Y (1992) Evidence for supporting cell mitosis in response to acoustic trauma in the avian inner ear. J Neurocytol 21:663–671.

Raphael Y (1993) Reorganization of the chick basilar papilla after acoustic trauma. J Comp Neurol 330:521–532.

Raphael Y, Adler HJ, Wang Y, Finger PA (1994) Cell cycle of transdifferentiating supporting cells in the basilar papilla. Hear Res 80:53–63.

Rask-Andersen H, Bostrom M, Gerdin B, Kinnefors A, Nyberg G, Engstrand T, Miller JM, Lindholm D (2005) Regeneration of human auditory nerve. In vitro/in video demonstration of neural progenitor cells in adult human and guinea pig spiral ganglion. Hear Res 203:180–191.

Rivolta M, Li H, Heller S (2006) Generation of inner ear cell types from embryonic stem cells. In: Turksen K (ed) Embryonic Stem Cell Protocols, Volume II: Differentiation, Models. Totowa, NJ: Humana Press, pp. 71–92.

Roberson DW, Kreig CS, Rubel EW (1996) Light microscopic evidence that direct transdifferentiation gives rise to new hair cells in regenerating avian auditory epithelium. Audit Neurosci 2:195–205.

Ryals BM, Rubel EW (1988) Hair cell regeneration after acoustic trauma in adult Coturnix quail. Science 240:1774–1776.

Sage C, Huang M, Karimi K, Gutierrez G, Vollrath MA, Zhang DS, Garcia-Anoveros J, Hinds PW, Corwin JT, Corey DP, Chen ZY (2005) Proliferation of functional hair cells in vivo in the absence of the retinoblastoma protein. Science 307:1114–1118.

Sen A, Kallos MS, Behie LA (2002) Passaging protocols for mammalian neural stem cells in suspension bioreactors. Biotechnol Prog 18:337–345.

Sherr CJ, Roberts JM (1999) CDK inhibitors: positive and negative regulators of G1-phase progression. Genes Dev 13:1501–1512.

Shou J, Zheng JL, Gao WQ (2003) Robust generation of new hair cells in the mature mammalian inner ear by adenoviral expression of Hath1. Mol Cell Neurosci 23: 169–179.

Staecker H, Gabaizadeh R, Federoff H, Van De Water TR (1998) Brain-derived neurotrophic factor gene therapy prevents spiral ganglion degeneration after hair cell loss. Otolaryngol Head Neck Surg 119:7–13.

Stone JS, Cotanche DA (1994) Identification of the timing of S phase and the patterns of cell proliferation during hair cell regeneration in the chick cochlea. J Comp Neurol 341:50–67.

Stone JS, Shang JL, Tomarev S (2004) cProx1 immunoreactivity distinguishes progenitor cells and predicts hair cell fate during avian hair cell regeneration. Dev Dyn 230: 597–614.

Stöver T, Yagi M, Raphael Y (1999) Cochlear gene transfer: round window versus cochleostomy inoculation. Hear Res 136:124–130.

Svendsen CN, ter Borg MG, Armstrong RJ, Rosser AE, Chandran S, Ostenfeld T, Caldwell MA (1998) A new method for the rapid and long term growth of human neural precursor cells. J Neurosci Methods 85:141–152.

Tateya I, Nakagawa T, Iguchi F, Kim TS, Endo T, Yamada S, Kageyama R, Naito Y, Ito J (2003) Fate of neural stem cells grafted into injured inner ears of mice. NeuroReport 14:1677–1681.

Torres M, Giraldez F (1998) The development of the vertebrate inner ear. Mech Dev 71:5–21.

Tsue TT, Oesterle EC, Rubel EW (1994) Hair cell regeneration in the inner ear. Otolaryngol Head Neck Surg 111:281–301.

Tucci DL, Rubel EW (1990) Physiologic status of regenerated hair cells in the avian inner ear following aminoglycoside ototoxicity. Otolaryngol Head Neck Surg 103:443–450.

Warchol ME (2002) Cell density and N-cadherin interactions regulate cell proliferation in the sensory epithelia of the inner ear. J Neurosci 22:2607–2616.

Warchol ME, Lambert PR, Goldstein BJ, Forge A, Corwin JT (1993) Regenerative proliferation in inner ear sensory epithelia from adult guinea pigs and humans. Science 259:1619–1622.

Wilkins HR, Presson JC, Popper AN, Ryals BM, Dooling RJ (2001) Hair cell death in a hearing-deficient canary. J Assoc Res Otolaryngol 2:79–86.

Witte MC, Montcouquiol M, Corwin JT (2001) Regeneration in avian hair cell epithelia: identification of intracellular signals required for S-phase entry. Eur J Neurosci 14: 829–838.

Woods C, Montcouquiol M, Kelley MW (2004) Math1 regulates development of the sensory epithelium in the mammalian cochlea. Nat Neurosci 7:1310–1318.

Yagi M, Magal E, Sheng Z, Ang KA, Raphael Y (1999) Hair cell protection from aminoglycoside ototoxicity by adenovirus-mediated overexpression of glial cell line-derived neurotrophic factor. Hum Gene Ther 10:813–823.

Zheng JL, Gao WQ (2000) Overexpression of Math1 induces robust production of extra hair cells in postnatal rat inner ears. Nat Neurosci 3:580–586.

Zine A (2003) Molecular mechanisms that regulate auditory hair-cell differentiation in the mammalian cochlea. Mol Neurobiol 27:223–238.

Index

For more information about the series, please visit www.springer-ny.com/shar.

Printed in the United States of America